IET COMPUTING SERIES 68

Explainable Artificial Intelligence for Trustworthy Internet of Things

IET Book Series on Big Data – Call for Authors

Editor-in-Chief: Professor Albert Y. Zomaya, University of Sydney, Australia

The topic of Big Data has emerged as a revolutionary theme that cuts across many technologies and application domains. This new book series brings together topics within the myriad research activities in many areas that analyse, compute, store, manage, and transport massive amounts of data, such as algorithm design, data mining and search, processor architectures, databases, infrastructure development, service and data discovery, networking and mobile computing, cloud computing, high-performance computing, privacy and security, storage and visualization.

Topics considered include (but not restricted to) IoT and Internet computing; cloud computing; peer-to-peer computing; autonomic computing; data centre computing; multi-core and many-core computing; parallel, distributed and high-performance computing; scalable databases; mobile computing and sensor networking; green computing; service computing; networking infrastructures; cyberinfrastructures; e-Science; smart cities; analytics and data mining; Big Data applications and more.

Proposals for coherently integrated International co-edited or co-authored handbooks and research monographs will be considered for this book series. Each proposal will be reviewed by the editor-in-chief and some board members, with additional external reviews from independent reviewers. Please email your book proposal for the IET Book Series on Big Data to: Professor Albert Y. Zomaya at albert.zomaya@sydney.edu.au or to the IET at author_support@theiet.org.

Explainable Artificial Intelligence for Trustworthy Internet of Things

Mohamed Abdel-Basset, Nour Moustafa,
Hossam Hawash and Albert Y. Zomaya

The Institution of Engineering and Technology

Published by The Institution of Engineering and Technology, London, United Kingdom

The Institution of Engineering and Technology is registered as a Charity in England & Wales (no. 211014) and Scotland (no. SC038698).

First published 2024

The Institution of Engineering and Technology
Futures Place
Kings Way, Stevenage
Hertfordshire, SG1 2UA, United Kingdom

www.theiet.org

British Library Cataloguing in Publication Data
A catalogue record for this product is available from the British Library

ISBN 978-1-83953-802-5 (hardback)
ISBN 978-1-83953-803-2 (PDF)

Typeset in India by MPS Limited

Cover Image: simon2579 / DigitalVision Vector via Getty Image

Contents

About the authors **xiii**

1 Explaining AI for safeguarding and securing Internet of Things (IoT) systems—an introduction **1**
1.1 Introduction to the book 1
1.2 The IoT revolution 2
1.3 Emergence of AI in IoT 3
1.4 Need for safeguarding and security 4
1.5 Understanding XAI 4
1.6 XAI in IoT: enhancing security 5
1.7 Building trust and compliance 6
1.8 The human–machine interface 7
References 7

2 Securing the Internet of Things: architectures and designs **11**
2.1 Introduction 11
2.2 IoT fundamentals 12
2.3 IoT architecture 14
2.3.1 Three-layer IoT architecture 14
2.3.2 Four-layer IoT architecture 21
2.3.3 Five-layer IoT architecture 23
2.3.4 Seven-layer IoT architecture 26
2.4 Security challenges in IoT 32
References 37

3 Convergence of Internet of Things and computing technologies **39**
3.1 Introduction 39
3.2 Centralized computing 40
3.3 Distributed computing 41
3.3.1 Client–server architecture 42
3.3.2 Cloud computing 43
3.3.3 Edge computing 43
3.3.4 Fog computing 45
3.3.5 Quantum computing 46
3.4 Advanced distributed computing paradigms 47
3.4.1 Cloudlet computing 48

3.4.2 Swarm computing 49
3.4.3 Dew computing 50
3.4.4 Mist computing 50
3.5 Emerging technologies and innovations 52
3.5.1 Fabric computing 52
3.5.2 Multi-access edge computing (MEC) 53
3.5.3 Mobile computing 55
References 55

4 Security vulnerabilities in Internet of Things: attack surfaces, threats, and defense 59
4.1 Introduction 59
4.2 IoT attack surfaces 60
4.2.1 Network-based attack surfaces 60
4.2.2 Physical attack surfaces 64
4.2.3 Software attack surfaces 69
4.2.4 Data attack surfaces 72
4.2.5 Cloud and backend attack surfaces 73
4.3 Common IoT attacks 78
4.3.1 Physical security attacks 79
4.3.2 Data privacy attacks 80
4.3.3 Network attacks 81
4.3.4 Firmware and software attacks 84
4.4 Defensive strategies and mitigation techniques 84
4.4.1 Secure design principles 84
4.4.2 Encryption and authentication 85
4.4.3 Network security measures 86
4.4.4 Secure software development 87
4.4.5 Monitoring and incident response 88
4.5 Regulatory frameworks and standards 89
References 92

5 Black-box machine learning for IoT security 95
5.1 Introduction 95
5.2 Fundamentals of black-box ML 96
5.3 Classical ML 98
5.4 Deep learning 105
5.4.1 Feedforward neural networks 105
5.4.2 Recurrent neural networks 106
5.4.3 Convolutional neural networks 110
5.4.4 Generative adversarial networks 112
5.4.5 Autoencoders 114
5.4.6 Transformers 116
5.4.7 Memory-augmented neural networks 119
5.4.8 Spiking neural networks 120

5.4.9 Graph neural networks 122
5.5 Reinforcement learning 122
5.6 Meta-learning 126
References 129

6 Explainable artificial intelligence for safeguarding IoT 131
6.1 Introduction 131
6.2 Principles of XAI 132
6.3 The need for explainability in IoT systems 136
6.4 Dimensions of explainability in IoT security 140
6.4.1 Algorithmic explainability 140
6.4.2 Data explainability 141
6.4.3 Contextual explainability 142
6.4.4 Human-centric explainability 143
6.5 Regulatory landscape and standards 144
6.6 Frameworks and tools for implementing explainable AI in IoT security 145
References 147

7 Explainability methods in explainable security intelligence: fine-grained taxonomy 151
7.1 Introduction 151
7.2 Model-agnostic versus model-specific XAI methods 152
7.3 Post-hoc versus intrinsic XAI methods 154
7.4 Global versus local interpretability 155
7.5 Complexity and granularity 157
7.6 Human-understandable versus machine-oriented explanations 159
7.7 Domain-specific XAI methods 160
7.8 Quantitative versus qualitative explanation 161
7.9 Interpretability type 163
References 167

8 Intrinsically explainable security intelligence 169
8.1 Introduction 169
8.2 Linear models 170
8.2.1 Linear regression 170
8.2.2 Logistic regression 174
8.2.3 Generalized linear model 177
8.3 Graphical models 179
8.3.1 Naive Bayes 179
8.3.2 Bayesian network 181
8.4 Tree-based models 182
8.4.1 Decision tree 183
8.4.2 Decision rules 186

8.5 K-nearest neighbors 189
References 193

9 Model-agnostic methods for globally interpretable machine learning 195
9.1 Introduction 195
9.2 PDP 196
9.3 ICE plots 198
9.4 Automated dependence plots 201
9.5 Accumulated local effect plots 202
9.6 Feature interaction 205
9.7 Functional decomposition 206
9.8 Permutation feature importance 208
9.9 Global surrogate models 209
9.10 Prototypes and criticisms 211
References 212

10 Model-agnostic methods for locally explainable AI to secure IoT system 215
10.1 Introduction 215
10.2 ICE plots 217
10.3 Local interpretable model-agnostic explanations 220
10.4 SHapley Additive exPlanations 222
10.5 PyExplainer: a local rule-based model-agnostic technique 224
10.6 Contrastive explanations 227
10.7 Counterfactual explanations 229
10.8 Anchor 232
References 234

11 Explainability evaluation metrics for explainable security intelligence in the Internet of Things 237
11.1 Introduction 237
11.2 Foundations of evaluation metrics 238
11.3 Fairness metrics 241
11.3.1 Group unawareness metric 241
11.3.2 Demographic parity metric 242
11.3.3 Disparate impact metric 243
11.3.4 Equal opportunity metric 244
11.3.5 Equal odds metric 245
11.3.6 PPV-parity metric 245
11.3.7 FPR-parity metric 246
11.4 Faithfulness metrics 247
11.4.1 Faithfulness correlation metric 248
11.4.2 Faithfulness estimate metric 249
11.4.3 Pixel-flipping metric 250

11.4.4 Monotonicity metric 251
11.4.5 Monotonicity correlation metric 252
11.4.6 Selectivity metric 253
11.4.7 SensitivityN metric 254
11.4.8 Infidelity metric 256
11.5 Robustness metrics 257
11.5.1 Continuity test 257
11.5.2 Local Lipschitz estimate 258
11.5.3 Max-sensitivity metric 259
11.5.4 Average sensitivity (Avg Sensitivity) metric 260
11.5.5 Consistency metric 262
11.5.6 Relative input stability metric 263
11.5.7 Relative output stability metric 264
11.6 Localisation metrics 265
11.6.1 Pointing game metric 265
11.6.2 Top-*K* intersection metric 266
11.6.3 Relevance mass accuracy metric 267
11.6.4 Relevance rank accuracy metric 269
11.7 Axiomatic metrics 270
11.7.1 Completeness metric 270
11.7.2 Non-sensitivity metric 271
11.7.3 Input invariance metric 272
References 273

12 Adversarial attacks and defense in explainable security intelligence 277
12.1 Introduction 277
12.2 Adversarial attacks 278
12.2.1 Knowledge of adversarial attacks 279
12.2.2 One-time versus iterative attacks 281
12.2.3 Training versus inference attacks 283
12.2.4 Targeted versus non-targeted attacks 284
12.3 Adversarial perturbation techniques 288
12.3.1 Fast gradient sign method 288
12.3.2 Projected gradient descent 289
12.3.3 Carlini–Wagner attack 289
12.4 Zeroth-order optimization attacks 290
12.5 Evaluating adversarial attacks 291
12.6 Adversarial defense 293
12.6.1 Preprocessing techniques 294
12.6.2 Robust model architectures 297
12.6.3 Gradient masking and obscuring 300
12.7 Adversarial attacks on XAI systems 301

12.8 Explainability and robustness: a balancing act 303
12.9 Explainable adversarial attacks 305
12.10 Explainable defense mechanisms 308
References 309

13 Federated learning meets explainable AI at the edge of things 313
13.1 Introduction 313
13.2 Fundamentals FL 315
13.2.1 Definition and motivation 315
13.2.2 Key concepts in FL 316
13.2.3 Advantages and challenges 319
13.3 Taxonomy of FL methods 321
13.3.1 Network topology 321
13.3.2 Communication mechanism 325
13.3.3 Aggregation methods 329
13.3.4 Model architecture 333
13.3.5 Participant selection 334
13.3.6 Convergence techniques 336
13.3.7 Failure tolerance 339
13.3.8 Resource allocation 340
13.4 Intersection of FL and XAI 341
13.4.1 Challenges in achieving explainability in FL 341
13.4.2 Techniques for explainable FL 342
13.5 Accountable FL 344
13.5.1 Accountability in FL 345
13.5.2 Regulatory compliance 346
13.6 Trustworthy FL 348
13.6.1 Trust challenges in FL 348
13.6.2 Security and privacy measures 349
References 352

14 Explainable security intelligence for zero-trust IoT 355
14.1 Introduction 355
14.2 Background of history of zero-trust security 356
14.3 Principles of zero-trust IoT security 360
14.3.1 Primary principles 360
14.3.2 Secondary principles 364
14.4 Zero-trust architecture 366
14.5 Synergies between zero trust and XAI 369
14.5.1 Explainability in zero-trust systems 369
14.5.2 Integration of XAI in zero trust 369
14.5.3 Enhancing trust through XAI 371
14.5.4 Challenges and opportunities 372
References 374

15 Explainable security intelligence in IoT applications **377**
15.1 Introduction 377
15.2 Smart healthcare: leveraging XAI for enhanced services 378
15.3 Smart transportation: optimizing systems with XAI 381
15.4 Smart manufacturing: implementing XAI for improved processes 387
15.5 Smart agriculture: XAI for precision farming 390
15.6 Smart grid: enhancing efficiency through XAI 395
15.7 Smart retail: personalization and efficiency with XAI 400
15.8 Smart building: optimizing spaces with XAI 404
15.9 Nanotechnology: advancements through XAI 407
References 412

Index **415**

About the authors

Mohamed Abdel-Basset is an associate professor with the Faculty of Computers and Informatics, Zagazig University, Egypt. His research interests include AI, machine and deep learning, computational intelligence, security intelligence, IoT data mining, and applied statistics. He has been program chair for many conferences in the fields of AI optimization and complexity. He is also an editor and a reviewer for international journals and conferences. He holds a PhD from the Faculty of Computers and Informatics, Menoufia University, Egypt.

Nour Moustafa is a senior lecturer and leader of intelligent security at the School of Systems & Computing, University of New South Wales (UNSW Canberra), Australia. His areas of research interest include cyber security and artificial intelligence, in particular network security, IoT security, intrusion detection, and machine and deep learning. He is associate editor of *IEEE Systems*, *IEEE Transactions on Industrial Informatics*, *IEEE IoT Journal*, *IEEE Access*, and *Ad Hoc Networks*. He is vice-chair, session chair, technical program committee (TPC) member and proceedings chair for leading conferences. He has been awarded the 2020 prestigious ARC DECRA Fellowship and Australian Spitfire Memorial Defence Fellowship Awards. He is a senior IEEE member, ACM distinguished speaker, and Spitfire fellow. He received his PhD degree in computing and cyber security from the UNSW Canberra, Australia.

Hossam Hawash is a research assistant and an lecturer with the Department of Computer Science, Zagazig University, Egypt. His research interests include explainable artificial intelligence, machine and deep learning, the Internet of Things (IoT), cyber security, and fuzzy learning. He holds an MSc from the Department of Computer Science, Faculty of Computers and Informatics, Zagazig University, Egypt.

Albert Y. Zomaya is Peter Nicol Russell Chair Professor of Computing, Director of the Centre for Distributed and High-Performance Computing, and Australian Research Council professorial fellow in the School of Computer Science, the University of Sydney, Australia. He has published over 30 book titles and is chief editor for the IET Book Series on Big Data. He is a fellow of the Australian Academy of Science, American Association for the Advancement of Science (AAAS), a distinguished member of the Association for Computing Machinery (ACM), fellow of the Institute of Electrical and Electronic Engineers (IEEE, USA), fellow of the Institution of Engineering and Technology and IET chartered engineer (CEng). He received the 2021 IEEE Computer Society's Technical Committee on Cloud Computing Research Innovation Award. He holds a PhD from the Department of Automatic Control and Systems Engineering, Sheffield University, UK.

Chapter 1

Explaining AI for safeguarding and securing Internet of Things (IoT) systems—an introduction

1.1 Introduction to the book

In the contemporary digital era, the fusion of explainable artificial intelligence (XAI) and Internet of Things (IoT) systems represents a pioneering endeavor at the intersection of science, technology, and security. This introductory chapter serves as a portal into the realm of our textbook, where we embark on a journey to comprehend, dissect, and harness the transformative potential of these two groundbreaking domains [1]. The IoT refers to the network of interconnected physical devices and sensors that collect and exchange data, enabling smart, efficient, and automated decision-making. The IoT, a paradigm shift in connectivity, is reshaping industries, businesses, and our daily lives. IoT devices have permeated homes, factories, healthcare, transportation, and cities, facilitating unprecedented of data-driven decision-making and automation. Nevertheless, this interconnected utopia is paradoxically characterized by increased vulnerability. The very interconnectedness that empowers IoT ecosystems also exposes them to diverse security threats, making robust security measures a paramount concern [2,3].

XAI is a pioneering concept that transcends AI algorithms' typical "black-box" nature. XAI advocates for transparent, interpretable AI models, thereby facilitating human comprehension of AI-driven decisions. When incorporated into IoT systems, XAI is poised to revolutionize the security landscape. By demystifying AI-driven actions and ensuring accountability, XAI enhances security and builds trust among users, regulators, and stakeholders. Our journey commences with exploring the IoT revolution, dissecting its far-reaching impact across industries. We delve into the emergence of AI as an enabler of IoT's transformative potential, elucidating its role in predictive analytics, automation, and real-time decision-making within IoT ecosystems.

We explore the critical vulnerabilities and security challenges facing IoT systems, underscoring the urgent need for robust safeguards to protect these interconnected networks. Real-world case studies underscore the urgency of addressing these concerns. In elucidating the foundation of our discourse, we scrutinize XAI—its definition, significance, and ethical underpinnings [4,5]. By demystifying AI, XAI introduces transparency, accountability, and comprehensibility to AI systems, laying the groundwork for their fusion with IoT. As we progress through this textbook, we will examine the techniques, applications, and strategies that integrate XAI with IoT security. This include broad range of XAI methodologies, illustrating their adaptability to IoT ecosystems [6–8].

Moreover, this journey extends beyond technicalities. We will explore the human-machine interface of IoT systems, highlighting the critical role of effective communication and the bridge that XAI provides between technical experts and non-technical stakeholders [9]. Our mission is to equip you, the reader, with a profound understanding of this evolving landscape. The following chapters will dissect challenges, anticipate future trends, and elucidate the path forward. By the time you reach the final page, you will be well-versed in the symbiotic relationship between XAI and IoT security, poised to advance secure, transparent, and trustworthy IoT systems.

1.2 The IoT revolution

In the annals of technological progress, the Internet of Things, or IoT, stands as a monumental paradigm shift. At its core, IoT is an ecosystem where everyday objects, devices, and machines are imbued with the ability to connect, communicate, and share data over the Internet [10–13]. This Internet transcends traditional computing devices, extending its reach to encompass almost anything imaginable: from smart thermostats optimizing energy consumption in homes to wearable health monitors tracking vital signs and industrial sensors overseeing factory operations to autonomous vehicles navigating our roads. IoT thrives on the concept of data exchange between devices, where information is collected, processed, and often acted upon in real-time [14]. The hallmark of IoT is its capacity to transform inert, inanimate objects into intelligent entities capable of enhancing efficiency, productivity, and the quality of life for individuals and entire industries.

IoT is not a singular technology but an ecosystem characterized by its ubiquity. It has swiftly permeated an array of sectors, sparking transformative change. In healthcare, IoT has given rise to a new era of telemedicine, where wearable devices remotely monitor patients' health and transmit data to healthcare professionals, potentially saving lives through early intervention. In agriculture, precision farming harnesses IoT's power to optimize crop management, reduce waste, and enhance yield. Smart cities leverage IoT for traffic management, environmental monitoring, and improved public services, making urban life more sustainable and efficient. Industries as diverse as logistics, manufacturing, energy, and retail have all experienced substantial enhancements through IoT adoption. This ubiquity stems from IoT's unique ability to extract meaningful insights from vast data streams, revolutionizing decision-making and enabling predictive analytics. It has ushered in a new era of hyper-connectivity, where previously siloed systems now communicate seamlessly, propelling society into an era of unprecedented efficiency and innovation [15].

The proliferation of IoT devices and systems has been nothing short of meteoric. As the world becomes increasingly interconnected, IoT devices have exploded in number, far surpassing traditional computing devices. It is estimated that billions of devices are currently part of the IoT ecosystem, and this number continues to grow exponentially. However, this rapid expansion has ushered in a new era of security challenges. The very connectivity that empowers IoT also exposes vulnerabilities, creating a vast attack surface for cybercriminals. IoT devices, often resource-constrained, are susceptible to a various security threats, from malware and data breaches to physical attacks [16–19]. Moreover, their sheer diversity, from consumer gadgets to critical infrastructure components, complicates security efforts. The potential consequences of a security breach in an IoT context are staggering, ranging from compromised personal privacy to disruptions in essential services, underscoring the urgency of robust security measures. As we delve deeper into this textbook, we will explore these challenges and unravel XAI's critical role in mitigating them, ensuring the security and trustworthiness of our interconnected world [20].

1.3 Emergence of AI in IoT

The seamless integration of AI into the IoT marks a pivotal moment in technological evolution. This convergence empowers otherwise static and inert IoT devices with intelligence, enabling them to not only collect data but also analyze, interpret, and act upon it. AI algorithms are embedded into the very fabric of IoT systems, facilitating real-time data processing, decision-making, and adaptive responses [21]. Whether it is the intuitive control of smart homes, the autonomous navigation of vehicles, or the optimization of industrial processes through Industrial IoT (IIoT), the union of AI and IoT has unleashed a new era of efficiency and convenience. These smart systems learn and adapt, making sense of the ever-increasing volumes of data generated by IoT devices, transforming how we interact with our environments, and catalyzing innovations across diverse sectors.

The synergistic relationship between AI and IoT promises a myriad of benefits that extend far beyond the sum of their parts. At its core, AI augments IoT by imparting it with the ability to make sense of the deluge of data generated by IoT devices. By harnessing the power of AI-driven analytics, organizations can extract valuable insights, enabling informed decision-making, predictive maintenance, and proactive responses to emerging trends. Additionally, AI enhances the security posture of IoT systems, effectively identifying anomalies and potential threats in real-time. This amalgamation of technologies is poised to increase efficiency and drive resource optimization, energy conservation, and cost-effectiveness. In the larger context, it redefines the very nature of connectivity, promising smarter cities, safer transportation, and more responsive healthcare, ultimately enriching our lives and reimagining industries across the globe [22].

As the bond between AI and IoT strengthens, it inevitably brings forth its own set of challenges and considerations. One of the foremost challenges lies in adapting AI algorithms to operate effectively within the resource-constrained environment of IoT devices. These devices often have limited computational power and memory, necessitating the development of lightweight and efficient AI models. Additionally, integrating AI into IoT raises important ethical concerns surrounding data privacy, transparency, and bias. The vast amounts of data collected by IoT devices require careful handling to ensure that individual privacy rights are preserved [23]. Moreover, the opacity of certain AI algorithms challenges the notion of accountability and fairness. Finally, the growing interconnectedness of AI-enhanced IoT systems calls for a robust regulatory framework to address potential issues related to safety, security, and compliance with evolving standards. Balancing the transformative potential of AI in IoT with these considerations is pivotal as we navigate this dynamic technological landscape [24].

1.4 Need for safeguarding and security

The interconnected nature of IoT devices, while offering unparalleled convenience and efficiency, presents a complex web of vulnerabilities and security threats. As these devices communicate with each other and with central systems over networks, they create numerous potential entry points for cyberattacks. The very diversity of IoT devices, ranging from smart thermostats to industrial sensors, means that security risks vary widely. Common vulnerabilities include weak or default passwords, lack of regular security updates, insufficient encryption, and insecure communication protocols [15]. Moreover, IoT devices often collect and transmit sensitive data, making them prime targets for data breaches. Beyond data breaches, IoT devices can be exploited to launch distributed denial of service (DDoS) attacks, infiltrate networks, or compromise critical infrastructure. These vulnerabilities necessitate robust security measures to safeguard the interconnected IoT ecosystem, a topic we will delve deeper into in the subsequent chapters.

Real-world instances of IoT security breaches serve as stark reminders of the profound implications of inadequate security within the IoT landscape. One notable example is the Mirai botnet attack in 2016, where compromised IoT devices, including cameras and routers, were harnessed to launch massive DDoS attacks, disrupting major online services. This incident underscored the vulnerabilities of unsecured IoT devices and their potential to wreak havoc on the Internet's infrastructure. Another instance is the Stuxnet worm, which targeted industrial control systems, emphasizing the critical importance of securing IoT devices in critical infrastructure [25]. IoT security breaches have also led to privacy infringements, with unauthorized access to smart home cameras and voice assistants raising concerns about data privacy and surveillance. These real-world breaches illuminate the pressing need for proactive security measures and the critical role of XAI in detecting, mitigating, and providing insights into such security incidents. Understanding these implications is paramount as we explore how XAI can address the evolving challenges of IoT security [22].

1.5 Understanding XAI

XAI stands as a pivotal advancement in AI, driven by the necessity to unravel the decision-making processes of increasingly complex AI models. XAI is a framework that emphasizes the development of AI systems capable of providing understandable and coherent explanations for their actions and predictions. This transparency is paramount, especially in the context of IoT systems, where the stakes are high and the potential consequences of AI-driven decisions are profound [7]. XAI not only facilitates a deeper comprehension of AI outputs but also ensures accountability by allowing users and stakeholders to trace the logic and reasoning behind these outputs. In a world increasingly reliant on AI-driven decisions, XAI is the beacon of trust, as it instills confidence in the technology's reliability and fairness. It plays a pivotal role in enhancing the ethical, legal, and practical aspects of AI deployment within IoT ecosystems, ensuring that decisions made by AI systems align with human values and ethical norms [10–12].

The pervasive use of complex, "black-box" AI models in IoT systems introduces a litany of challenges, primarily centered on their lack of interpretability and explainability. These models, like deep neural networks, can produce highly accurate results but often do so without offering insights into their decision-making processes. This opacity raises concerns related to accountability, trustworthiness, and bias, especially when AI systems influence critical decisions in IoT contexts [3]. Interpretable AI models, in contrast, provide clarity by allowing users to understand the rationale behind AI-driven actions. This interpretability fosters trust among users, regulators, and stakeholders, particularly in scenarios such as autonomous vehicles, healthcare diagnostics, and smart grids, where AI plays a pivotal role. Moreover, interpretable models enable debugging and auditing, making it easier to identify and rectify errors or biases. The use of interpretable AI models in IoT systems not only enhances transparency but also contributes to ethical and regulatory compliance, thereby ensuring that these interconnected systems are not only efficient but also reliable, fair, and accountable [20].

1.6 XAI in IoT: enhancing security

The application of XAI techniques within IoT systems opens a fascinating avenue for comprehending the often intricate and opaque world of AI-driven decisions. By integrating XAI, IoT systems can provide profound insights into their decision-making processes, allowing users, operators, and stakeholders to gain a deep understanding of why a particular decision was made. This transparency is achieved through a variety of interpretable AI techniques such as rule-based systems, decision trees, local interpretable model-agnostic explanations (LIME), and SHapley Additive exPlanations (SHAP). These techniques, when deployed in IoT systems, transform complex algorithms into comprehensible decision pathways [11]. This not only fosters trust and accountability but also empowers users to make more informed decisions based on the AI's recommendations. By demystifying AI-driven decisions, XAI bridges the gap between technical intricacy and user comprehension, offering a valuable layer of transparency in IoT systems [3].

XAI stands as a formidable ally in the battle against cybersecurity threats within IoT systems. Its potential becomes particularly evident when applied to the detection of anomalies and the identification of security breaches. XAI models can scrutinize the vast streams of data generated by IoT devices, discerning patterns and anomalies that may elude traditional security measures. When an anomaly is detected, XAI can not only flag it but also provide clear explanations for why it is considered an anomaly. This explanation is pivotal for security analysts and stakeholders as it offers insight into the root cause of the issue, aiding in quicker and more precise remediation efforts. Moreover, in the unfortunate event of a security breach, XAI can play a critical role in post-incident analysis by elucidating the sequence of events and decisions that led to the breach, thereby facilitating the development of strategies to prevent future attacks. In essence, XAI serves as a vigilant guardian within IoT systems, identifying security threats and acting as a transparent sentinel, ready to explain and fortify defenses when necessary [19].

1.7 Building trust and compliance

XAI plays a pivotal role in building user trust within the context of IoT systems. IoT devices and systems increasingly influence various aspects of our lives, from healthcare and transportation to home automation and industrial processes. In this interconnected world, users rely on AI-driven decisions made by these systems, often without a deep understanding of the complex algorithms at work. XAI addresses this trust deficit by providing understandable explanations for AI-driven actions. When users can clearly see why an IoT system made a particular decision or recommendation, they are more likely to trust and embrace the technology [16]. This trust is critical for the widespread adoption and acceptance of AI-powered IoT systems. Furthermore, in scenarios where safety and reliability are paramount, such as autonomous vehicles or healthcare diagnostics, user trust is not just a convenience but a fundamental necessity. XAI, by making AI-driven actions transparent and interpretable, paves the way for users to confidently embrace the benefits of IoT technology while understanding the rationale behind the decisions made by these systems [25].

The regulatory and ethical dimensions of IoT security are of paramount importance in our increasingly connected world. As IoT systems become integral to critical infrastructure, healthcare, transportation, and more, there is a pressing need for comprehensive regulations and ethical considerations to safeguard individuals and society as a whole. Regulatory frameworks are evolving to address issues related to data privacy, security standards, and responsible AI use. In this context, XAI is vital in ensuring compliance with these regulations. By providing transparent explanations for AI-driven actions, XAI helps organizations and IoT system operators demonstrate adherence to regulatory requirements [21]. Moreover, XAI assists in mitigating ethical concerns surrounding AI, such as algorithmic bias and discrimination. By making the decision-making processes of AI models clear and interpretable, XAI allows organizations to assess and rectify any ethical issues in their IoT systems. This proactive approach to regulatory compliance and ethics minimizes legal risks and contributes to responsible and trustworthy AI deployments within the IoT landscape. In essence, XAI becomes an indispensable tool in ensuring that IoT systems align with evolving regulatory standards and ethical best practices, ultimately benefiting both users and society [14].

1.8 The human–machine interface

The interaction between humans and AI-powered IoT systems is a multifaceted dynamic that necessitates clear and transparent communication. In the IoT landscape, AI often plays a pivotal role in making complex decisions, be it optimizing energy consumption in a smart home or managing traffic flow in a smart city. For these systems to be effective and trustworthy, humans need to comprehend the outcomes and the underlying reasoning of AI-driven actions. Clear communication is vital because it builds trust and enables users and stakeholders to make informed decisions based on AI recommendations. Without transparent communication, there is a risk that users might hesitate to fully trust AI systems, leading to missed opportunities and underutilization of the technology's potential [6]. Therefore, the need for explicit and understandable communication between humans and AI-powered IoT systems goes beyond a mere technical requirement; it is essential for ensuring that IoT technologies truly enhance our lives [7].

XAI plays a pivotal role in bridging the gap between the highly technical world of AI development and the broader audience of non-technical stakeholders in IoT systems. In many IoT projects, there is a significant disparity in technical knowledge between those who design and implement AI algorithms (the technical experts) and those who use, regulate, or oversee these systems (the non-technical stakeholders). XAI serves as a translator, converting AI algorithms' often complex, "black-box" outputs into clear and understandable explanations. This translation is essential for several reasons. First, it enables non-technical stakeholders, such as business leaders, policymakers, and end-users, to make informed decisions regarding adopting and using AI-powered IoT systems [16–18]. Second, it facilitates collaboration, as technical experts can effectively communicate the benefits and limitations of AI models to a broader audience. Third, XAI aids in compliance and ethical considerations, allowing regulators to assess and enforce responsible AI use within IoT ecosystems. In essence, XAI acts as a conduit for democratizing AI insights, ensuring that the benefits of AI within IoT systems are accessible and comprehensible to all, regardless of their technical expertise. This not only promotes transparency but also empowers a diverse range of stakeholders to engage meaningfully with AI technologies and contribute to their responsible development and deployment.

References

[1] Kök, I., Okay, F. Y., Muyanlı, Ö., and Özdemir, S. (2023). Explainable artificial intelligence (xai) for internet of things: a survey. *IEEE Internet of Things Journal*, *10*(16), 14764–14779.

[2] Zaman, S., Alhazmi, K., Aseeri, M. A., *et al.* (2021). Security threats and artificial intelligence based countermeasures for internet of things networks: a comprehensive survey. *IEEE Access*, 9, 94668–94690.

[3] Meneghello, F., Calore, M., Zucchetto, D., Polese, M., and Zanella, A. (2019). IoT: Internet of threats? A survey of practical security vulnerabilities in real IoT devices. *IEEE Internet of Things Journal*, 6(5), 8182–8201.

[4] Arrieta, A. B., Díaz-Rodríguez, N., Del Ser, J., *et al.* (2020). Explainable Artificial Intelligence (XAI): concepts, taxonomies, opportunities and challenges toward responsible AI. *Information fusion*, *58*, 82–115.

[5] Adadi, A., and Berrada, M. (2018). Peeking inside the black-box: a survey on explainable artificial intelligence (XAI). *IEEE Access*, *6*, 52138–52160.

[6] Jagatheesaperumal, S. K., Pham, Q. V., Ruby, R., Yang, Z., Xu, C., and Zhang, Z. (2022). Explainable AI over the Internet of Things (IoT): overview, state-of-the-art and future directions. *IEEE Open Journal of the Communications Society*, *3*, 2106–2136.

[7] Garg, S., Kaur, K., Aujla, G. S., Kaddoum, G., Garigipati, P., and Guizani, M. (2023). Trusted explainable AI for 6G-enabled edge cloud ecosystem. *IEEE Wireless Communications*, *30*(3), 163–170.

[8] Nwakanma, C. I., Ahakonye, L. A. C., Njoku, *et al.* (2023). Explainable artificial intelligence (xai) for intrusion detection and mitigation in intelligent connected vehicles: A review. *Applied Sciences*, *13*(3), 1252.

[9] Haque, A. B., Islam, A. N., and Mikalef, P. (2023). Explainable Artificial Intelligence (XAI) from a user perspective: a synthesis of prior literature and problematizing avenues for future research. *Technological Forecasting and Social Change*, *186*, 122120.

[10] Li, S., Xu, L. D., and Zhao, S. (2015). The internet of things: a survey. *Information Systems Frontiers*, *17*, 243–259.

[11] Atzori, L., Iera, A., and Morabito, G. (2010). The internet of things: a survey. *Computer Networks*, *54*(15), 2787–2805.

[12] Da Xu, L., He, W., and Li, S. (2014). Internet of things in industries: a survey. *IEEE Transactions on Industrial Informatics*, *10*(4), 2233–2243.

[13] Ray, P. P. (2018). A survey on Internet of Things architectures. *Journal of King Saud University-Computer and Information Sciences*, *30*(3), 291–319.

[14] Ahmed, M. S. (2021). Designing of internet of things for real time system. *Materials Today: Proceedings*, *87*, 2214–7853.

[15] Vermesan, O., and Friess, P. (2015). *Building the Hyperconnected Society - Internet of Things Research and Innovation Value Chains, Ecosystems and Markets* (p. 332). Taylor & Francis.

[16] Thomas, K., Li, F., Zand, A., *et al.* (2017, October). Data breaches, phishing, or malware? Understanding the risks of stolen credentials. *In Proceedings of the 2017 ACM SIGSAC Conference on Computer and Communications Security* (pp. 1421–1434).

[17] Alharbi, F. S. (2020). Dealing with data breaches amidst changes in technology. *International Journal of Computer Science and Security (IJCSS)*, *14* (3), 108–115.

[18] Bhadouria, A. S. (2022). Study of: impact of malicious attacks and data breach on the growth and performance of the company and few of the world's biggest data breaches. *Int. J. Sci. Res. Publ.*, DOI:10.29322/IJSRP.X.2022.p091095.

[19] Makhdoom, I., Abolhasan, M., Lipman, J., Liu, R. P., and Ni, W. (2018). Anatomy of threats to the internet of things. *IEEE Communications Surveys & Tutorials*, *21*(2), 1636–1675.

[20] Masud, M. T., Keshk, M., Moustafa, N., Linkov, I., and Emge, D. K. (2024). Explainable artificial intelligence for resilient security applications in the Internet of Things. *IEEE Open Journal of the Communications Society*, 99, 1.

[21] Andronie, M., Lăzăroiu, G., Iatagan, M., Uță, C., Ștefănescu, R., and Cocoșatu, M. (2021). Artificial intelligence-based decision-making algorithms, internet of things sensing networks, and deep learning-assisted smart process management in cyber-physical production systems. *Electronics*, *10* (20), 2497.

[22] Umair, M., Cheema, M. A., Cheema, O., Li, H., and Lu, H. (2021). Impact of COVID-19 on IoT adoption in healthcare, smart homes, smart buildings, smart cities, transportation and industrial IoT. *Sensors*, *21*(11), 3838.

[23] Kalyani, G., and Chaudhari, S. (2022). Data privacy preservation in MAC aware Internet of things with optimized key generation. *Journal of King Saud University-Computer and Information Sciences*, *34*(5), 2062–2071.

[24] Vermesan, O., Eisenhauer, M., Sundmaeker, H., et al. (2022). Internet of things cognitive transformation technology research trends and applications. *Cognitive Hyperconnected Digital Transformation*, 17–95.

[25] Jagatheesaperumal, S. K., Pham, Q. V., Ruby, R., Yang, Z., Xu, C., and Zhang, Z. (2022). Explainable AI over the Internet of Things (IoT): overview, state-of-the-art and future directions. *IEEE Open Journal of the Communications Society*, *3*, 2106–2136.

Chapter 2

Securing the Internet of Things: architectures and designs

2.1 Introduction

The Internet of Things (IoT) represents a seismic shift in how we interact with technology and the world around us. In the IoT paradigm, everyday objects, from household appliances to industrial machinery, are equipped with sensors, connectivity, and computing power, allowing them to communicate and share data over the Internet. This interconnected web of devices forms a dynamic ecosystem that has penetrated virtually every sector, from healthcare and transportation to agriculture and manufacturing [1]. One of the defining characteristics of IoT is its ability to capture vast amounts of data in real time. These data streams provide invaluable insights, enabling us to make informed decisions, automate processes, and enhance efficiency. For example, IoT sensors in agriculture collect data on soil moisture and weather conditions, helping farmers optimize irrigation and crop yield. In healthcare, wearable IoT devices monitor vital signs, enabling remote patient monitoring and early disease detection [2].

The proliferation of IoT devices and applications has been nothing short of staggering. Over the past decade, we have witnessed an exponential increase in the number of IoT devices connected to the Internet. This growth has been driven by a confluence of factors, including advancements in sensor technology, the availability of low-cost connectivity options, and the insatiable demand for innovation [3]. From smart thermostats that learn our heating and cooling preferences to connected cars that can navigate traffic and provide real-time diagnostics, IoT has permeated our daily lives. Industries have also embraced IoT to enhance productivity and reduce operational costs. Smart factories now use IoT-enabled sensors and analytics to monitor machine health, predict maintenance needs, and optimize production schedules.

As the IoT landscape expands, the issue of security becomes paramount. Many IoT devices are designed with a focus on functionality and cost-effectiveness, often at the expense of robust security measures. This vulnerability has not gone unnoticed by cybercriminals and malicious actors who see IoT as a potential goldmine for exploits. Security breaches in the IoT ecosystem can have profound consequences. Compromised devices can be used as entry points into larger networks, leading to data breaches, privacy violations, and even threats to physical safety.

Protecting the integrity of IoT systems is not just about safeguarding data; it is about ensuring the reliability and safety of interconnected devices that increasingly govern our environments [4,5].

This chapter systematically explores the multifaceted world of IoT security and architecture. Our approach is designed to equip readers with a comprehensive understanding of the challenges and requirements inherent in securing the IoT ecosystem. We commence with a deep dive into IoT fundamentals, elucidating its core principles and highlighting the diverse domains where IoT is making a transformative impact. Subsequently, we embark on an insightful journey through various IoT architecture models, including the three-layer, five-layer, six-layer, and seven-layer models, providing a clear understanding of their components and functions [6]. Following this, perform a comparative analysis of these architectural models, empowering readers to make informed decisions when selecting the most appropriate framework for their projects. To address the pressing issue of security, we delve into the intricacies of IoT security challenges, offering insights into scalability, resource constraints, data privacy, authentication, and more. Finally, we outline the security requirements at each layer of the IoT architecture, ensuring readers are well-prepared to mitigate risks effectively. By the end of this chapter, you will possess a solid foundation for comprehending and addressing the multifaceted security challenges within the IoT ecosystem [7].

2.2 IoT fundamentals

The IoT, commonly called IoT, is a transformative technological paradigm that has redefined how we interact with our environment and the digital world. At its core, IoT is the interconnection of everyday objects, devices, and systems to the Internet, enabling them to collect, exchange, and utilize data autonomously. This connectivity allows previously inert objects to become "smart," facilitating enhanced functionality, real-time data analysis, and remote control. IoT encompasses a vast and diverse ecosystem of "things," including anything from household appliances like refrigerators and thermostats to industrial machinery, vehicles, and even wearable devices [8]. The basic principles of the IoT encompass the fundamental concepts that underpin this transformative technological paradigm. These principles collectively define the core tenets that guide IoT systems' design, implementation, and operation, facilitating their integration into diverse industries and applications [9].

Connectivity: At its core, IoT relies on connectivity to establish communication channels among devices, systems, and the Internet. This connectivity enables devices to share data, receive commands, and interact with each other and centralized control platforms. The range of connectivity options varies from traditional Wi-Fi and cellular networks to low-power options like Bluetooth Low Energy (BLE) and low-power wide area network (LPWAN) technologies. These diverse connectivity solutions are chosen based on factors such as range, power consumption, and data transfer rates.

Data acquisition: IoT devices are equipped with various sensors and data collection mechanisms to capture information from their environment. These sensors can include temperature sensors, humidity sensors, motion detectors, cameras, GPS modules, and more. Data acquisition is not limited to environmental data; it can also involve capturing user behavior, machine performance, or other relevant information. This continuous stream of data forms the foundation for IoT's real-time insights and decision-making capabilities [10].

Data analysis: IoT systems often involve real-time or near-real-time data analysis. This analysis can occur at different levels within the IoT ecosystem. Embedded processors and microcontrollers can perform initial data processing and filtering at the device level. Edge computing nodes are another layer where data can be analyzed and filtered closer to the data source, reducing latency and bandwidth usage. Finally, cloud-based platforms are utilized for more extensive data analytics, offering scalability and the ability to derive valuable insights from large datasets [11]. Data analysis in IoT encompasses various techniques, including statistical analysis, machine learning, and predictive modeling, to extract meaningful information and facilitate automated decision-making.

Interoperability: Interoperability is a critical principle in IoT because it ensures that devices and systems from different manufacturers can work together seamlessly. Standardization of communication protocols, data formats, and interfaces is essential to achieving interoperability. Open standards and industry alliances are vital in defining common frameworks for IoT devices. For example, Message Queuing Telemetry Transport (MQTT) and Hypertext Transfer Protocol (HTTP) are commonly used IoT communication protocols that enable interoperability between devices and IoT platforms. Interoperability fosters a cohesive IoT ecosystem where devices can easily share data and collaborate to deliver more advanced services and functionalities [12].

Security: IoT security is paramount due to the sensitive nature of the data involved and the potential for cyber threats. The principle of security encompasses various measures, including authentication, encryption, access control, and secure bootstrapping. Devices must be equipped with secure identity mechanisms to ensure that only authorized entities can access them. Data transmission between devices and backend systems should be encrypted to prevent eavesdropping and tampering. Regular software updates and security patches are crucial to address vulnerabilities. Furthermore, device manufacturers and IoT service providers must adopt best practices in security to safeguard data, devices, and networks from unauthorized access, breaches, and potential attacks [13].

Scalability: IoT deployments can vary significantly in size, from a few devices in a smart home to millions in a smart city infrastructure. Scalability is a foundational principle that accommodates the growth of IoT ecosystems while maintaining performance and reliability. Scalability considerations encompass both hardware and software aspects. Hardware should be designed to support increased device counts, data volumes, and network traffic. On the software side,

IoT solutions must be architected to scale horizontally and vertically to meet the demands of larger deployments. Cloud-based IoT platforms often provide scalability options by automatically provisioning additional resources as needed [14]. **Energy efficiency:** Energy efficiency is crucial in IoT, particularly for devices operating on battery power or in remote locations with limited energy sources. IoT devices are often designed to minimize power consumption, employing techniques such as low-power processors, sleep modes, and optimized data transmission. Additionally, energy harvesting technologies, such as solar panels and kinetic energy converters, can be integrated into IoT devices to extend their operational lifespan without requiring frequent battery replacements [15].

2.3 IoT architecture

In the realm of the IoT, where billions of devices converge to reshape the fabric of our digital and physical existence, architecture plays a pivotal role. IoT architecture serves as the blueprint that defines how devices, networks, and systems are structured, interconnected, and orchestrated to deliver seamless functionality and data flow. IoT architectures serve as the structural blueprints that underpin the deployment of interconnected devices, networks, and data-processing systems. In this section, we embark on an analytical journey into the diverse architectural models forming the IoT landscape's backbone. Through rigorous examination and comparison, we seek to elucidate these architectures' distinct characteristics, functionalities, and applicability. Up on interpreting the intricacies of IoT architectures, we lay the groundwork for informed decision-making in IoT deployments, ensuring not only functionality but also resilience, scalability, and security within this dynamic ecosystem.

2.3.1 Three-layer IoT architecture

Within the domain of the IoT, the three-layer architecture represents a foundational structural model that orchestrates the interplay of devices, networks, and applications. This architectural framework, grounded in rigorous design principles, delineates the core functionalities of an IoT ecosystem (refer to Figure 2.1). At its essence, the three-layer IoT architecture exhibits a hierarchical structure comprising three primary layers: the perception network, and application layers [16].

2.3.1.1 Perception layer (sensing layer)

The perception layer, constituting the foundational tier of the three-layer IoT architecture, serves as the interface between the physical world and the digital realm. This section meticulously examines the intricacies of the perception layer, elucidating its paramount role in data acquisition and the transformation of real-world phenomena into digital information comprehensible to the broader IoT system. At the core of the perception layer lies an array of sensors and actuators, meticulously designed to capture and respond to environmental stimuli. Sensors

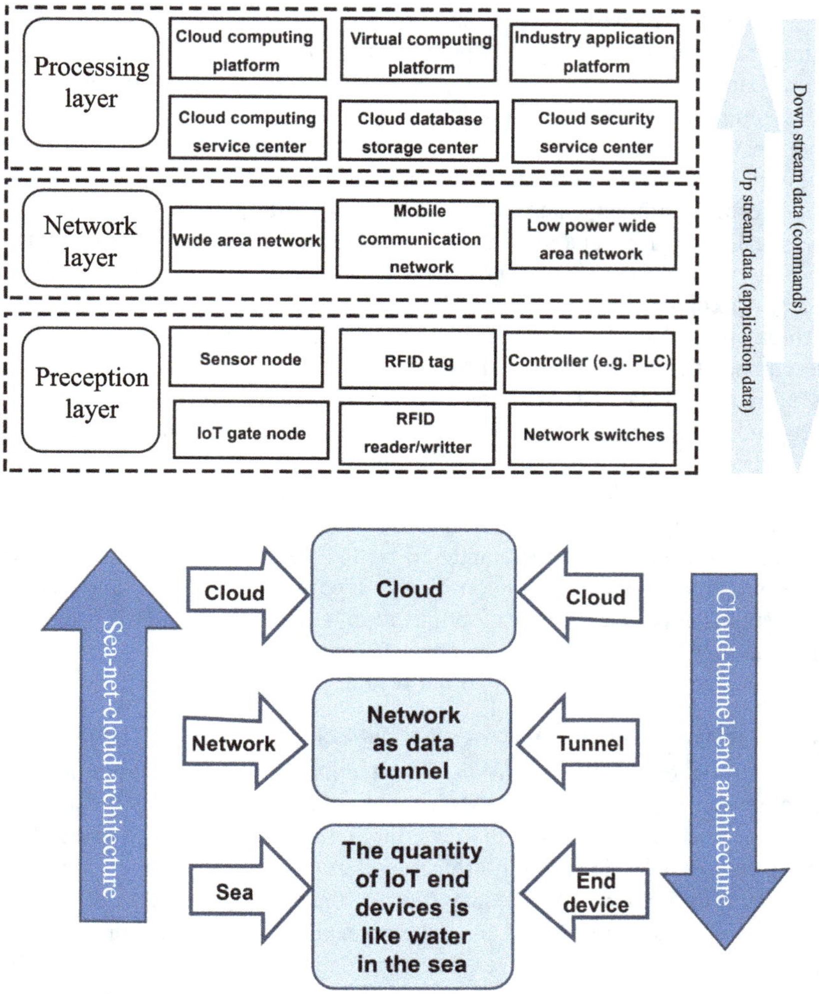

Figure 2.1 Three-layer IoT architecture

are the sensory organs of the IoT ecosystem, ranging from commonplace temperature and humidity sensors to more sophisticated devices such as accelerometers and image sensors. These sensors translate physical attributes into electrical signals, initiating the first step in the data acquisition process. Simultaneously, actuators, the responsive counterparts to sensors, execute actions based on the information received, enabling real-time control and interaction with the physical environment [17].

The primary function of the perception layer is to collect data emanating from the diverse array of sensors embedded within it. This data, representative of the nuances of the environment, undergoes a meticulous encoding process to ensure compatibility with subsequent stages of the IoT architecture. Common encoding formats include JavaScript Object Notation (JSON) and eXtensible Markup Language (XML), fostering a standardized data representation for seamless processing and transmission. Given the time-sensitive nature of many IoT applications, the perception layer often incorporates elements of real-time processing. Microcontrollers and edge computing devices situated within this layer perform preliminary data processing, facilitating rapid decision-making at the device level. This capability is particularly crucial in scenarios demanding immediate responses, such as emergency notifications, autonomous vehicle control, or predictive maintenance [18].

An emerging facet of the perception layer is the integration of edge analytics, where computational capabilities are pushed closer to the data source. Edge analytics allows for localized data processing, reducing the burden on centralized systems and minimizing latency. This not only enhances efficiency but also supports applications requiring real-time insights, such as anomaly detection in industrial settings or adaptive traffic management in smart cities. An inherent challenge in designing the perception layer lies in balancing power efficiency with the resource constraints often encountered by IoT devices. Many sensors operate within environments where continuous power supply may be impractical. As such, the perception layer necessitates optimization strategies, including low-power design, intermittent sensing, and energy-efficient communication protocols, to ensure prolonged device operation within resource limitations [3].

2.3.1.2 Network layer (intermediate layer)

The network layer, positioned as the intermediary tier in the three-layer IoT architecture, emerges as a critical data transportation and management orchestrator. This section rigorously dissects the intricacies of the network layer, unveiling its pivotal role in facilitating seamless communication, data aggregation, and routing within the IoT ecosystem. At the crux of the network layer lies the facilitation of data movement from the perception layer to the upper echelons of the architecture. This involves intricate data routing mechanisms, where information collected by sensors is efficiently directed toward its intended destination. Simultaneously, the network layer assumes the responsibility of aggregating data streams from multiple devices, optimizing bandwidth utilization and minimizing latency. This process is instrumental in ensuring that the vast volumes of data generated within the perception layer are streamlined for further analysis and utilization [19].

A cornerstone of the network layer is the deployment of robust communication protocols that govern the interaction between devices. The selection and implementation of these protocols profoundly influence the efficiency, security, and scalability of IoT deployments. The network layer thus serves as a nexus where the diverse languages of interconnected devices converge and harmonize. The network layer involves diverse arrays of communication protocols that govern data exchange, shaping the efficiency, security, and interoperability of the IoT ecosystem [8–10].

Internet protocol (IP): Fundamental to the fabric of the Internet, the IP is a cornerstone of IoT communication within the network layer. IPv4 and the increasingly adopted IPv6 provide a standardized addressing system, enabling devices to locate and communicate with each other across the vast expanse of the Internet. IP serves as the foundational protocol upon which many other IoT-specific communication protocols are built, forming an essential layer for global connectivity.

Message queuing telemetry transport (MQTT): In scenarios where low bandwidth and intermittent connectivity are prevalent, MQTT emerges as a lightweight and efficient communication protocol within the network layer. Following a publish-subscribe model, MQTT facilitates communication between devices through a broker, allowing for real-time data transmission. Its adaptability to constrained environments and support for quality of service (QoS) levels make it a preferred choice for IoT deployments, particularly in scenarios where resource efficiency is paramount.

Constrained application protocol (CoAP): Tailored for resource-constrained devices in the IoT ecosystem, CoAP operates within the network layer to enable efficient communication. Employing a client-server model, CoAP integrates seamlessly with HTTP, providing a RESTful communication paradigm. Its simplicity, low overhead, and support for user datagram protocol (UDP) make it an apt choice for applications with constrained resources and intermittent connectivity, such as in smart cities and industrial IoT.

Advanced message queuing protocol (AMQP): Designed to address the challenges of interoperability and data exchange in complex IoT environments, AMQP operates at the network layer to enable efficient and secure messaging between devices. Its support for diverse message patterns, including publish-subscribe and point-to-point, makes it versatile in scenarios where heterogeneity and scalability are crucial, such as in industrial automation and smart grids.

Coexistence of protocols: The network layer often accommodates a coexistence of communication protocols within IoT deployments. This flexibility allows devices with varying capabilities and requirements to interact seamlessly. For instance, while resource-efficient protocols like MQTT and CoAP may be suitable for constrained devices, more robust protocols like HTTP or AMQP may be employed in scenarios demanding higher reliability and advanced features.

Security considerations: Ensuring secure communication is imperative within the network layer, and communication protocols play a pivotal role in achieving this goal. Many protocols integrate security features such as encryption, authentication, and secure channels to safeguard data during transmission. Selecting a communication protocol within the network layer involves carefully considering these security measures to mitigate potential threats and vulnerabilities.

Protocol interoperability: Interoperability is a key consideration in the network layer, where devices from diverse manufacturers must seamlessly communicate. Protocols prioritizing interoperability facilitate a cohesive IoT ecosystem, allowing devices with different communication protocols to understand and exchange data effectively. Standardization bodies and industry alliances play a crucial role in developing and promoting interoperable communication standards within the network layer.

Data distribution service (DDS): Data distribution service, often referred to as DDS, is an advanced communication protocol designed to address the requirements of real-time and mission-critical IoT applications. DDS operates within the network layer and follows a publish-subscribe model, allowing devices to efficiently share and consume data. With features such as QoS controls, DDS is adept at ensuring timely and reliable data delivery, making it well suited for applications like industrial automation, healthcare, and autonomous systems.

Extensible messaging and presence protocol (XMPP): Originally developed for instant messaging, XMPP has found its application in IoT communication within the network layer. XMPP operates on an open standard, enabling real-time communication and collaboration among devices. Its extensibility and support for a wide range of applications make it suitable for scenarios where dynamic, near-instantaneous interactions are crucial, such as in smart homes and wearable devices.

Lightweight machine-to-machine (LwM2M): Specifically crafted for machine-to-machine (M2M) communication within the network layer, LwM2M is an advanced protocol that addresses the unique challenges of IoT deployments. LwM2M is designed for resource-constrained devices and focuses on efficient communication and device management. With firmware updates, device registration, and efficient data reporting capabilities, LwM2M is well suited for smart agriculture, logistics, and remote monitoring applications.

Advanced security protocols: As security remains a paramount concern in IoT communication, advanced security protocols within the network layer play a pivotal role. Protocols like datagram transport layer security (DTLS) and transport layer security (TLS) ensure secure communication by encrypting data and authenticating devices. Their incorporation is crucial in safeguarding sensitive information exchanged between devices in applications like healthcare, finance, and smart cities.

Time-sensitive networking (TSN): In applications demanding precise and synchronized communication, TSN protocols operate within the network layer to ensure deterministic data delivery. TSN enhances standard Ethernet networks to provide low-latency and jitter, making it suitable for applications like industrial control systems, where real-time synchronization is critical.

Advanced QoS protocols: To cater to the varied QoS requirements of IoT applications, advanced QoS protocols within the network layer offer enhanced control over factors like reliability, latency, and bandwidth.

Protocols like differentiated services (DiffServ) and resource reservation protocol (RSVP) enable fine-grained control over network resources, ensuring that critical applications receive the necessary prioritization and performance.

The network layer, traditionally constituted by wide area networks (WANs), establishes the backbone for long-distance data transmission. Wired Internet connections, forming a part of this layer, have played a pivotal role in connecting disparate regions. The advent of wireless WANs has also introduced a paradigm shift, enabling more flexible and mobile communication over vast geographical areas [20]. These networks, whether wired or wireless networks, provide the foundational infrastructure for the seamless flow of data within the IoT ecosystem. Recent advancements in wireless technology have given rise to a specialized category within the network layer known as LPWANs. Engineered to cater specifically to the low-power and long-range requirements of IoT applications, LPWANs have witnessed notable developments. Networks such as narrowband IoT (NB-IoT), long range (LoRa), and SigFox have emerged as prominent players in this domain. Their deployment is characterized by commercialized operations, offering extended coverage, energy efficiency, and cost-effectiveness. LPWANs have become instrumental in connecting a myriad of IoT devices spread across diverse environments.

The network layer embodies a diversity of network architectures, often operating collaboratively to ensure robust and reliable data transmission. A notable example is the seamless integration of multiple networks within the same layer. Data collected from IoT devices may embark on a journey traversing different network infrastructures. Initially transmitted through mobile communication networks for their mobility and accessibility, the data may subsequently transition to the wired Internet for broader connectivity and reliable data transport. This multi-network approach optimizes data transmission based on specific network characteristics, ensuring efficiency and adaptability to varying communication needs. The advent of LPWANs, tailored specifically for IoT applications, signifies a transformative shift in the network layer. These networks leverage technologies like NB-IoT, LoRa, and SigFox to provide not only extended coverage but also address the unique challenges posed by IoT devices, such as low power consumption and long-range communication. The commercialization of these wireless WANs underscores their viability and scalability for widespread IoT deployment, spanning industries from agriculture and healthcare to smart cities and industrial automation [18–20]. As the network layer accommodates diverse network architectures and technologies, security and scalability considerations become paramount. Ensuring the confidentiality and integrity of data during its journey across networks is essential. Additionally, scalability considerations address the growing demand for accommodating an increasing number of IoT devices within the network, necessitating robust infrastructure and optimized protocols.

2.3.1.3 Application layer (business layer)

The pinnacle tier of the three-layer IoT architecture, the application layer, represents the culmination of data's transformative journey. This section meticulously dissects the intricacies of the application layer, where data harvested from the perception layer and traversing the network layer is refined into actionable insights and value, fostering innovation and efficiency. The reason for being of the application layer lies in its ability to harness the raw data acquired from the perception layer and refined in the network layer, converting it into actionable insights and informed decision-making. Here, data is processed, analyzed, and interpreted to extract meaningful information that serves as the foundation for applications and services. This layer acts as the catalyst for translating data into real-world value, driving the efficiency and effectiveness of IoT deployments. At the core of the application layer lies the development and deployment of applications that cater to specific use cases and end-user requirements. Application developers leverage the processed data to create solutions that range from smart home automation and predictive maintenance to healthcare monitoring and industrial optimization. These applications, whether residing on edge devices, cloud platforms, or distributed across the network, play a pivotal role in realizing the potential of IoT data in diverse domains.

The application layer interfaces directly with end-users through user interfaces and interactive systems. These interfaces, whether graphical or through other sensory modalities, allow users to interact with and derive value from the IoT ecosystem. From mobile applications that allow users to monitor and control smart devices to web-based dashboards that offer comprehensive data visualization, the user interfaces within the application layer are designed to enhance user experience and foster intuitive engagement. Embedded within the application layer is the business logic that defines how data is processed, interpreted, and utilized to achieve specific objectives. This logic encompasses algorithms, rules, and workflows that govern the behavior of applications, facilitating tasks such as automation, optimization, and anomaly detection. Additionally, services offered within this layer extend beyond individual applications, providing a broader ecosystem where businesses can derive value and insights from the aggregated data [5]. The application layer often interfaces with external systems, allowing for integration with existing enterprise software, cloud services, and other third-party applications. This integration facilitates a holistic approach to data utilization, enabling organizations to leverage IoT insights within their broader operational frameworks. Such integration enhances the scalability and interoperability of IoT applications, fostering a seamless exchange of information across diverse systems. In scenarios demanding real-time responsiveness, the application layer incorporates mechanisms to ensure timely and adaptive actions. Whether it be instantaneously adjusting environmental conditions in a smart building or triggering preventive maintenance in an industrial setting, the real-time capabilities of this layer are essential for applications that require immediate and dynamic responses to changing conditions. Security considerations within the application layer are paramount to safeguard sensitive data and ensure the integrity of applications and services. Authentication mechanisms, encryption protocols, and access controls are implemented to mitigate potential threats and unauthorized access. As applications within this layer often handle critical data, robust security measures are essential to instill confidence in the integrity and confidentiality of the IoT ecosystem [6].

2.3.2 Four-layer IoT architecture

The evolution of IoT architectures has led to the emergence of more intricate models to address the diverse needs of applications and industries. The four-layer IoT architecture represents a refined and extended framework, introducing additional layers to enhance functionality and scalability. This section scrutinizes the intricacies of the four-layer IoT architecture, shedding light on its composition and the nuanced roles each layer plays in the overall system (refer to Figure 2.2).

2.3.2.1 Perception layer (sensing layer)

The foundational tier of the four-layer IoT architecture remains the perception layer, akin to its counterpart in the three-layer model. This layer houses sensors and actuators interacting with the physical environment, capturing data and

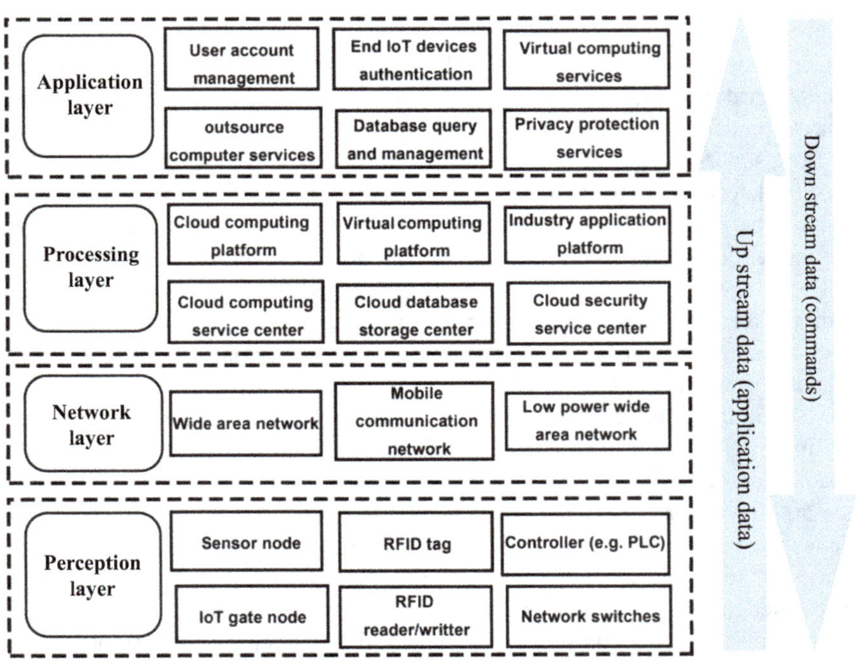

Figure 2.2 Four-layer IoT architecture

initiating actions. The key enhancement in the four-layer model lies in the increased sophistication and diversity of sensors, enabling more nuanced data collection. The perception layer is the gateway for real-world data to enter the IoT ecosystem.

2.3.2.2 Network layer (transport layer)

The network layer, analogous to the three-layer architecture, is responsible for data transportation, aggregation, and routing. In the four-layer model, this layer may exhibit heightened capabilities to accommodate the increasing volume and diversity of data generated by an expanding array of sensors. It leverages traditional WANs, wireless WANs, and LPWANs to establish robust connectivity. Additionally, the network layer facilitates seamless communication between devices, ensuring efficient data flow within the IoT ecosystem.

2.3.2.3 Processing layer

A pivotal addition to the four-layer IoT architecture, the processing layer introduces a sophisticated stratum dedicated to advanced data processing and analytics. The processing layer serves as the crucible where raw data from the perception layer undergoes intricate processing methodologies. Beyond simple aggregation and routing, this layer engages in advanced data processing techniques, including filtering, normalization, and transformation. These processes ensure that data entering the layer is refined, standardized, and prepared for more intricate analyses, laying the groundwork for a comprehensive understanding of the information at hand.

Central to the processing layer is the integration of analytical algorithms and machine-learning models. These computational mechanisms decipher complex patterns, trends, and anomalies within the data. Machine-learning algorithms, ranging from supervised learning for predictive analytics to unsupervised learning for anomaly detection, contribute to the layer's capability to derive meaningful insights. The iterative learning process of these algorithms enables the system to adapt and evolve in response to changing data patterns over time.

The processing layer accommodates both real-time and batch processing paradigms to cater to diverse IoT application requirements. Real-time processing allows instantaneous analysis and response to data streams, crucial in scenarios demanding immediate actions, such as healthcare monitoring or autonomous systems. Simultaneously, batch processing facilitates the analysis of large datasets over time, contributing to historical trend analysis, optimization, and long-term decision-making. Beyond processing data generated within the IoT ecosystem, the processing layer seamlessly integrates with external data sources. This integration enriches the analytics process by incorporating contextual information from diverse origins. For instance, weather data, social media feeds, or market trends may be incorporated, providing a more holistic understanding of the factors influencing the IoT system. Such integration amplifies the depth and accuracy of insights derived within the processing layer [15–17].

An emerging facet of the processing layer is its ability to leverage edge computing resources. By pushing computational capabilities closer to the data source, edge computing reduces latency, conserves bandwidth, and enhances real-time responsiveness. This is particularly valuable in applications where immediate actions based on locally processed data are essential, such as in industrial automation or autonomous vehicles. As IoT deployments scale, the processing layer exhibits robust scalability mechanisms. Efficient resource management ensures optimal utilization of computational resources, whether distributed across edge devices, cloud platforms, or a combination of both. Scalability considerations encompass load balancing, parallel processing, and adaptive resource allocation, guaranteeing the processing layer's ability to handle the increasing volume and complexity of data generated by a burgeoning number of IoT devices.

Security within the processing layer is paramount, given its role in handling and analyzing sensitive data. Encryption, access controls, and secure communication protocols are implemented to fortify the layer against potential threats. Additionally, anomaly detection algorithms contribute to identifying irregular patterns that may indicate security breaches, bolstering the overall security posture of the IoT ecosystem [1].

2.3.2.4 Application layer

The apex of the four-layer IoT architecture, the application layer, remains akin to its counterpart in the three-layer model but now interacts directly with the processing layer. Applications within this layer leverage the enriched and refined data to provide sophisticated services, automate processes, and drive innovation. The expanded capabilities of the four-layer model enable more complex applications, catering to diverse domains such as healthcare, smart cities, and industrial automation. User interfaces, business logic, and integration with external systems are all orchestrated within the application layer [2].

2.3.3 Five-layer IoT architecture

Analogous to its counterparts in the three- and four-layer models, the perception layer remains the foundational tier of the five-layer IoT architecture. Here, sensors and actuators interact with the physical environment, collecting an extensive array of data (refer to Figure 2.3). The unique strength of the perception layer within the five-layer model lies in the enhanced diversity and precision of sensors, facilitating more nuanced data acquisition for a broader spectrum of applications [3].

2.3.3.1 Perception layer (sensing layer)

Analogous to its counterparts in the three- and four-layer models, the perception layer remains the foundational tier of the five-layer IoT architecture. Here, sensors and actuators interact with the physical environment, collecting an extensive array of data. The unique strength of the perception layer within the five-layer model lies in the enhanced diversity and precision of sensors, facilitating more nuanced data acquisition for a broader spectrum of applications.

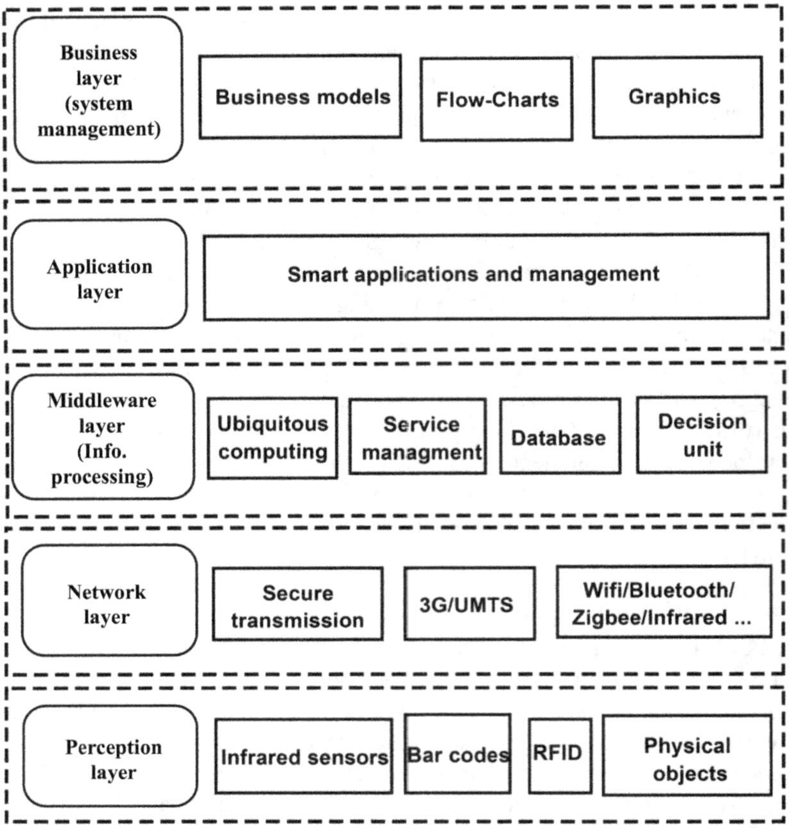

Figure 2.3 Five-layer IoT architecture

2.3.3.2 Network layer (transport layer)

Like previous models, the network layer retains its pivotal role in data transportation, aggregation, and routing. However, within the five-layer architecture, this layer may exhibit heightened capabilities to accommodate the increasingly complex data generated by an expansive array of sensors. Traditional WANs, wireless WANs, and LPWANs contribute to robust connectivity, ensuring efficient data flow within the IoT ecosystem.

2.3.3.3 Middleware layer

A distinctive addition to the five-layer IoT architecture, the middleware layer is an intermediary, orchestrating seamless communication and data management between the network and application layers. This section scrutinizes the nuanced intricacies of the middleware layer, elucidating its critical functions and the transformative impact it brings to the orchestration of the IoT ecosystem.

At the heart of the middleware layer lies its ability to harmonize and integrate diverse data streams originating from the perception layer. This layer serves as a convergence point for heterogeneous data formats, protocols, and structures, facilitating a cohesive flow of information within the IoT ecosystem. Through advanced data integration and aggregation mechanisms, the middleware layer ensures that data from various sources is efficiently processed and ready for analysis in higher layers. A fundamental responsibility of the middleware layer is to manage the multitude of devices populating the IoT landscape. Device management encompasses tasks such as device registration, authentication, and configuration. Furthermore, the middleware layer governs communication protocols, ensuring standardized and secure interactions between devices. By providing a centralized hub for device management, this layer contributes to the efficient orchestration of the IoT infrastructure [7].

Given the diversity of devices and communication protocols in the IoT ecosystem, the middleware layer excels in protocol translation and interoperability. It acts as a translator, enabling devices with different communication languages to communicate seamlessly. This functionality is crucial for fostering interoperability, allowing devices from different manufacturers and with varying communication protocols to work cohesively within the IoT framework. Within the middleware layer, data undergoes transformation and enrichment processes to enhance its quality and relevance. This involves converting data into standardized formats, enriching it with additional context, and ensuring consistency across the IoT ecosystem. These processes lay the groundwork for more meaningful analyses and insights in the higher layers of the architecture, particularly in the processing and application layers [1].

As the IoT ecosystem expands, the middleware layer showcases robust scalability mechanisms to accommodate a growing number of devices and data streams. Efficient resource optimization ensures that computational and storage resources are allocated judiciously, preventing bottlenecks and ensuring the smooth functioning of the entire system. This scalability is vital for adapting to the dynamic nature of IoT deployments and varying workloads. Given its central role in managing device interactions and data flow, the middleware layer is equipped with robust security measures. Authentication protocols, encryption mechanisms, and access controls are implemented to safeguard against unauthorized access and data breaches. Security within the middleware layer is crucial for maintaining the integrity and confidentiality of data as it traverses between devices and applications. The middleware layer often incorporates real-time processing capabilities to address time-sensitive requirements in the IoT ecosystem. Real-time analytics and response mechanisms enable immediate actions based on incoming data, enhancing the overall responsiveness of the system. This is particularly essential in applications where timely decisions and actions are critical, such as in smart cities or healthcare monitoring [11–14].

2.3.3.4 Processing layer

Retaining its role from the four-layer model, the processing layer within the five-layer architecture engages in advanced data processing and analytics. Here, data undergoes complex computations, machine-learning algorithms, and statistical analyses. The processing layer extracts valuable insights, identifies patterns, and prepares refined data for consumption by higher-level applications, enhancing the system's analytical capabilities.

2.3.3.5 Application layer

The apex of the five-layer IoT architecture, the application layer, remains similar to its counterparts in previous models but now interacts directly with both the middleware and processing layers. Applications within this layer leverage the enriched and refined data to provide sophisticated services, automate processes, and drive innovation. The expanded capabilities of the five-layer model enable more complex applications, catering to diverse domains such as healthcare, smart cities, and industrial automation.

2.3.4 Seven-layer IoT architecture

The seven-layer IoT architecture represents a comprehensive framework designed to address the intricacies of modern IoT applications. This section scrutinizes each layer, elucidating the specific functions and interactions that contribute to the seamless operation of the IoT ecosystem (refer to Figure 2.4).

2.3.4.1 Physical layer (perception layer)

At the foundation of the seven-layer IoT architecture lies the perception layer. This layer comprises sensors and actuators that interact with the physical environment, capturing data and initiating actions. The sensors within this layer are diversified and sophisticated, catering to a wide array of applications. From temperature and humidity sensors in smart homes to industrial sensors monitoring machinery performance, the perception layer serves as the entry point for real-world data into the IoT ecosystem.

2.3.4.2 Connectivity layer

The connectivity/transport layer serves as the conduit for data transmission between devices within the IoT ecosystem. This layer incorporates traditional wide-area networks (WANs) and wireless WANs, ensuring reliable and efficient communication over long distances. As data is collected from the perception layer, the connectivity/transport layer facilitates its seamless transfer to subsequent layers for further processing and analysis.

2.3.4.3 Edge/fog computing layer

The edge/fog computing layer stands as a pivotal stratum within the seven-layer IoT architecture, designed to cater to the speed, security, and scalability

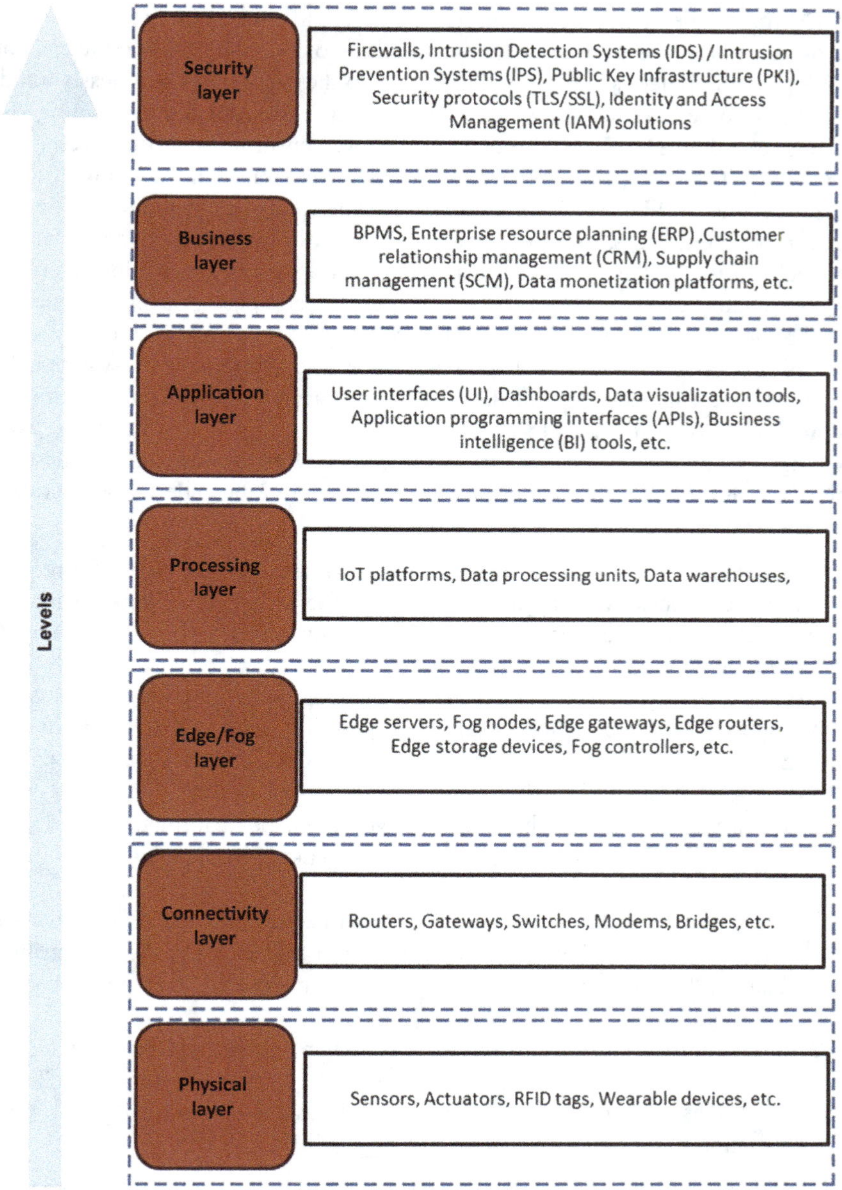

Figure 2.4 Seven-layer IoT architecture

requirements ushered in by the advent of the fifth-generation (5G) mobile network. This section meticulously dissects the functionalities and significance of the edge/fog computing layer, elucidating its role in processing and managing data as close to the source as possible [17].

Central to the edge/fog computing layer's significance is its role in aligning IoT systems with the transformative capabilities of 5G. The 5G mobile network promises unparalleled speeds, lower latency, and the capacity to seamlessly handle a multitude of connected devices. By leveraging edge and fog computing, IoT systems can harness the full potential of 5G, ensuring real-time responses, enhanced performance, and a level of connectivity that transcends the limitations of previous standards like fourth generation [11]. The crux of edge and fog computing lies in processing and storing information as close to its sources as possible. This approach mitigates the need to transport all data to centralized cloud services, saving valuable time and resources. The edge/fog computing layer becomes the locus for analyzing and transforming high volumes of real-time data locally, at the edge of the networks. This localized data processing capability is instrumental in reducing system latency, paving the way for real-time responses, and optimizing overall system performance [15].

Edge computing unfolds on gateways, local servers, or other edge nodes strategically scattered across the network. At this level, data undergoes a series of crucial functions:

- *Evaluation for further processing*: Data is evaluated to determine if it requires additional processing at higher levels within the architecture. This ensures that only relevant and actionable information is transmitted, optimizing bandwidth and computational resources.
- *Formatting for further processing*: The edge/fog computing layer formats data for subsequent processing, ensuring compatibility with higher-level analytics and applications. This step streamlines the flow of information and prepares it for more sophisticated analyses.
- *Decoding*: In scenarios where data arrives in encoded formats, the edge/fog computing layer decodes the information, making it accessible for further interpretation and processing.
- *Filtering*: Unnecessary or redundant data is filtered out at the edge, preventing the transmission of superfluous information to higher layers. This contributes to bandwidth efficiency and reduces the computational load on subsequent processing stages.
- *Redirection to additional destinations*: Based on specific criteria or predefined rules, data may be redirected to additional destinations within the IoT ecosystem. This dynamic routing capability allows for optimized data flow and distribution across the network.

The localized processing capabilities of the edge/fog computing layer lead to reduced system latency, enabling real-time responses in dynamic IoT environments. By minimizing the time it takes for data to travel to centralized cloud services, the edge/fog computing layer significantly enhances overall system performance. This is particularly crucial in applications where immediate actions based on real-time data are imperative, such as autonomous vehicles, healthcare monitoring, or industrial automation [9].

2.3.4.4 Processing layer

Situated above the edge/fog computing layer, the processing layer engages in sophisticated data processing and analytics. Complex computations, machine-learning algorithms, and statistical analyses are employed to extract meaningful insights from the data. Positioned strategically within the seven-layer IoT architecture, the processing layer emerges as the nexus where data from preceding layers converges, undergoes intricate transformations, and is prepared for consumption by higher-level applications. This section meticulously dissects the functionalities and critical stages of the processing layer, elucidating its role in accumulating, storing, and abstracting data to unveil actionable insights.

At the core of the processing layer lies the data accumulation stage, a vital component that orchestrates the initial handling of real-time data. This stage acts as a transit hub, seamlessly capturing real-time data via APIs and placing it at rest to meet the varied requirements of both real-time and non-real-time applications. Its key functions include the following:

- **Transit hub for data:** Serving as a transit hub, the data accumulation stage bridges the gap between event-based data generation and query-based data consumption. It ensures a smooth transition from the immediacy of data generation to the structured storage that aligns with the needs of diverse applications.
- **Relevance determination:** The stage plays a crucial role in determining the relevance of data to business requirements. Through sophisticated algorithms, it evaluates incoming data streams, deciding where each piece of data should be placed for optimal storage efficiency. This includes storing data in diverse storage solutions, ranging from data lakes capable of handling unstructured data like images and video streams to event stores and telemetry databases.
- **Efficient data storage:** The ultimate goal of the data accumulation stage is to sift through vast and diverse data, storing it in the most efficient way possible. Whether dealing with images, video streams, or telemetry data, this stage ensures that data is organized and accessible for subsequent processing stages.

Following the data accumulation stage, the data abstraction stage takes center stage, finalizing data preparation for consumption by consumer applications. This stage involves a multifaceted process:

- **Combining data from different sources:** The data abstraction stage excels in combining data from diverse sources, not limited to IoT alone. It reconciles information from enterprise resource management (ERM), enterprise resource planning (ERP), customer relationship management (CRM) systems, and other non-IoT sources, creating a unified data landscape.
- **Reconciling multiple data formats:** Given the heterogeneity of data sources, the stage reconciles various data formats, ensuring compatibility and consistency. This reconciliation process is pivotal in creating a seamless and standardized dataset that applications can readily consume.

- **Aggregating data for insights:** Aggregation is a vital facet of the data abstraction stage, where data is aggregated in one place or made universally accessible through data virtualization. This step ensures that insights derived from the combined dataset are coherent, providing a holistic view regardless of the original data's location. Furthermore, the reformatted data is prepared for transmission to the physical level, ensuring compatibility with devices.

A distinctive feature of the data accumulation and abstraction stages is their role in veiling hardware intricacies. By abstracting away details of the underlying hardware, these stages enhance the interoperability of smart devices. This abstraction allows software developers to concentrate on solving specific business tasks without delving into the intricacies of devices from different vendors. The focus shifts from hardware specifications to the creation of innovative applications and solutions, fostering a dynamic and collaborative IoT ecosystem.

2.3.4.5 Application layer

The application layer remains the apex where data is transformed into tangible outcomes. Applications within this layer leverage the insights generated by the processing layer to provide services, automate processes, and drive innovation. User interfaces, business logic, and integration with external systems are orchestrated within the application layer, serving as the interface between the IoT ecosystem and end-users or other systems.

2.3.4.6 Business layer

Elevated above the intricacies of application-specific layers, the business layer emerges as the orchestrator of value creation within the seven-layer IoT architecture. The linchpin of the business layer lies in its ability to derive tangible value from the information generated in preceding layers. When transformed into meaningful insights, data, attains significance only when it catalyzes problem-solving solutions and aligns with broader business goals. The business layer acts as the nexus where raw data evolves into actionable intelligence, initiating collaborative efforts among stakeholders to introduce new processes that enhance productivity and drive innovation. A hallmark of the business layer is its role in fostering collaboration among diverse stakeholders. The information flowing through the IoT architecture often involves multiple actors, each contributing unique perspectives and expertise. The business layer becomes the stage where stakeholders converge, leveraging insights from the lower layers to collectively strategize and make informed decisions. This collaborative approach ensures that the full spectrum of expertise is harnessed to benefit the entire ecosystem [11–15].

Recognizing the complexity of decision-making in the IoT landscape, the business layer acknowledges that decisions typically involve more than one person working with multiple software solutions. This layer is designed to facilitate decision-making processes that are inherently collaborative and multifaceted. By accommodating the interplay of diverse perspectives and integrating insights from various sources, the business layer ensures that decisions are comprehensive and aligned with the overarching goals of the enterprise. The business layer introduces new processes that are informed by the insights gleaned from the lower layers. These processes are not static; instead, they are dynamic and responsive to the evolving nature of IoT data. As new information emerges, the business layer catalyzes the introduction of innovative processes that enhance productivity, streamline operations, and contribute to the efficiency of the entire business ecosystem. This adaptability is crucial in the ever-changing landscape of IoT applications.

Given the diversity of data sources and the multifaceted nature of decision-making, the business layer seamlessly integrates with multiple software solutions. This integration ensures that data from various sources is harmonized, providing a unified view for decision-makers. Whether leveraging analytics platforms, ERP systems, or industry-specific applications, the business layer orchestrates a cohesive integration that supports holistic decision-making. At its core, the business layer is driven by a commitment to achieving holistic business objectives. Beyond the realm of individual applications, this layer aligns IoT insights with broader organizational strategies. Whether the goal is to optimize operational efficiency, enhance customer experiences, or drive innovation, the business layer ensures that the insights derived from the lower layers are strategically channeled to contribute to the overarching success of the enterprise.

2.3.4.7 Security layer

Undoubtedly, a paramount consideration is the security layer in the intricate landscape of the seven-layer IoT architecture. Encompassing all layers discussed above, this section meticulously examines the comprehensive role of the security layer, acknowledging its broad scope and the critical imperative it holds in safeguarding the entire IoT ecosystem. At the heart of the security layer is the recognition that security concerns permeate every facet of the IoT landscape. Whether in the perception layer with sensors and actuators, the connectivity/transport layer ensuring seamless communication, the edge/fog computing layer orchestrating real-time processing, or the business layer aligning with overarching goals—security is omnipresent. Recognizing this ubiquity, the security layer stands as the guardian, ensuring the data confidentiality, integrity, and availability across all layers. IoT security is not a one-size-fits-all proposition but demands a nuanced and layered approach. The security layer is designed to implement a spectrum of security measures, acknowledging the diversity of devices, data types, and communication protocols within the IoT ecosystem. From encryption mechanisms safeguarding data in transit to robust access controls governing device interactions, the security layer orchestrates a holistic array of measures tailored to the intricacies of each layer [21].

Central to IoT security is the rigorous authentication and authorization of devices. The security layer ensures that only legitimate and authorized devices gain access to the IoT ecosystem. This involves the implementation of robust authentication protocols, secure key management, and access controls that delineate the permissions granted to each device. The meticulous verification of device identity mitigates the risk of unauthorized access and potential malicious activities. Given the sensitive nature of IoT data, especially in applications like healthcare, smart homes, or industrial control, the security layer prioritizes data encryption and confidentiality. End-to-end encryption mechanisms are implemented to safeguard data in transit, preventing eavesdropping and unauthorized access. Moreover, the security layer ensures that data at rest is stored in encrypted formats, fortifying against potential breaches and data compromises.

IoT ecosystems are susceptible to an evolving landscape of cyber threats. The security layer proactively addresses these challenges through advanced threat detection mechanisms and intrusion prevention systems. Machine-learning algorithms and anomaly detection techniques are employed to identify abnormal patterns or activities within the network, enabling swift responses to potential security incidents. Security is an ongoing process, and the security layer incorporates continuous monitoring and auditing to assess the health of the IoT ecosystem. The security layer identifies deviations from normal behavior through real-time monitoring of device activities, data flows, and network communications. Regular security configurations and policy audits ensure that the IoT infrastructure remains resilient against emerging threats and vulnerabilities. In an era of heightened awareness regarding data privacy and regulatory compliance, the security layer takes a proactive stance. It ensures that IoT deployments adhere to relevant privacy regulations and industry standards. This involves transparent data handling practices, informed consent mechanisms, and incorporating privacy-by-design principles. The security layer serves as a bulwark, ensuring that IoT applications meet the ethical and legal expectations regarding user privacy and data protection. Recognizing that human factors play a pivotal role in security, the security layer extends its reach to include educational and awareness initiatives. From end-users to developers, stakeholders, are provided with resources and information to foster a security-conscious mindset. Training programs, best practices dissemination, and regular updates contribute to a security culture that permeates every level of the IoT ecosystem [19–21].

2.4 Security challenges in IoT

Navigating the vast and interconnected landscape of the IoT is not without its challenges, particularly when it comes to security. This section delves into the multifaceted aspects that render security in the IoT ecosystem a challenging and complex endeavor. Understanding these challenges is essential for developing robust security measures that safeguard the integrity and trustworthiness of IoT deployments.

One of the primary challenges in IoT security stems from the sheer scale and proliferation of devices within the ecosystem. As IoT devices continue to multiply exponentially, ranging from sensors and actuators to connected appliances and industrial machines, managing the security of each device becomes a monumental task. The diversity in device types, manufacturers, and communication protocols adds layers of complexity, making it challenging to enforce uniform security standards across the entire spectrum of IoT deployments. In addition, many IoT devices operate under stringent resource constraints, including limited processing power, memory, and energy. These resource limitations pose a significant challenge to implementing robust security measures. Traditional security protocols that may be feasible for more powerful devices are often impractical for resource-constrained IoT devices. Striking a balance between security and resource efficiency becomes a delicate undertaking, requiring innovative solutions tailored to the constraints of each device category [7–9]. The heterogeneity of the IoT ecosystem introduces another layer of complexity in terms of security. IoT applications span diverse domains, from smart homes and healthcare to industrial automation and smart cities. Each domain comes with its unique set of requirements, standards, and regulatory frameworks. Securing this heterogeneous landscape necessitates flexible and adaptive security measures that can accommodate the diverse needs and characteristics of each IoT application.

The vast amounts of data generated by IoT devices, often of a sensitive or personal nature, raise significant concerns regarding data privacy and confidentiality. Ensuring that this data is handled securely, transmitted without interception, and stored with robust encryption becomes paramount. Privacy breaches pose a threat to individual users and erode trust in the entire IoT ecosystem, making it imperative to implement stringent measures to safeguard the privacy of IoT-generated data. Establishing secure authentication and authorization mechanisms in the IoT ecosystem presents inherent challenges. Traditional username/password models may be inadequate for IoT devices, and the scale of device interactions makes manual provisioning impractical. Implementing effective and scalable authentication, where devices can reliably verify each other's identity, is complex. Furthermore, defining and managing authorization policies across a myriad of interconnected devices require careful consideration to prevent unauthorized access and potential malicious activities [18].

Understanding and mitigating security threats within the IoT ecosystem are imperative for ensuring the integrity, confidentiality, and availability of data. Table 2.1 outlines the multifaceted security requirements across various layers of the IoT architecture, addressing both the technical implementation details and the essential monitoring and compliance measures. Each layer faces distinct threats, ranging from unauthorized access and data tampering at the perception layer to sophisticated attacks such as advanced persistent threats at the security layer. By adopting a proactive approach outlined in this comprehensive table, organizations can fortify their IoT deployments against a spectrum of potential threats. The implementation details offer practical insights into securing devices, networks, and applications, while continuous monitoring and compliance assessments act as vigilant guardians against evolving security challenges. As the IoT landscape expands, this table serves as a strategic guide for devising robust security strategies that safeguard the interconnected fabric of IoT systems [14–16].

Table 2.1 Comprehensive overview of IoT architecture security requirements

Layer	Security requirements	Implementation details	Monitoring and compliance	Threats
Perception layer (sensing layer)	Device authentication: ensure only legitimate devices can access the network	Implement strong device authentication mechanisms	Monitor device authentication logs	Unauthorized device access
	Data integrity: implement mechanisms to prevent tampering or manipulation of sensor data	Use cryptographic techniques for data integrity verification	Regularly audit sensor data for anomalies	Data tampering and manipulation
	Privacy controls: safeguard the privacy of data collected by sensors, adhering to applicable regulations	Apply anonymization or pseudonymization techniques to sensitive data	Periodically assess compliance with privacy regulations	Unauthorized access to sensitive user data
Network layer (connectivity layer)	Secure communication: encrypt data transmission to prevent eavesdropping and unauthorized access	Implement robust encryption protocols for data in transit	Monitor network traffic for encryption usage and anomalies	Man-in-the-middle attacks on data transmission
	Network segmentation: implement segmentation to isolate critical components and mitigate lateral movement	Define and enforce network segmentation policies	Regularly audit network segmentation configurations	Lateral movement within the network
	Traffic monitoring: continuously monitor network traffic for anomalies and potential security threats	Employ intrusion detection and prevention systems	Analyze network traffic patterns for unusual behavior	Network-based attacks, including DDoS attacks
Processing layer	Access controls: define and enforce access controls for data stored and processed within the layer	Implement role-based access controls for data processing resources	Regularly review and update access control policies	Unauthorized access to processed data
	Data encryption: apply encryption to data at rest and during processing to protect against unauthorized access	Utilize encryption algorithms for securing stored and processed data	Conduct periodic audits of encryption mechanisms	Data breaches leading to unauthorized data access
	Intrusion detection: deploy intrusion detection systems to identify and respond to unauthorized activities	Set up intrusion detection rules and alerts for anomalous activities	Conduct regular drills and simulations for incident response	Intrusion attempts and security policy violations

Platform/ middleware layer	Identity management: implement robust identity and access management systems for devices and users	Utilize secure identity verification methods such as multi-factor authentication	Regularly review and update user access privileges	Identity theft and unauthorized access to platform resources
	Secure APIs: ensure APIs used for data integration are secure and resistant to unauthorized access	Implement API security measures such as token-based authentication	Monitor API usage and access logs for security incidents	API-based attacks, including injection attacks
	Data cirtualization security: protect data virtualization processes against potential vulnerabilities	Apply encryption and access controls to virtualized data	Regularly assess the security posture of data virtualization processes	Data manipulation through virtualization vulnerabil-ities
Application layer	Role-based access control: implement role-based access controls for applications and services	Define roles and permissions based on job responsibilities	Regularly audit appli-cation access logs for unauthorized activities	Unauthorized access to sensitive application functionalities
	Secure application development: adhere to secure coding practices to prevent common vulnerabilities	Conduct secure code reviews and static code analysis during development	Implement automated security testing tools in the development pipeline	Exploitation of software vulnerabilities
	Data privacy measures: embed privacy controls into applications to safeguard user data	Implement privacy-by-design principles in application development	Regularly review and update privacy controls in applications	Privacy breaches through application vulnerabilities
Business layer	Collaboration security: ensure secure collaboration tools and protocols for stakeholders in decision-making	Implement end-to-end encryption for collaborative platforms	Monitor collaborative platforms for security incidents	Unauthorized access to collaborative tools and sensitive discussions
	Data governance: enforce data governance policies to maintain the quality and security of business data	Establish data ownership, quality, and retention policies	Periodically audit adherence to data governance policies	Data quality issues, unauthorized data modifications
	Regulatory compliance: align with industry regulations and standards relevant to the specific business domain	Conduct regular assessments to ensure compliance with relevant regulations	Maintain a compliance register and update it as regulations evolve	Non-compliance with industry regulations lead-ing to legal consequences

(Continues)

Table 2.1 (Continued)

Layer	Security requirements	Implementation details	Monitoring and compliance	Threats
Edge/fog computing layer	Edge device authentication: implement secure authentication for devices at the edge	Utilize strong device authentication protocols at the edge	Monitor edge device authentication logs	Unauthorized access to edge devices
	Localized data encryption: apply encryption for data processed at the edge to protect against local threats	Implement encryption mechanisms for data processed locally	Periodically assess the effectiveness of edge data encryption	Data breaches at the edge leading to unauthorized access
	Real-time threat detection: deploy real-time threat detection mechanisms to identify and respond to local threats	Employ real-time threat detection algorithms and anomaly detection	Regularly review threat detection logs for incidents at the edge	Localized threats, including malware and intrusions
Security layer	Continuous monitoring: implement continuous monitoring of the entire IoT ecosystem for security incidents	Employ security information and event management (SIEM) systems	Establish real-time dashboards for security monitoring	Advanced persistent threats and ongoing security breaches
	Incident response plan: develop and maintain an incident response plan to address security breaches promptly	Establish incident response teams and procedures	Conduct regular drills and simulations of incident response scenarios	Delayed or inadequate response to security incidents
	Security awareness training: conduct regular training for stakeholders to enhance awareness of security best practices	Provide training programs on security best practices and threat awareness	Monitor user compliance with security training requirements	Insider threats and human errors leading to security incidents

References

[1] Mrabet, H., Belguith, S., Alhomoud, A., and Jemai, A. (2020). A survey of IoT security based on a layered architecture of sensing and data analysis. *Sensors*, *20*(13), 3625.

[2] Kashani, M. H., Madanipour, M., Nikravan, M., Asghari, P., and Mahdipour, E. (2021). A systematic review of IoT in healthcare: applications, techniques, and trends. *Journal of Network and Computer Applications*, *192*, 103164.

[3] Kassab, W. A., and Darabkh, K. A. (2020). A–Z survey of Internet of Things: Architectures, protocols, applications, recent advances, future directions and recommendations. *Journal of Network and Computer Applications*, *163*, 102663.

[4] Obaidat, M. A., Obeidat, S., Holst, J., Al Hayajneh, A., and Brown, J. (2020). A comprehensive and systematic survey on the internet of things: Security and privacy challenges, security frameworks, enabling technologies, threats, vulnerabilities and countermeasures. *Computers*, *9*(2), 44.

[5] Donta, P. K., Srirama, S. N., Amgoth, T., and Annavarapu, C. S. R. (2022). Survey on recent advances in IoT application layer protocols and machine learning scope for research directions. *Digital Communications and Networks*, *8*(5), 727–744.

[6] Nebbione, G., and Calzarossa, M. C. (2020). Security of IoT application layer protocols: Challenges and findings. *Future Internet*, *12*(3), 55.

[7] Zhang, J., Ma, M., Wang, P., and Sun, X. D. (2021). Middleware for the Internet of Things: A survey on requirements, enabling technologies, and solutions. *Journal of Systems Architecture*, *117*, 102098.

[8] Firouzi, F., Farahani, B., Weinberger, M., DePace, G., and Aliee, F. S. (2020). IOT fundamentals: Definitions, architectures, challenges, and promises. *Intelligent Internet of Things: From Device to Fog and Cloud*, 3–50.

[9] Laghari, A. A., Wu, K., Laghari, R. A., Ali, M., and Khan, A. A. (2021). A review and state of art of Internet of Things (IoT). *Archives of Computational Methods in Engineering*, 1–19.

[10] Gupta, V., Sharma, M., Pachauri, R. K., and Babu, K. D. (2021). A low-cost real-time IOT enabled data acquisition system for monitoring of PV system. *Energy Sources, Part A: Recovery, Utilization, and Environmental Effects*, *43*(20), 2529–2543.

[11] Bansal, M., Chana, I., and Clarke, S. (2020). A survey on IOT big data: Current status, 13 v's challenges, and future directions. *ACM Computing Surveys (CSUR)*, *53*(6), 1–59.

[12] Lee, E., Seo, Y. D., Oh, S. R., and Kim, Y. G. (2021). A Survey on Standards for Interoperability and Security in the Internet of Things. *IEEE Communications Surveys & Tutorials*, *23*(2), 1020–1047.

[13] Schiller, E., Aidoo, A., Fuhrer, J., Stahl, J., Ziörjen, M., and Stiller, B. (2022). Landscape of IoT security. *Computer Science Review*, *44*, 100467.

[14] Jouhari, M., Saeed, N., Alouini, M. S., and Amhoud, E. M. (2023). A survey on scalable LoRaWAN for massive IoT: Recent advances, potentials, and challenges. *IEEE Communications Surveys & Tutorials*, *25*(3), 1841–1876.

[15] Rana, B., Singh, Y., and Singh, P. K. (2021). A systematic survey on internet of things: Energy efficiency and interoperability perspective. *Transactions on Emerging Telecommunications Technologies*, *32*(8), e4166.

[16] Pukkasenung, P., and Lilakiatsakun, W. (2021). Improved Generic Layer Model for IoT Architecture. *Journal of Information Science & Technology*, *11*(1).

[17] Aarika, K., Bouhlal, M., Abdelouahid, R. A., Elfilali, S., and Benlahmar, E. (2020). Perception layer security in the internet of things. *Procedia Computer Science*, *175*, 591–596.

[18] Luis, Á., Casares, P., Cuadrado-Gallego, J. J., and Patricio, M. A. (2021). PSON: A serialization format for IoT sensor networks. *Sensors*, *21*(13), 4559.

[19] Tewari, A., and Gupta, B. B. (2020). Security, privacy and trust of different layers in Internet-of-Things (IoTs) framework. *Future Generation Computer Systems*, *108*, 909–920.

[20] Burhan, M., Rehman, R. S., Khan, B., and Byung-Seo Kim. IoT elements, layered architectures and security issues: A comprehensive survey. *Sensors 18.9* (2018): 2796.

[21] Yousuf, O., and Mir, R. N. (2019). A survey on the Internet of Things security: State-of-art, architecture, issues and countermeasures. *Information & Computer Security*, *27*(2), 292–323.

Chapter 3

Convergence of Internet of Things and computing technologies

3.1 Introduction

The advent of the Internet of Things (IoT) has sparked a transformative revolution in the way we perceive and interact with technology. It represents a connected ecosystem where a myriad of devices, sensors, and systems communicate seamlessly, generating a wealth of data that propels innovation across industries. At the core of this paradigm shift lies the convergence of IoT with various computing paradigms, each offering unique attributes and capabilities that contribute significantly to the evolution and security of IoT ecosystems [1].

The IoT landscape encompasses a vast array of interconnected devices, ranging from household appliances and wearable gadgets to industrial sensors and smart infrastructure components. These devices collect, transmit, and process data, fostering a networked environment that enables automation, efficiency improvements, and data-driven decision-making. The convergence of IoT with diverse computing paradigms has been instrumental in shaping the landscape of connected technologies. Traditional centralized computing models have paved the way for distributed computing architectures, offering scalability, efficiency, and resilience.

The remarkable progression of technology has been punctuated by a series of transformative computing paradigms that have shaped the way we perceive, process, and utilize information. Understanding the historical evolution of these paradigms provides a lens through which we can appreciate the trajectory leading to the convergence of computing technologies with the IoT. The genesis of computing paradigms traces back to the era of centralized computing. Initially, computing power and data storage were centralized in mainframe computers, marking the inception of a technological landscape characterized by central control and processing [2].

The emergence of distributed computing paradigms marked a pivotal shift from the centralized model. Client–server architectures evolved, enabling multiple devices to interact with a centralized server, distributing computational tasks across networks. One of the most influential milestones in computing history was the advent of cloud computing. This paradigm introduced a scalable, on-demand model for delivering computing services over the Internet. It revolutionized data storage, processing, and accessibility, laying the groundwork for scalable and flexible computing infrastructures. As the IoT ecosystem burgeoned, demands for faster processing, reduced latency, and enhanced security became apparent. Edge computing surfaced as a solution, empowering devices to process data closer to the source and reducing reliance on distant data centers. Fog computing extended this concept by integrating edge devices with cloud resources, bridging the gap between local and cloud-based computing [3].

Additionally, the prospect of quantum computing, although in its nascent stages, holds promise for revolutionizing computational capabilities, potentially disrupting existing encryption methods and introducing new dimensions to data processing. The amalgamation of these historical computing paradigms with contemporary advancements in IoT technologies has given rise to a convergence that transcends conventional boundaries. This convergence heralds an era where distributed computing paradigms, from edge to cloud, are seamlessly interwoven with the fabric of IoT, transforming the technological landscape.

The significance of this convergence lies in the amalgamation of these diverse computing paradigms with IoT, offering a symbiotic relationship that leverages the strengths of each paradigm. This union unlocks unparalleled opportunities for enhancing the functionality, efficiency, and security of IoT ecosystems. While this convergence promises immense innovation, it also presents multifaceted challenges, particularly in terms of security. The distributed nature of IoT, combined with diverse computing architectures, introduces complexities and vulnerabilities that demand robust solutions to ensure the integrity, confidentiality, and resilience of IoT networks [4].

This chapter aims to delve deeper into this convergence, exploring the nuances of various computing paradigms and their role in securing IoT. By examining these paradigms and their integration with IoT, we aim to decipher how their amalgamation contributes to technological evolution and the imperative need for ensuring a secure IoT landscape.

3.2 Centralized computing

Centralized computing stands as a cornerstone in the historical evolution of computing paradigms, defining an epoch where computational power and data storage were concentrated within monolithic mainframe systems. At its essence, this paradigm embodies the consolidation of control and processing capabilities in a singular, centralized infrastructure. In this section, we delve into the fundamental principles and historical significance of centralized computing, exploring its role in shaping technological landscapes and laying the groundwork for subsequent computing models. This paradigm not only established the initial framework for computational processes but also contributed significantly to the emergence of distributed and decentralized computing architectures that followed [1–6].

Mainframe computing marks a significant milestone in the annals of technological evolution. Its origins can be traced back to the mid-twentieth century when computing power was centralized within massive, room-sized machines. The advent of mainframe computers, typified by iconic systems such as the IBM System/360, heralded a new era in computational capabilities. These behemoth machines were designed to execute vast computational tasks, handle large volumes of data, and cater to the computational needs of governments, enterprises, and scientific communities.

During the 1960s and 1970s, mainframes were the backbone of computing infrastructure, facilitating groundbreaking technological advancements. They played an instrumental role in various sectors, including banking, telecommunications, aerospace, and scientific research, handling critical operations with reliability and efficiency. Mainframes were characterized by their robustness, high processing power, and exceptional reliability. They featured a hierarchical architecture that facilitated multitasking and concurrent user access, catering to the computational demands of multiple users simultaneously. The systems boasted impressive storage capacities and computational speed, setting them apart as the pinnacle of computing technology during their prime.

The significance of mainframe computing transcends its technological prowess; it laid the foundational concepts for centralized computing paradigms. Its centralized architecture and emphasis on data integrity and reliability became the bedrock upon which subsequent computing models, including distributed architectures, were built. Mainframes were the precursor to the centralized control and processing that underpinned computing infrastructures for decades. Despite the evolution toward distributed and decentralized computing models, mainframe computing continues to persist in critical sectors where reliability, security, and high-throughput processing are paramount. Its legacy endures, serving as a testament to the enduring significance of centralized computing paradigms in shaping the trajectory of technological innovation [5]. Mainframe computing represents a pivotal chapter in the history of computing paradigms, laying the groundwork for centralized computing architectures that have underpinned technological progress for decades. Its robustness, reliability, and massive computational capabilities have left an indelible mark on the technological landscape, influencing subsequent paradigms and contributing to the foundational principles of modern computing infrastructures [6].

3.3 Distributed computing

Distributed computing heralds a paradigm shift from the centralized models of computing, embracing a decentralized architecture that empowers networks of interconnected devices to collaboratively tackle computational tasks. This section ventures into the multifaceted realm of distributed computing, a domain encompassing diverse architectures and models designed to distribute processing power, enhance scalability, and foster collaboration among nodes within a network. Exploring the evolution and principles of distributed computing, we unravel the intricacies of various models such as client-server architecture, cloud computing, edge computing, fog computing, and even the nascent implications of quantum computing. This section endeavors to elucidate the significance of distributed computing paradigms in reshaping the technological landscape and their profound impact on securing the expansive realm of the IoT [7].

3.3.1 Client–server architecture

Client-server architecture represents a computing model that distributes tasks or processes between servers (centralized systems managing resources) and clients (devices or systems that request services from servers). It operates on the modularization principle, where clients make requests for data, resources, or services, and servers fulfill these requests by processing them and delivering the required information back to the clients. This architectural model fosters a separation of concerns, facilitating efficient resource management and scalability. Within the realm of client-server architecture, three primary service models have emerged:

1. **Infrastructure as a Service (IaaS):** IaaS provides virtualized computing resources over the Internet, offering scalable and flexible infrastructure resources such as virtual machines, storage, and networking. It allows users to rent these resources on a pay-as-you-go basis, enabling greater control over the infrastructure without managing physical hardware.
2. **Platform as a Service (PaaS):** PaaS offers a platform allowing developers to build, deploy, and manage applications without dealing with the underlying infrastructure complexities. It provides tools, development frameworks, and deployment environments, streamlining the application development process.
3. **Software as a Service (SaaS):** SaaS delivers software applications over the Internet on a subscription basis. Users can access and use these applications through web browsers without needing to install or maintain software locally. Common examples include email services, office suites, and customer relationship management (CRM) tools.

The introduction of client-server architecture offered many primary benefits, which are summarized as follows:

- **Scalability:** Client-server architectures allow for scalable systems where servers can handle increased client demands by adding resources.
- **Centralized management:** Centralized servers enable efficient resource management, data storage, and security implementations.
- **Resource sharing:** Servers facilitate resource sharing among clients, reducing redundancy and improving efficiency.
- **Flexible service models:** The various service models (IaaS, PaaS, and SaaS) offer flexibility and cost-effectiveness based on user needs.

Despite the success of client-server architecture, it has come with many limitations:

- **Single point of failure:** Dependency on a centralized server creates vulnerabilities; failure of the server can disrupt services.

- **Potential bottlenecks:** Heavy reliance on servers for processing can lead to performance bottlenecks during peak usage.
- **Scalability challenges:** Scaling may become complex due to the need for server upgrades or load-balancing configurations.

Client-server architecture, with its modular approach and service models, remains a fundamental computing paradigm, offering versatility, scalability, and efficient resource utilization. However, while it provides numerous benefits, considerations regarding potential limitations underline the necessity of balancing centralized processing with distributed approaches for robust and secure IoT ecosystems [7,8].

3.3.2 Cloud computing

Cloud computing refers to the delivery of computing services, including servers, storage, databases, networking, software, and analytics, over the Internet ("the cloud"). It allows users to access and utilize these resources on-demand, without the need for direct management of physical infrastructure. The essence of cloud computing lies in its ability to provide scalable, flexible, and cost-effective services, catering to diverse computing needs.

The underlying principles of cloud computing are based on delivering services through a shared pool of configurable resources. It operates on the concepts of virtualization, where resources are abstracted from the underlying hardware, allowing for flexible allocation and usage. Additionally, self-service provisioning enables users to request and manage resources autonomously, while elasticity ensures that resources can be scaled up or down dynamically based on demand. Moreover, pay-as-you-go pricing models offer cost-effectiveness by charging users based on their actual resource consumption.

In the realm of IoT, cloud computing plays a pivotal role in addressing IoT devices' computational and storage demands. IoT generates vast amounts of data, requiring robust processing and storage capabilities, which cloud infrastructure readily provides. Cloud platforms act as repositories for IoT-generated data, enabling analysis, machine learning, and actionable insights. Moreover, by offloading computational tasks to the cloud, IoT devices with limited processing capabilities can leverage the vast computational resources available in the cloud, enhancing their functionality and capabilities.

3.3.3 Edge computing

Edge computing is a distributed computing paradigm that involves processing data closer to the source of generation, i.e., at the edge of the network, rather than relying solely on centralized data centers or the cloud. It brings computational resources and data storage closer to where the data is generated, reducing latency, enhancing real-time processing capabilities, and enabling quicker decision-making [9,10]. The foundational principles of edge computing revolve around proximity, scalability, and distributed processing. By deploying computational resources at or near the data source, edge computing minimizes the distance data must traverse, thereby reducing latency and improving response times for critical applications. This paradigm operates on the principle of scalability, allowing for the deployment of edge devices and servers in a distributed manner, enabling the network to handle varying workloads and demands efficiently. Furthermore, edge computing emphasizes distributed processing, allowing the data to be processed and acted upon locally, thus reducing the need for constant communication with centralized cloud servers.

Edge computing presents a spectrum of benefits, primarily characterized by enhanced performance, improved efficiency, and increased autonomy. Processing data locally significantly reduces latency, making it ideal for applications requiring real-time or low-latency responses, such as industrial automation, autonomous vehicles, and IoT-enabled healthcare systems. Moreover, the decentralized nature of edge computing contributes to reduced bandwidth usage, as only pertinent data is transmitted to centralized servers, optimizing network traffic and minimizing costs. Additionally, the increased autonomy afforded by edge computing enables devices to operate independently, even in scenarios with limited or intermittent connectivity (Figure 3.1).

In the realm of the IoT, edge computing assumes a pivotal role in addressing the challenges posed by massive data volumes, stringent latency requirements, and diverse application scenarios. IoT devices generate copious amounts of data, and by leveraging edge computing, this data can be processed locally, enabling faster

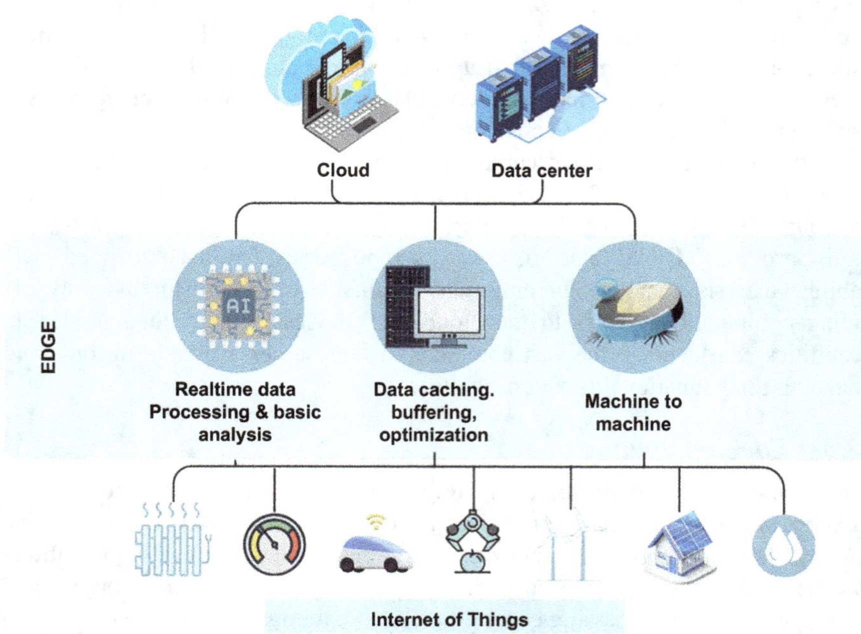

Figure 3.1 The role of edge computing in IoT ecosystems for low-latency data processing

decision-making and reducing the strain on centralized cloud infrastructure. Edge computing is instrumental in IoT applications spanning various domains, including smart cities, industrial IoT, healthcare monitoring, and smart homes, where real-time insights and rapid responses are critical for optimal functionality and user experience.

3.3.4 Fog computing

Fog computing is a distributed computing paradigm that extends cloud computing capabilities to the network's edge. It aims to bring computing resources and services closer to the data source, providing real-time data processing, storage, and analysis at intermediate points between the edge devices and centralized cloud servers. Fog computing complements edge computing by adding a hierarchical layer of intermediate fog nodes, enhancing the capabilities of edge devices and reducing latency for critical applications [11] (Figure 3.2).

Fog computing introduces intermediate fog nodes between edge devices and centralized cloud servers, creating a hierarchical architecture that enables processing tasks at different levels of the network. Fog nodes supplement edge devices by providing additional computational resources, storage, and services to handle more complex tasks that exceed the capabilities of edge devices alone. By leveraging proximity to edge devices, fog computing enables the provision of real-time services, analytics, and responses, enhancing the efficiency of IoT applications and critical operations.

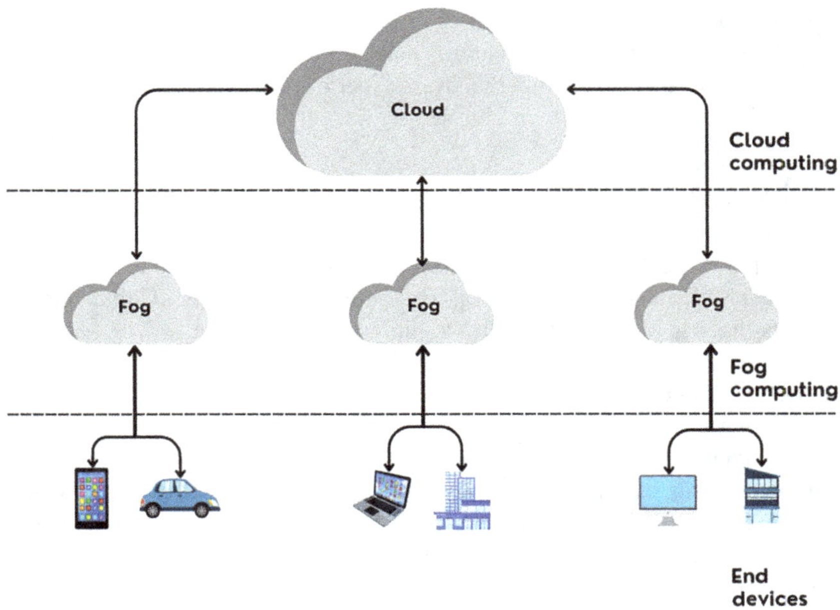

Figure 3.2 Illustrating fog computing's role in IoT for distributed edge processing

The principles underpinning fog computing emphasize decentralized processing and storage, leveraging edge devices and fog nodes to handle data closer to the source. This model involves a hierarchical structure where edge devices interact with intermediary fog nodes, which, in turn, communicate with cloud resources. Fog computing operates on the premise of distributing computational tasks across this hierarchy, optimizing data processing, and reducing latency by bringing computing resources closer to where data is generated. The principles of fog computing thus align with the fundamental aspects of edge computing, emphasizing the importance of proximity to data sources for efficient processing.

The benefits of fog computing are multi-fold, particularly in the context of IoT. By facilitating distributed data processing and analysis at the network's edge, fog computing significantly reduces latency, enhancing real-time application performance. Moreover, it minimizes the amount of data that needs to be transmitted to centralized cloud servers, thereby reducing network congestion and bandwidth usage. This localized approach to data processing also enhances privacy and security by mitigating the risks associated with transmitting sensitive information over networks. Additionally, fog computing enhances the scalability and flexibility of IoT ecosystems, enabling seamless integration and management of many devices and services.

In the realm of IoT, fog computing plays a pivotal role in addressing the evolving demands for efficient data processing, low-latency analytics, and real-time decision-making. Its proximity-based approach to processing data at the network's edge complements IoT applications that require immediate responses, such as smart cities, autonomous vehicles, and industrial automation. By enabling distributed computation and storage closer to IoT devices, fog computing empowers IoT systems to manage and analyze data efficiently, facilitating intelligent decision-making and enhancing the overall performance and reliability of IoT applications [12].

3.3.5 Quantum computing

Quantum computing represents a revolutionary paradigm in computing that leverages the principles of quantum mechanics to perform complex calculations and solve problems beyond classical computers' capabilities. Unlike classical computers that use bits (binary digits) to represent information as 0 or 1, quantum computers use quantum bits or qubits, which can exist simultaneously in multiple states due to quantum superposition and entanglement, vastly increasing computational power and capability [13].

The principles underlying quantum computing encompass key quantum phenomena, including superposition, where qubits can exist in multiple states simultaneously, and entanglement, wherein the states of qubits become interdependent regardless of distance. These principles pave the way for quantum algorithms to execute tasks such as Shor's algorithm for integer factorization and Grover's algorithm for unstructured search, offering exponential speedups for specific problems that are classically challenging. The potential benefits of quantum computing

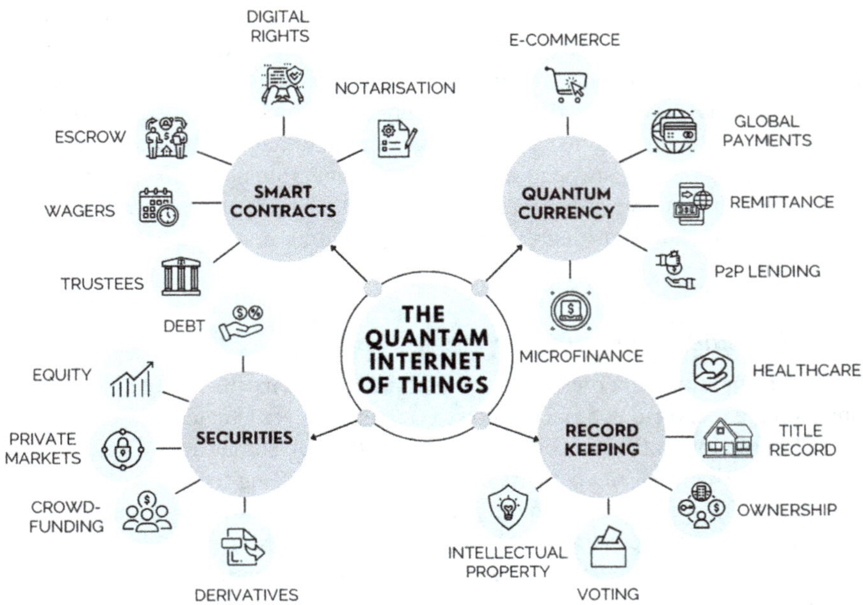

Figure 3.3 Illustrating quantum computing's potential in IoT for complex problem solving

span various domains, promising advancements in fields such as cryptography, optimization, drug discovery, and material science. Quantum computers possess the capability to break conventional cryptographic protocols through Shor's algorithm, compelling the development of quantum-resistant encryption methods. Furthermore, in optimization tasks, quantum algorithms could significantly enhance the efficiency of complex computations, revolutionizing supply chain management, logistics, and financial modeling.

In the realm of the IoT, quantum computing holds the promise of addressing specific challenges faced by IoT devices and networks. The massive amounts of data IoT devices generate demand for robust computational power for real-time analytics, security protocols, and pattern recognition. Quantum computing's ability to process and analyze large datasets more efficiently could bolster IoT systems by enabling faster processing of complex algorithms, enhancing data security through quantum-resistant encryption, and optimizing resource allocation for IoT networks (Figure 3.3).

3.4 Advanced distributed computing paradigms

The landscape of distributed computing continues to evolve, ushering in a realm of advanced paradigms that extend the boundaries of traditional distributed architectures. Section 3.4 explores these innovative paradigms, unveiling a spectrum of models that transcend conventional distributed computing approaches. From cloudlet computing to swarm computing, dew computing, and mist computing, this section navigates each paradigm's intricacies and distinctive attributes. By delving into these advanced distributed computing models, we unravel their transformative potential, unique contributions to enhancing computational capabilities, and pivotal role in reshaping the landscape of distributed systems. Join us on this journey as we unravel the nuances and implications of these cutting-edge paradigms, poised to redefine the very fabric of distributed computing in an era of rapid technological evolution [3].

3.4.1 Cloudlet computing

Cloudlet computing emerges as a pivotal paradigm in distributed computing, representing a localized and intermediary computing layer situated at the network's edge. This section delves into the essence of cloudlet computing, elucidating its fundamental principles, the benefits it offers, and discerning its role within the IoT ecosystem [10]. Cloudlet computing embodies an architectural concept where small-scale cloud servers or clusters are positioned at the network's edge, proximate to end-users or devices. These cloudlets provide computational resources, storage, and services, offering a localized platform for offloading tasks from resource-constrained devices. This decentralized approach facilitates quicker data processing and access to cloud services, minimizing latency and enhancing the efficiency of edge-based applications.

The core principles underlying cloudlet computing revolve around proximity, resource augmentation, and hierarchical computing. Cloudlets are strategically positioned in close proximity to end-users or devices, enabling swift data processing without relying solely on distant data centers. These cloudlets augment the computational capabilities of edge devices, alleviating their processing burdens by offloading complex tasks to these intermediary servers. Cloudlets also establish a hierarchical computing model, complementing edge computing and cloud computing infrastructures by providing localized services while retaining connectivity to broader cloud resources. Cloudlet computing presents an array of advantages, encompassing reduced latency, enhanced scalability, improved reliability, and efficient resource utilization. By minimizing data transit to distant servers, cloudlets significantly reduce latency, ensuring swift response times for time-sensitive applications. Moreover, these intermediary servers offer scalability by providing additional computational resources, bolstering the efficiency of edge devices. The localized nature of cloudlet computing enhances reliability by reducing dependency on distant servers and offers optimized resource utilization by efficiently distributing computational tasks.

In the context of the IoT, cloudlet computing plays a pivotal role in optimizing IoT ecosystem functionality. Cloudlets empower IoT devices to efficiently execute tasks that demand immediate responses or intensive computations by enabling localized data processing and services. This local offloading of tasks to cloudlets reduces the need for constant reliance on cloud servers, preserving network bandwidth, and enhancing the overall responsiveness and resilience of IoT applications.

Cloudlet computing, thus, emerges as a critical facilitator in addressing latency concerns and optimizing resource utilization within IoT frameworks.

3.4.2 Swarm computing

Swarm computing embodies a decentralized approach to problem-solving and computation, drawing inspiration from the collective behaviors observed in natural swarms of organisms, such as ants, bees, and flocking birds. At its core, swarm computing represents a computational paradigm where many simple entities, often called agents, interact locally and autonomously to achieve complex tasks without centralized control. These agents follow simple rules and interact with their local environment and neighboring agents, collectively exhibiting emergent behavior that solves complex problems beyond the capabilities of individual entities [12].

The principles governing swarm computing revolve around self-organization, decentralized control, and emergence. In this paradigm, agents exhibit limited capabilities and communicate and coordinate with nearby agents based on local information. Through interactions and simple rules governing their behavior, these agents collectively exhibit emergent behaviors, such as self-optimization, adaptive responses to changes in the environment, and robustness against failures. Swarm computing emphasizes the autonomy and flexibility of agents, fostering a system where individual entities' actions contribute to the collective intelligence of the swarm (Figure 3.4).

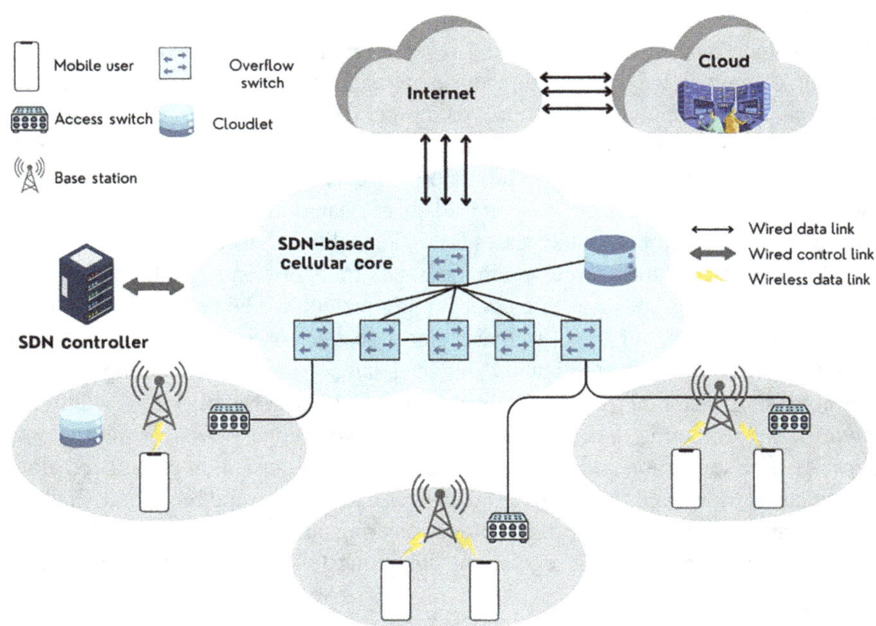

Figure 3.4 Illustrating cloudlet system's role in IoT, enabling proximity-based processing

Swarm computing offers a spectrum of benefits, primarily stemming from its decentralized and self-organizing nature. The inherent redundancy and fault tolerance within swarm systems enable robustness against individual failures or disruptions, ensuring the system's resilience. Additionally, the adaptability of swarm systems to dynamic environments allows for flexible responses to changing conditions. Moreover, the scalability and parallelism inherent in swarm computing empower these systems to handle complex tasks efficiently, making them suitable for solving optimization problems, task allocation, and pattern recognition. In the realm of the IoT, swarm computing emerges as a compelling paradigm, offering solutions to various IoT challenges. The decentralized nature of swarm systems aligns with the distributed and heterogeneous nature of IoT environments. Swarm algorithms and strategies find applications in IoT scenarios, such as distributed sensor networks, autonomous vehicles' coordination, smart grid management, and collaborative robotics. Swarm computing's ability to handle decentralized decision-making and adapt to dynamic IoT environments positions it as a promising approach to address coordination, optimization, and scalability challenges prevalent in complex IoT ecosystems.

3.4.3 Dew computing

Dew computing represents a paradigm that extends cloud and edge computing by emphasizing the utilization of resources available at both the edge and within the cloud infrastructure. It aims to leverage resources ranging from edge devices to the cloud, incorporating a spectrum of intermediate computing layers, to provide efficient and flexible solutions for data processing, storage, and computation [9–13]. The principles underpinning Dew computing revolve around the distribution and optimization of computational tasks within the network peripheries. Emphasizing decentralization, Dew computing orchestrates data processing and storage tasks across dew nodes, enabling collaborative efforts among edge devices. These nodes operate autonomously yet collaboratively, embracing the concept of self-organizing systems that dynamically allocate resources and adapt to changing computational demands. Additionally, Dew computing advocates for data locality, promoting the processing of critical tasks closer to the source, minimizing data transmission to centralized clouds, and thereby reducing latency while conserving bandwidth. The advent of Dew computing heralds a suite of benefits, particularly in the context of IoT ecosystems. Notably, Dew computing significantly mitigates latency issues by fostering localized data processing, enabling quicker responses for time-sensitive applications within IoT networks. By promoting data processing at the network edges, Dew computing optimizes resource utilization and bandwidth, alleviating the burden on centralized cloud infrastructure. Moreover, this paradigm enhances data privacy and security by limiting the transmission of sensitive information, fostering an environment conducive to compliance with stringent data regulations governing IoT deployments (Figure 3.5).

3.4.4 Mist computing

Mist computing, also known as fog-to-edge computing, extends the capabilities of fog computing by emphasizing the integration of edge devices with more localized

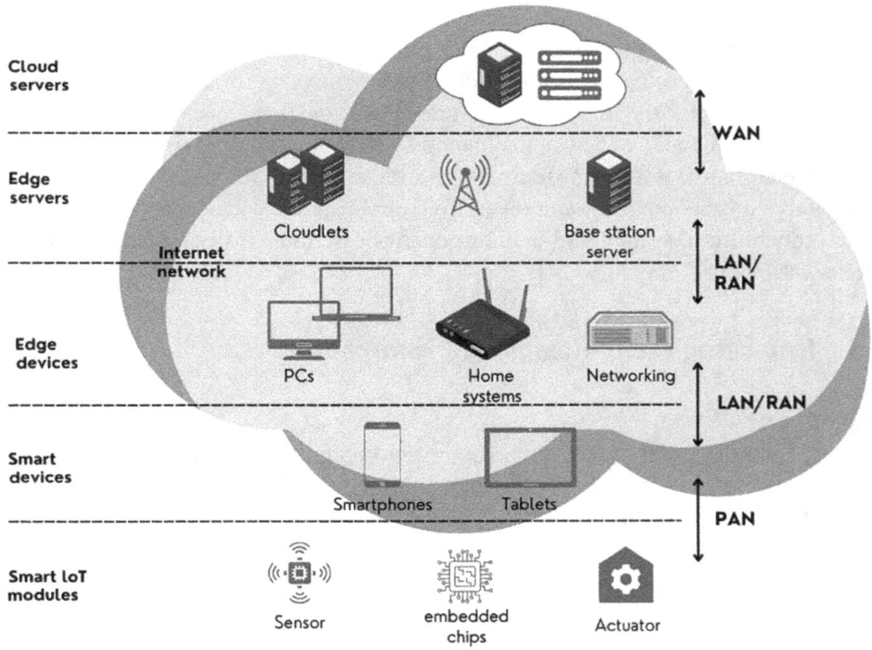

Figure 3.5 Exploring Dew computing's significance in IoT for integrated edge-cloud solutions

computing resources, often at the sensor or device level. It focuses on distributing computational tasks across a hierarchical network that includes sensors, actuators, edge devices, and fog nodes, enabling localized data processing, analysis, and decision-making closer to the source of data generation [14–16].

At its core, mist computing operates on fundamental principles that encompass decentralization, adaptability, and resource optimization. This paradigm embraces decentralized architectures, enabling distributed data processing and services among fog nodes and edge devices. It emphasizes adaptability by dynamically adjusting computational tasks and resource allocation based on evolving network conditions. Moreover, mist computing prioritizes resource optimization by orchestrating tasks across fog nodes and edge devices, aiming for efficient utilization of computational resources while minimizing latency in data processing.

Mist computing offers a spectrum of advantages within IoT ecosystems. By facilitating localized data processing and analytics, it significantly reduces latency, ensuring real-time responsiveness for critical IoT applications. This paradigm enhances system scalability and resilience by distributing computational tasks across fog nodes and edge devices, mitigating the risk of single-point failures. Additionally, mist computing bolsters data privacy and security by processing sensitive information closer to its source, minimizing exposure during data transmission and storage.

In the context of IoT, mist computing plays a pivotal role in orchestrating distributed computing tasks and optimizing resource utilization. It acts as a bridge between edge devices and centralized cloud infrastructures, offering a middle layer for computational offloading, data aggregation, and preprocessing. This intermediary role enhances IoT systems' capabilities by facilitating efficient data processing, enabling faster response times, and reducing the burden on cloud servers. Ultimately, mist computing empowers IoT applications with enhanced computational capabilities, scalability, and responsiveness, thus fostering advancements across various IoT domains [17].

3.5 Emerging technologies and innovations

3.5.1 Fabric computing

Fabric computing refers to an architecture that integrates and orchestrates various computing resources (processing, storage, and networking) into a unified, flexible, and scalable infrastructure. It emphasizes the creation of a dynamic fabric or mesh-like structure that interconnects resources across data centers, edge devices, and cloud environments, allowing for seamless resource utilization and management [18,19] (Figure 3.6).

Fabric computing aims to create a unified and abstracted infrastructure where diverse resources are seamlessly interconnected and managed. It emphasizes

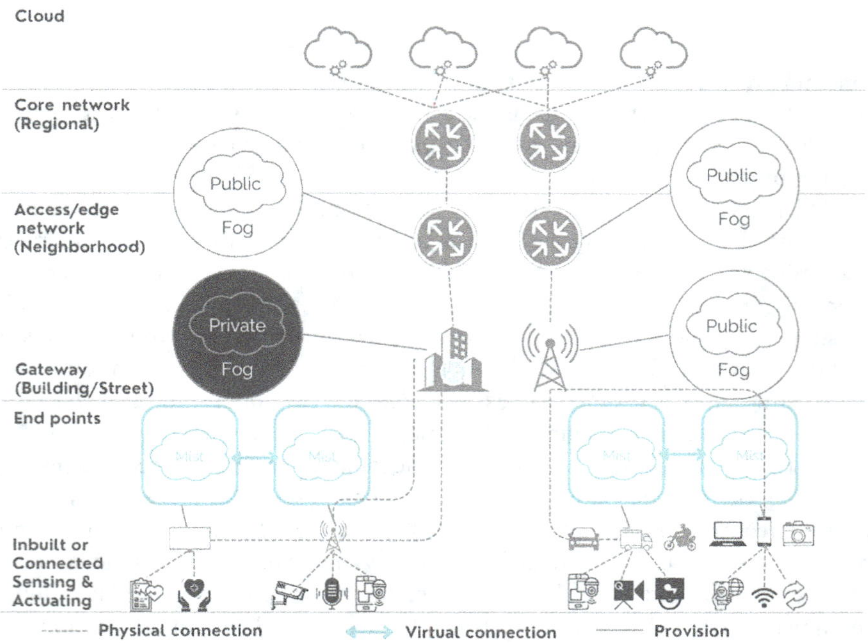

Figure 3.6 Illustrating MIST computing system's role in IoT

resource virtualization, enabling the abstraction of physical resources into a pool of virtualized resources that can be dynamically allocated based on demand. Fabric computing systems are designed to be highly scalable and elastic, allowing them to efficiently scale up or down to accommodate changing workloads. The adoption of fabric computing offers several key benefits. First, fabric computing optimizes resource utilization by allowing dynamic resource allocation based on application needs, thereby improving overall system efficiency. Second, the flexible and resilient nature of fabric computing architectures enhances system robustness by enabling resource reallocation in response to failures or changing demands. Third, fabric computing's scalability and adaptability cater to the diverse and evolving requirements of IoT applications, ensuring they can handle varying workloads and emerging challenges.

In IoT environments, fabric computing orchestrates computing resources across edge devices, cloud servers, and intermediate layers, ensuring efficient utilization and management of resources. Fabric computing enables IoT systems to dynamically adapt to changing environmental conditions or operational requirements, ensuring continuous and optimized performance.

3.5.2 Multi-access edge computing

Multi-access edge computing (MEC) refers to a distributed computing architecture that brings computational resources and services closer to the edge of the network, specifically at the edge of the radio access network (RAN). MEC aims to process and store data closer to end-users, devices, or sensors, reducing latency and enabling faster data processing, real-time analytics, and content delivery [19,20].

MEC represents a distributed computing architecture that brings computational capabilities, storage, and networking resources closer to the network edge, typically at base stations or access points. MEC aims to reduce latency and improve bandwidth efficiency by enabling computation and storage resources to be positioned closer to end-users and IoT devices. This proximity facilitates the efficient processing of data, enabling real-time applications and services, while also offloading tasks from centralized cloud servers. The core principle of MEC lies in providing computing capabilities at the edge of the network, enhancing the responsiveness and efficiency of applications and services. MEC operates on the principles of proximity, low latency, and distributed computing. It leverages the physical proximity of edge servers or access points to end-users and IoT devices, reducing the distance data needs to travel and minimizing latency. Additionally, MEC facilitates distributed computing by decentralizing computational tasks, enabling devices to process data locally or within nearby edge servers. This principle enhances the responsiveness of applications by reducing the reliance on distant cloud servers and optimizing the flow of data between devices and computational resources.

The significance of MEC lies in its ability to support emerging applications that demand ultra-low latency and high bandwidth, such as augmented reality (AR), virtual reality (VR), connected vehicles, industrial automation, and smart cities.

The architecture of MEC typically comprises edge nodes or servers deployed at the edge of the network, strategically positioned to serve nearby end-users and devices. These edge nodes are equipped with computational resources, storage, and network functionalities to process and respond to data in close proximity, thereby enhancing application performance and user experience. The benefits of MEC encompass reduced latency, improved network efficiency, enhanced scalability, and the ability to support emerging low-latency applications. In the realm of IoT, MEC plays a crucial role by enabling real-time data processing, analysis, and decision-making at the edge of the network. This capability is particularly vital for IoT applications that require immediate responses, such as smart healthcare, autonomous vehicles, and industrial IoT. MEC facilitates efficient IoT deployments by enabling devices to process data locally, reducing reliance on centralized cloud infrastructure, and improving overall system responsiveness and reliability.

3.5.2.1 Cloud of Things

Cloud of Things (CoT) represents an integrated system where the IoT devices, sensors, and applications are interconnected and seamlessly integrated with cloud computing infrastructure and services. It leverages cloud resources to manage, analyze, and store the vast amounts of data generated by IoT devices, enabling scalable and efficient data processing and management [18]. The significance of CoT lies in its ability to address inherent challenges in IoT ecosystems, primarily concerning data processing, scalability, and resource limitations. Its architecture typically comprises multiple layers, including IoT devices/sensors at the edge, gateway devices for data aggregation and preprocessing, and cloud-based platforms for data analysis and storage. This architecture enables the seamless transmission of sensor data to cloud services, where sophisticated analytics, machine-learning algorithms, and large-scale data processing can be executed. The centralized architecture ensures efficient utilization of cloud resources, accommodating the ever-growing volume of data generated by IoT devices while offering scalable and robust computational capabilities.

The adoption of CoT presents numerous benefits for IoT applications. First, it offers scalable and flexible computing resources, allowing for the handling of vast amounts of data generated by IoT devices. Second, it enables advanced analytics and machine-learning algorithms to derive actionable insights from the accumulated data, facilitating data-driven decision-making. Moreover, CoT enhances interoperability among diverse IoT devices by providing standardized interfaces for seamless integration with cloud services. Additionally, it fosters improved security and privacy measures by centralizing data management and implementing robust access control mechanisms [19].

CoT plays a pivotal role in the IoT landscape by serving as a unifying platform that harmonizes edge devices with cloud computing resources. Its role encompasses facilitating efficient data management, enabling advanced analytics, and supporting real-time decision-making processes for various IoT applications. CoT empowers IoT deployments by offering scalable and dynamic infrastructures, enabling seamless communication, and enhancing the overall performance of IoT ecosystems, thereby driving innovation across diverse industry verticals.

3.5.3 Mobile computing

Mobile computing refers to the utilization of computational resources and services on mobile devices, enabling users to access and interact with data, applications, and services while on the move. It encompasses a wide range of technologies and services that facilitate connectivity, data transmission, and access to resources on portable devices such as smartphones, tablets, and wearables [10,15]. The principles underpinning mobile computing revolve around portability, connectivity, and resource optimization. Portability denotes the capability of devices to be easily transported, ensuring computational resources are available regardless of location. Connectivity emphasizes continuous access to networks, enabling communication and data exchange. Resource optimization involves efficient utilization of hardware and software resources to ensure functionality within mobile devices despite their limitations compared to traditional computing systems.

Mobile computing holds paramount importance in contemporary technological landscapes. Its architecture comprises hardware components (e.g., smartphones and tablets) integrated with operating systems and mobile applications. These devices connect to various networks, including Wi-Fi, cellular, and Bluetooth, ensuring seamless communication. The architecture also incorporates cloud-based services, allowing offloading of complex computational tasks to remote servers for enhanced processing power and storage capabilities. The benefits of mobile computing lie in its ability to empower users with ubiquitous access to information, services, and applications. Its role within IoT ecosystems is multifaceted, offering a conduit for IoT devices to connect and communicate effectively. Mobile devices serve as gateways for data collection, transmission, and control, enabling real-time monitoring and management of IoT devices. Additionally, mobile applications facilitate user interaction with IoT systems, providing interfaces for data visualization, remote control, and analysis, enhancing the user experience and fostering the integration of IoT technologies into everyday life [20].

References

[1] Ammar, M., Russello, G., and Crispo, B. (2018). Internet of Things: A survey on the security of IoT frameworks. *Journal of Information Security and Applications*, *38*, 8–27.

[2] Ceruzzi, P. E. (2012). *Computing: A Concise History*. MIT press. DOI:-https://doi.org/10.7551/mitpress/9426.001.0001

[3] Aazam, M., Zeadally, S., and Harras, K. A. (2018). Fog computing architecture, evaluation, and future research directions. *IEEE Communications Magazine*, *56*(5), 46–52.

[4] Anand, P., Singh, Y., Selwal, A., Alazab, M., Tanwar, S., and Kumar, N. (2020). IoT vulnerability assessment for sustainable computing: threats, current solutions, and open challenges. *IEEE Access*, *8*, 168825–168853.

[5] Oktian, Y. E., Witanto, E. N., and Lee, S. G. (2021). A conceptual architecture in decentralizing computing, storage, and networking aspect of IoT infrastructure. *IoT*, *2*(2), 205–221.

[6] Lindsay, D., Gill, S. S., Smirnova, D., and Garraghan, P. (2021). The evolution of distributed computing systems: from fundamental to new frontiers. *Computing*, *103*(8), 1859–1878.

[7] Chang, C., Srirama, S. N., and Buyya, R. (2019). Internet of Things (IoT) and new computing paradigms. *Fog and Edge Computing: Principles and Paradigms*, *6*, 1–23.

[8] Jing, J., Helal, A. S., and Elmagarmid, A. (1999). Client-server computing in mobile environments. *ACM Computing Surveys (CSUR)*, *31*(2), 117–157.

[9] Yu, W., Liang, F., He, X., Hatcher, W. G., Lu, C., Lin, J., and Yang, X. (2017). A survey on the edge computing for the Internet of Things. *IEEE Access*, *6*, 6900–6919.

[10] Kong, L., Tan, J., Huang, J., *et al.* (2022). Edge-computing-driven internet of things: A survey. *ACM Computing Surveys*, *55*(8), 1–41.

[11] Mouradian, C., Naboulsi, D., Yangui, S., Glitho, R. H., Morrow, M. J., and Polakos, P. A. (2017). A comprehensive survey on fog computing: State-of-the-art and research challenges. *IEEE Communications Surveys & Tutorials*, *20*(1), 416–464.

[12] Puliafito, C., Mingozzi, E., Longo, F., Puliafito, A., and Rana, O. (2019). Fog computing for the internet of things: A survey. *ACM Transactions on Internet Technology (TOIT)*, *19*(2), 1–41.

[13] Caleffi, M., Amoretti, M., Ferrari, D., Illiano, J., Manzalini, A., and Cacciapuoti, A. S. (2024). Distributed quantum computing: a survey. *Computer Networks*, *254*, 110672.

[14] Vasconcelos, D. R., Andrade, R. M. C., Severino, V., and Souza, J. D. (2019). Cloud, fog, or mist in IoT? That is the question. *ACM Transactions on Internet Technology (TOIT)*, *19*(2), 1–20.

[15] Ketu, S., and Mishra, P. K. (2022). Cloud, fog and mist computing in IoT: An indication of emerging opportunities. *IETE Technical Review*, *39*(3), 713–724.

[16] López Escobar, J. J., Díaz Redondo, R. P., and Gil-Castiñeira, F. (2022). In-depth analysis and open challenges of Mist Computing. *Journal of Cloud Computing*, *11*(1), 81.

[17] Fazel, E., Najafabadi, H. E., Rezaei, M., and Leung, H. (2023). Unlocking the power of mist computing through clustering techniques in IoT networks. *Internet of Things*, *22*, 100710.

[18] Xhafa, F. (2020). The Vision of Edges of Internet as a Computing Fabric. *In Advances in Edge Computing: Massive Parallel Processing and Applications* (pp. 1–21). IOS Press.

[19] Chen, M., Liu, J., Li, P., *et al.* (2022). Fabric computing: Concepts, opportunities, and challenges. *The Innovation*, *3*(6).

[20] Porambage, P., Okwuibe, J., Liyanage, M., Ylianttila, M., and Taleb, T. (2018). Survey on multi-access edge computing for internet of things realization. *IEEE Communications Surveys & Tutorials*, *20*(4), 2961–2991.

Chapter 4

Security vulnerabilities in Internet of Things: attack surfaces, threats, and defense

4.1 Introduction

The rapid evolution of technology has ushered in an era of unprecedented connectivity and interactivity, epitomized by the paradigm-shifting concept of the Internet of Things (IoT). At its core, IoT represents a vast network of interconnected devices, systems, and objects, seamlessly communicating and sharing data to streamline processes, enhance efficiency, and revolutionize various industries. From smart home devices and wearable gadgets to industrial sensors and autonomous vehicles, the pervasive influence of IoT has transformed the way we interact with our surroundings, bridging the physical and digital realms.

While the proliferation of IoT brings forth a multitude of opportunities and advancements, it equally introduces a myriad of security challenges and vulnerabilities that demand meticulous attention. The sheer interconnectedness and ubiquity of IoT devices create an intricate web of vulnerabilities, elevating the stakes for ensuring robust security measures. This chapter delves deep into the security intricacies of IoT systems, illuminating the critical significance of safeguarding these interconnected environments from potential threats, breaches, and malicious activities. The exploration of IoT security becomes paramount in preserving data integrity, protecting user privacy, thwarting cyber threats, and maintaining the reliability of interconnected ecosystems.

The cornerstone of this chapter lies in unraveling the multifaceted landscape of IoT security, dissecting its intricacies, and elucidating comprehensive strategies to fortify these interconnected environments. From understanding the attack surfaces and vulnerabilities inherent in IoT networks to elucidating the significance of regulatory compliance and industry standards, this chapter offers a panoramic view of the challenges and defenses surrounding IoT security. By elucidating the definition, significance, and vulnerabilities inherent in IoT, this chapter aims to provide a holistic overview of the security intricacies within the expansive realm of interconnected devices, while offering actionable insights into securing these dynamic ecosystems.

4.2 IoT attack surfaces

The rapid proliferation of interconnected devices within the IoT landscape has revolutionized the way we interact with technology, offering unparalleled convenience and functionality across various domains. However, this interconnectedness also introduces a myriad of vulnerabilities, collectively referred to as IoT attack surfaces, which pose significant security risks. Attack surfaces in IoT encompass a diverse array of potential entry points and weaknesses within the intricate network of interconnected devices, interfaces, and systems [1]. These vulnerabilities span across hardware, software, networks, data, and physical components, providing malicious actors with opportunities to exploit weaknesses, compromise integrity, and infiltrate systems. Understanding these attack surfaces is essential to fortify the defenses of IoT ecosystems, mitigate risks, and safeguard against evolving threats. In this section, we delve into an exploration of these attack surfaces, examining their intricacies, vulnerabilities, and implications within the context of IoT security, aiming to equip stakeholders with insights and strategies to proactively address these critical security challenges [2].

4.2.1 Network-based attack surfaces

Network-based attack surfaces within the realm of IoT encompass the vulnerabilities and entry points that arise from the interconnectedness and communication protocols utilized by IoT devices. These attack surfaces constitute the various weaknesses present in the networks through which devices exchange data, interact with backend systems, and connect with other devices. They include vulnerabilities in communication protocols, inadequate encryption methods, unsecured network interfaces, and potential points of entry for unauthorized access or interception. Essentially, network-based attack surfaces in IoT represent the susceptibility of the communication infrastructure and interfaces to exploitation, manipulation, or infiltration by malicious actors aiming to disrupt data integrity, intercept sensitive information, or gain unauthorized control over interconnected devices. Understanding and securing these network-based attack surfaces is paramount to fortifying the overall security posture of IoT ecosystems and ensuring the confidentiality, integrity, and availability of transmitted data and interactions among IoT devices and systems [3].

The vulnerabilities associated with wireless communication protocols such as Wi-Fi, Bluetooth, and Zigbee present significant challenges in securing IoT devices and networks. These protocols, while enabling seamless connectivity and communication among devices, also introduce specific vulnerabilities that can be exploited by attackers. Here are some key vulnerabilities associated with these wireless communication protocols:

Wi-Fi vulnerabilities: Wi-Fi, being one of the most prevalent wireless communication standards in IoT devices, presents its own set of vulnerabilities. Weaknesses in Wi-Fi security protocols, such as WEP (Wired Equivalent Privacy) or WPA (Wi-Fi Protected Access), can be exploited by attackers. For instance, WEP encryption is susceptible to brute force attacks due to its easily crackable keys. Additionally, Wi-Fi networks are prone to rogue access points, allowing attackers to set up fake hotspots to intercept data or launch man-in-the-middle (MitM) attacks. The lack of timely security updates for older Wi-Fi-enabled IoT devices further exacerbates these vulnerabilities, leaving them exposed to known exploits and vulnerabilities.

Bluetooth vulnerabilities: Bluetooth technology, commonly used for short-range communication in IoT devices, faces security challenges as well. BlueBorne, a set of vulnerabilities discovered in the Bluetooth protocol, allows attackers to take control of devices, execute arbitrary code, and propagate malware without requiring any user interaction. Furthermore, Bluetooth devices may be susceptible to eavesdropping and unauthorized access due to weaknesses in pairing mechanisms or lack of proper encryption. Attackers exploiting these vulnerabilities can potentially gain access to sensitive data transmitted between devices or execute remote attacks to compromise device functionality.

Zigbee vulnerabilities: Zigbee, often employed in smart home automation and industrial IoT applications, also exhibits vulnerabilities that pose security risks. One critical concern is the lack of robust authentication and encryption mechanisms in some Zigbee implementations, making devices susceptible to unauthorized access or interception of communications. Inadequate encryption or weak key management practices within Zigbee networks can facilitate replay attacks or enable unauthorized devices to join the network, compromising its integrity and exposing sensitive data transmitted between devices (refer to Table 4.1).

Across these wireless communication protocols, common vulnerabilities include inadequate encryption methods, weak authentication mechanisms, and a lack of timely security patches. To mitigate these vulnerabilities, best practices such as implementing stronger encryption algorithms, enforcing strict authentication protocols, regularly updating firmware and software to patch known vulnerabilities, and conducting thorough security assessments and penetration testing are essential. Additionally, adopting newer, more secure versions of these protocols and ensuring proper configuration and management of network settings can significantly enhance the security posture of IoT devices leveraging wireless communication technologies [4].

The progression of cellular networks from 2G through to 5G has brought about transformative advancements in connectivity, offering faster speeds, lower latency, and expanded capabilities. However, this evolution has also introduced a spectrum of vulnerabilities, evolving alongside each generation of cellular technology [5].

Second generation (2G): This network, initially developed primarily for voice calls and text messaging, laid the groundwork for digital cellular communication. Despite technological advancements, 2G networks suffer from vulnerabilities such as the use of weak encryption algorithms like the A5/1 cipher, making them susceptible to eavesdropping and interception. Security weaknesses within the Signaling System 7 (SS7) protocol enabled attackers to perform location tracking, intercept calls, and conduct SMS spoofing attacks. Furthermore, the lack of mutual

Table 4.1 Summary of vulnerabilities and mitigation strategies across common communication protocols used in IoT devices

Communication protocol	Vulnerabilities	Examples of exploits	Potential impact	Mitigation strategies
Wi-Fi	– Weaknesses in WEP and early WPA encryption protocols – Vulnerability to rogue access point attacks – Lack of timely security updates for older devices	– Brute force attacks on weak encryption – Creation of rogue hotspots for data interception	– Unauthorized access to sensitive data – Network compromise – Data interception	– Upgrade to WPA3 – Regular firmware updates – Implementation of intrusion detection systems (IDSs)
Bluetooth	– BlueBorne vulnerabilities allowing remote code execution – Weaknesses in pairing mechanisms – Eavesdropping and unauthorized access risks	– BlueBorne exploit for remote device takeover – Sniffing attacks on unencrypted connections	– Compromise of device control – Data interception – Malware propagation	– Installation of security patches – Use of latest Bluetooth versions – Enforcing stronger pairing methods
Zigbee	– Lack of robust authentication and encryption in some implementations – Weaknesses in key management – Susceptibility to replay attacks	– Unauthorized device joins due to weak authentication – Replay of legitimate messages to disrupt the network	– Compromise of device integrity – Data interception – Network disruption	– Implementation of strong encryption – Regular key rotation – Use of trusted Zigbee stacks
Z-Wave	– Potential security gaps due to insufficient encryption – Vulnerability to certain key attacks	– Encryption key extraction attacks – Interception of communication for device control	– Compromise of device security – Unauthorized access to devices	– Use of stronger encryption methods – Regular updates and patches
NFC (near-field communication)	– Risks related to data interception during transmission – Possible relay attacks	– Intercepting sensitive data during transmission – Relay attacks for unauthorized access	– Unauthorized access to data – Unauthorized transactions	– Use of secure communication protocols – Implementing anti-relay measures

LoRaWAN	– Vulnerabilities in encryption and authentication mechanisms – Potential security gaps in data transmission	– Exploitation of encryption weaknesses for data interception – Sniffing attacks on unencrypted data	– Data interception and tampering – Compromise of device integrity	– Implementation of robust encryption – Regular monitoring for anomalies
Cellular (3G/4G/5G)	– Vulnerabilities in network infrastructure – Susceptibility to MitM attacks	– Interception and manipulation of transmitted data – Impersonation for unauthorized access	– Data tampering – Unauthorized access to networks	– Use of VPNs for secure connections – Implementation of strong authentication methods
Thread	– Weaknesses in security implementations leading to unauthorized access	– Unauthorized device joins the network	– Unauthorized access to devices – Network compromise	– Use of secure network keys – Regular security audits
MQTT (message queuing telemetry transport)	– Lack of encryption and authentication in some implementations	– Unauthorized access to MQTT brokers – Interception of messages	– Data interception – Unauthorized control of devices	– Implementation of TLS encryption – Use of authentication mechanisms
Ethernet	– Vulnerabilities in network switches or routers – Sniffing attacks on unencrypted data transmission	– Unauthorized access to network traffic – Injection of malicious packets	– Data interception – Unauthorized access to networks	– Implementation of VLANs for network segmentation – Encryption of data transmission

authentication between devices and the network facilitated the creation of rogue base stations, exposing users to potential MitM attacks.

Third generation (3G): With the advent of 3G networks, data transmission capabilities expanded, enabling faster Internet access and multimedia services. However, vulnerabilities persisted. The introduction of universal subscriber identity (USIM) cards aimed to bolster security through mutual authentication, yet certain flaws persisted. Vulnerabilities like the MitM attacks against mutual authentication and the KASUMI cipher used in 3G networks highlighted security concerns. Additionally, flaws in the mobile equipment identity (MEID) and international mobile equipment identity (IMEI) systems allowed for device tracking and manipulation.

Fourth generation (4G): The rollout of 4G networks aimed to provide significantly faster data transfer rates and enhanced security features. Despite these advancements, vulnerabilities such as attacks exploiting flaws in Diameter signaling protocols, known as Diameter-based attacks, emerged. These attacks targeted user authentication, subscription information, and session establishment, posing risks of service disruption, data interception, and unauthorized network access. Moreover, weaknesses in the LTE protocol's encryption and authentication mechanisms were identified, potentially leaving devices vulnerable to interception and hijacking of data.

Fifth generation (5G): The onset of 5G technology promises revolutionary advancements in connectivity, yet it introduces its own set of vulnerabilities. The transition to a software-defined network (SDN) and network function virtualization (NFV) raises concerns about potential security gaps and vulnerabilities in virtualized network functions. Additionally, the increased reliance on interconnected devices through the IoT in 5G networks amplifies the attack surface, potentially exposing critical infrastructure to diverse threats, including device compromise, data breaches, and network disruptions.

In each successive generation of cellular networks, while technological advancements have brought significant improvements in speed and capabilities, the concurrent evolution of vulnerabilities demands continuous vigilance, robust security measures, and ongoing advancements in network security protocols to mitigate risks and ensure the integrity and privacy of users' data and communications (refer to Table 4.2).

4.2.2 Physical attack surfaces

In the context of IoT security, physical attack surfaces represent the tangible vulnerabilities and potential entry points existing in the physical components and infrastructure of interconnected devices. These attack surfaces encompass weaknesses that malicious actors can exploit through direct physical access, manipulation, or tampering with hardware, supply chain elements, or the environment surrounding IoT devices. They include vulnerabilities in hardware design,

Table 4.2 Summary of vulnerabilities associated with various generations of cellular networks, including 2G through 6G

Cellular network generation	Vulnerabilities	Examples of exploits	Potential impact	Mitigation strategies
2G	– Weak encryption algorithms (e.g., A5/1) – SS7 vulnerabilities for call interception and tracking – Lack of mutual authentication	– Eavesdropping attacks on voice and text communications – Location tracking via SS7 exploits	– Data interception and manipulation – Call and SMS spoofing – User privacy breaches	– Migration to more secure encryption algorithms – Implementation of secure signaling protocols – Enhanced mutual authentication
3G	– Vulnerabilities in mutual authentication and KASUMI cipher – MEID and IMEI system flaws	– MitM attacks exploiting mutual authentication – Device tracking and manipulation via IMEI vulnerabilities	– Data interception and manipulation – Device tracking – Security breaches	– Strengthening encryption algorithms and protocols – Improved authentication mechanisms
4G	– Diameter signaling protocol vulnerabilities – LTE protocol weaknesses in encryption and authentication	– Diameter-based attacks targeting authentication and session establishment – LTE protocol vulnerabilities exploited for data interception	– Service disruption – Unauthorized network access – Data interception	– Patching and securing – Diameter protocol implementations – Enhanced encryption and authentication protocols
5G	– Potential security gaps in software-defined networking (SDN) and network function virtualization (NFV) – Increased attack surface due to IoT integration	– Exploitation of SDN/NFV vulnerabilities for network disruption – IoT-based attacks compromising critical infrastructure	– Network disruptions – Device compromise and data breaches	– Implementation of robust security measures for SDN/NFV – Strengthening IoT device security
6G	– As 6G technology is still in development, specific vulnerabilities may not be fully identified or standardized at this stage	—	—	—

exposed ports, inadequate tamper resistance, supply chain risks such as counterfeit components, vulnerabilities introduced during third-party manufacturing processes, and the lack of robust physical access controls. Understanding and addressing these physical attack surfaces are crucial as they can lead to unauthorized access, compromise of device integrity, data breaches, or manipulation of IoT devices, emphasizing the importance of implementing stringent physical security measures, supply chain audits, and access controls to mitigate potential risks effectively.

Physical vulnerabilities in IoT devices manifest across various categories, encompassing hardware weaknesses, supply chain risks, third-party manufacturing vulnerabilities, and concerns related to physical access and unauthorized handling. Hardware vulnerabilities encompass flaws in the design, such as exposed ports or inadequate tamper resistance, potentially leading to unauthorized access or manipulation of device functionality. Supply chain risks introduce vulnerabilities arising from counterfeit or compromised components, undermining the device's integrity or introducing backdoors. Third-party manufacturing risks involve vulnerabilities introduced during assembly or production, including the insertion of malicious hardware or firmware. Physical access and unauthorized handling vulnerabilities stem from inadequate access controls or insider threats, leading to unauthorized tampering, theft, or misuse of devices. Understanding these categories is pivotal in fortifying IoT security, necessitating robust measures such as tamper-resistant designs, stringent supply chain audits, manufacturing process validation, and strict physical access controls to mitigate potential risks effectively (refer to Table 4.3).

Hardware vulnerabilities in IoT devices encompass various aspects, including exposed ports and interfaces, which pose significant risks due to their susceptibility to unauthorized access or manipulation. These vulnerabilities arise from accessible physical ports such as sensors, actuators, or controllers, along with peripheral interfaces like USB ports or SD card slots. These entry points offer avenues for potential exploitation, enabling unauthorized entities to gain access to sensitive areas within the device or inject malicious code, ultimately compromising the functionality or integrity of the device. Addressing these vulnerabilities requires stringent measures, including implementing access controls, encryption protocols for data transfer through ports, and disabling unnecessary ports or interfaces when not in use to mitigate unauthorized access and manipulation risks.

Moreover, tamper resistance and physical security play a critical role in mitigating hardware vulnerabilities in IoT devices. Robust physical casing and tamper-resistant designs serve as essential barriers against unauthorized entry or manipulation of internal components. Effective physical security measures aim to prevent physical access to sensitive parts of the device, deterring attempts at tampering or extracting critical information. Implementing mechanisms such as secure enclosures, sealed casings, or anti-tamper seals help deter unauthorized physical access and tampering. Additionally, incorporating IDSs that trigger alerts upon detecting physical breaches can further fortify the device's physical security. Overall, a combination of robust physical casing, tamper-resistant designs, and vigilant monitoring of physical access

Table 4.3 Summary of physical vulnerabilities in IoT systems across four categories

Category	Vulnerabilities	Examples of exploits	Potential impact	Mitigation strategies
Hardware vulnerabilities	– Exposed ports and interfaces allowing unauthorized access – Lack of tamper-resistant casing	– Unauthorized access to internal components – Manipulation of device functionality	– Compromise of device integrity – Unauthorized data access	– Implementation of tamper-resistant designs – Secure casing and enclosure
Supply chain risks	– Inclusion of counterfeit or compromised components in devices – Lack of supply chain transparency	– Insertion of compromised hardware components – Use of counterfeit parts	– Introduction of backdoors or vulnerabilities – Device malfunctioning	– Rigorous supply chain audits – Use of trusted suppliers
Third-party manufacturing risks	– Vulnerabilities introduced during manufacturing or assembly processes	– Insertion of malicious hardware or firmware during assembly	– Compromise of device security and functionality	– Verification and validation of manufacturing processes
Physical access and unauthorized handling	– Lack of physical access controls or inadequate security measures – Insider threats due to unauthorized handling	– Unauthorized access to devices or components – Insider misuse of access	– Tampering, theft, or manipulation of devices – Data breaches	– Implementation of strict physical access controls – Employee training

forms the cornerstone of mitigating hardware vulnerabilities and safeguarding the integrity and functionality of IoT devices.

Supply chain risks present substantial threats to the security and integrity of IoT devices, encompassing concerns related to counterfeit components and vulnerabilities introduced during third-party manufacturing processes. Counterfeit components pose a significant hazard within the supply chain, potentially infiltrating IoT devices with compromised or fake hardware components. These components, if undetected, may introduce vulnerabilities or backdoors, undermining the device's functionality or providing avenues for unauthorized access. Detecting and mitigating these risks often involve rigorous supply chain audits, verifying the authenticity of components, and establishing relationships with trusted suppliers to ensure the procurement of genuine and secure hardware components for IoT devices.

Additionally, third-party manufacturing risks contribute to vulnerabilities as outsourcing manufacturing processes to external vendors increases the potential for compromises in device security. Trusting third-party vendors with sensitive manufacturing procedures might inadvertently introduce vulnerabilities during assembly or production. An oversight in the manufacturing phase can result in the insertion of malicious hardware or firmware into IoT devices, compromising their security. To mitigate these risks, stringent verification and validation processes throughout the manufacturing chain are crucial. Implementing thorough vetting processes for third-party vendors, conducting regular audits, and ensuring adherence to strict security protocols during the manufacturing phase are essential steps toward minimizing vulnerabilities introduced through third-party manufacturing practices in the supply chain. These proactive measures are vital in bolstering the security posture of IoT devices and mitigating risks originating from supply chain vulnerabilities.

Environmental risks pose a substantial threat to the reliability and functionality of IoT devices, particularly vulnerabilities arising from exposure to extreme environmental conditions. Devices deployed in harsh environments with extreme temperature fluctuations, high humidity, or physical impact face increased susceptibility to operational degradation or complete failure. These conditions can compromise the device's integrity, affecting its performance, accuracy, and overall functionality. Extreme temperatures, for instance, can lead to overheating or freezing of internal components, causing malfunctions or permanent damage. High humidity levels might trigger corrosion or short circuits within the device, resulting in erratic behavior or hardware failure. Physical impacts, such as shocks or vibrations, can disrupt sensitive electronic components, leading to operational inconsistencies or outright device breakdown. To mitigate these environmental risks, robust design considerations, such as ruggedized casings, protective coatings, or temperature-resistant materials, are essential to ensure device durability and reliability in challenging environmental conditions [6].

Moreover, proactive measures, including thorough testing and validation of IoT devices under simulated extreme environmental conditions, are imperative. Stress-testing devices in controlled environments that replicate extreme temperatures, humidity levels, or physical impacts help identify vulnerabilities early in the development phase. Implementing environmental sensors within devices can aid in monitoring and alerting users about potential environmental hazards, allowing for proactive measures or device relocations when conditions surpass operational thresholds. Additionally, incorporating fail-safes or shutdown mechanisms triggered by extreme environmental parameters can prevent further damage or data loss, enhancing the resilience of IoT devices against environmental risks. Overall, addressing environmental vulnerabilities necessitates a multi-faceted approach involving robust design considerations, rigorous testing, and responsive mechanisms to ensure device reliability and longevity in adverse environmental conditions.

Physical access and unauthorized handling present critical vulnerabilities in IoT devices, stemming from both inadequate physical access controls and potential insider threats. Insufficient controls over physical access to IoT devices create vulnerabilities, leaving them open to unauthorized handling, tampering, or theft. Without robust access controls in place, unauthorized individuals may gain physical access to devices, compromising their integrity, manipulating settings, or extracting sensitive data. Strengthening physical access controls involves implementing stringent measures such as biometric authentication, keycard access, or secure enclosures, limiting physical access to authorized personnel only. Additionally, monitoring and logging access attempts, conducting regular audits, and employing surveillance measures contribute to bolstering physical security and mitigating risks associated with unauthorized physical access to IoT devices.

Moreover, insider threats pose significant risks in the context of IoT security, wherein authorized personnel misuse their access privileges to compromise device integrity or pilfer sensitive data. Such threats could stem from employees, contractors, or individuals with legitimate access to devices, abusing their permissions for malicious intent. Insider misuse might involve tampering with device configurations, introducing vulnerabilities, or stealing confidential information. Addressing insider threats requires a comprehensive approach, including thorough employee training on security protocols, implementing least privilege access, and establishing strict access controls and monitoring mechanisms. Encouraging a culture of security awareness and enforcing stringent policies and procedures can minimize the likelihood of insider threats, mitigating risks arising from potential unauthorized handling or misuse of IoT devices by authorized personnel.

4.2.3 Software attack surfaces

Software attack surfaces within the domain of IoT refer to the vulnerable points, interfaces, or components within the software ecosystem of interconnected devices that can be exploited by malicious entities to compromise the security, integrity, or functionality of the system. These attack surfaces encompass the various entry points or weaknesses in software elements, including operating systems, firmware, applications, and communication protocols, which adversaries can target to gain unauthorized access, manipulate data, launch cyber-attacks, or disrupt normal operations. Understanding software attack surfaces involves recognizing the vulnerabilities within software components, their interconnectivity, and the potential exploitation points that adversaries can leverage to infiltrate or compromise the IoT ecosystem. Addressing these vulnerabilities requires comprehensive security measures, including robust software design, secure coding practices, regular updates and patches, and effective network security protocols to fortify the software elements and defend against potential threats.

Software vulnerabilities in IoT devices encompass various facets, including weaknesses in embedded operating systems, firmware, and web/mobile interfaces, each posing distinct risks to the security and integrity of interconnected devices. Embedded operating systems, foundational to IoT devices, often face vulnerabilities due to limited resources, outdated software, or insufficient security measures. These vulnerabilities can stem from unpatched software flaws, weak authentication mechanisms, or inadequate encryption protocols, allowing attackers to exploit these entry points for unauthorized access, data interception, or device manipulation. Firmware vulnerabilities present another critical aspect, as flaws within the firmware—such as outdated versions, unencrypted or weakly encrypted firmware, or lack of secure boot mechanisms—can expose devices to the risk of unauthorized firmware modifications, leading to device compromise or the introduction of backdoors. Addressing these vulnerabilities necessitates regular firmware updates, implementing secure boot mechanisms, and adopting encryption protocols to enhance the resilience of IoT devices against firmware-based attacks.

Furthermore, web and mobile interfaces used for managing or interacting with IoT devices serve as crucial attack surfaces susceptible to various security risks. Weak authentication mechanisms, insufficient input validation, or poor session management within web or mobile interfaces can expose vulnerabilities that adversaries may exploit to compromise user credentials, gain unauthorized access to devices, or execute remote code execution attacks. Insecure communication protocols or insufficient data encryption in these interfaces can lead to data breaches or interception, compromising user privacy or allowing attackers to manipulate device functionalities. Mitigating these vulnerabilities involves employing stringent authentication mechanisms, implementing robust encryption standards for data transmission, and conducting regular security assessments and audits to identify and address vulnerabilities within web and mobile interfaces, thus fortifying the overall security posture of IoT ecosystems.

Addressing software vulnerabilities in IoT necessitates a multi-layered approach encompassing embedded operating systems, firmware, and web/mobile interfaces. Strengthening security in embedded operating systems involves continuous monitoring, regular updates, and the implementation of security patches to mitigate known vulnerabilities. Additionally, enhancing firmware security by employing secure boot mechanisms, encryption protocols, and timely firmware updates aids in safeguarding devices against unauthorized modifications or exploitation. Similarly, securing web and mobile interfaces demands rigorous adherence to secure coding practices, robust authentication mechanisms, and encryption standards to thwart potential attacks and ensure the confidentiality, integrity, and availability of data exchanged between users and IoT devices. Implementing these measures collectively bolsters the resilience of IoT systems against software-based vulnerabilities and strengthens their overall security posture (refer to Table 4.4).

Table 4.4 Summary of different categories of software vulnerabilities in IoT systems

Software vulnerability category	Description	Examples	Potential impact	Mitigation strategies	Best practices
Embedded operating systems	Vulnerabilities within the underlying operating systems of IoT devices, including flaws in kernel or drivers	Unpatched software flaws Weak authentication mechanism Insecure default settings	Unauthorized access Data interception Device manipulation	Regular updates and patches Strong authentication Secure default configurations	Regular security audits Secure boot mechanism Least privilege access
Firmware vulnerabilities	Weaknesses within the firmware of IoT devices, including outdated versions, unencrypted firmware, or lack of secure boot mechanisms	Outdated firmware versions Unencrypted firmware Lack of secure boot	Unauthorized firmware modifications Device compromise Backdoor introduction	Regular firmware updates Implementation of secure boot mechanism Encryption protocols	Secure firmware signing Firmware integrity checks Continuous monitoring
Web and mobile interfaces	Vulnerabilities in interfaces used to manage or interact with IoT devices, such as weak authentication, insufficient input validation, or poor session management	Weak authentication mechanism Insufficient input validation Poor session management	Compromise of user credentials Unauthorized access to devices Data breaches	Robust authentication methods Encryption for data transmission Regular security assessments	Regular code reviews API security measures HTTPs implementation

4.2.4 Data attack surfaces

Data attack surfaces in the realm of IoT encompass the various vulnerabilities and entry points within the data ecosystem of interconnected devices that adversaries can exploit to compromise the confidentiality, integrity, or availability of data. These attack surfaces represent the susceptible points, interfaces, or data pathways where unauthorized access, interception, manipulation, or exfiltration of sensitive information can occur. Data attack surfaces may include vulnerabilities in data transmission protocols, inadequate encryption methods, weaknesses in data storage or handling mechanisms, or exploitation of APIs and data interfaces. Understanding and fortifying these attack surfaces involve identifying potential weaknesses across data channels, ensuring robust encryption practices, implementing access controls, and strengthening data storage mechanisms to safeguard against unauthorized access, data breaches, or tampering, thereby reinforcing the overall data security posture within IoT ecosystems.

Data attack surfaces within the context of IoT refer to the vulnerable points, interfaces, or pathways through which data is transmitted, stored, processed, or accessed within interconnected devices, presenting opportunities for unauthorized manipulation, interception, or compromise of data integrity or confidentiality. These attack surfaces encompass various vulnerabilities across the data lifecycle, such as weaknesses in data encryption protocols, insecure data storage methods, susceptibility in data transmission channels, or exploitable flaws in data handling mechanisms, including APIs or interfaces. Addressing data attack surfaces involves implementing robust encryption methods, ensuring secure data storage practices, scrutinizing data transmission protocols, and implementing stringent access controls to prevent unauthorized access, data breaches, or tampering, thereby bolstering the overall resilience and security of data within IoT infrastructures [7].

Data privacy vulnerabilities stemming from inadequate protection of sensitive data collected by IoT devices present substantial risks that can lead to privacy breaches and compromise user confidentiality. IoT devices often collect an extensive array of sensitive user data, including personal information, behavioral patterns, or location data, making them enticing targets for attackers if not adequately protected. Insufficient data encryption, weak authentication mechanisms, or insecure data storage practices within IoT devices can result in unauthorized access or data leaks, potentially exposing individuals to privacy infringements or identity theft. Furthermore, inadequate anonymization or pseudonymization of collected data exacerbates the risk, as unauthorized disclosure of personally identifiable information might lead to severe consequences, including financial losses or reputational damage for users. Addressing these vulnerabilities necessitates robust encryption protocols, implementation of stringent access controls, and adopting anonymization techniques to safeguard sensitive data, ensuring compliance with privacy regulations and bolstering user trust in IoT technologies.

Moreover, the proliferation of interconnected devices amplifies the complexity of data privacy vulnerabilities across ecosystems. Inter-device communication and data sharing within IoT networks pose additional risks, potentially leading to unintended data exposure or correlation attacks. Weak linkages or unsecured interfaces between devices can facilitate data leakage or unauthorized access across interconnected systems. Furthermore, the aggregation and analysis of data from multiple sources heighten the risk of privacy violations through data inference or profiling, potentially revealing sensitive user information. To mitigate these risks, comprehensive security measures, including implementing secure communication protocols between devices, segmenting IoT networks, and employing data anonymization or differential privacy techniques, are crucial. Ensuring data privacy within IoT environments demands a holistic approach that emphasizes robust encryption, secure data handling practices, and continuous monitoring to safeguard sensitive information and protect user privacy in an increasingly interconnected digital landscape.

Data integrity vulnerabilities stemming from the manipulation of data stored or transmitted by IoT devices pose significant risks, potentially leading to false readings, erroneous analytics, or critical system errors. As IoT devices generate and transmit vast volumes of data, ensuring the accuracy and reliability of this information is paramount. Manipulating data within IoT systems, either through unauthorized access or cyber-attacks, can compromise the trustworthiness and fidelity of the data, leading to severe consequences. For instance, tampering with sensor data measuring environmental factors, such as temperature or humidity, could yield inaccurate readings, leading to erroneous decision-making or actions within automated systems. Moreover, falsified data in critical infrastructure systems, such as energy grids or healthcare devices, may result in systemic failures, disruptions, or even endanger lives. Addressing data integrity vulnerabilities requires implementing robust authentication mechanisms, end-to-end encryption (E2EE), and implementing data validation protocols to detect and prevent unauthorized alterations or tampering of data.

Furthermore, the interconnected nature of IoT ecosystems amplifies the complexity of data integrity vulnerabilities across various interconnected devices and systems. Inter-device communication and data sharing within IoT networks open avenues for potential data manipulation across multiple endpoints. Compromising the integrity of data during transmission or at intermediary points within the network can lead to cascading effects, where manipulated data propagates throughout the system, influencing decision-making processes based on inaccurate or falsified information. Additionally, the use of centralized data storage or cloud-based platforms for IoT data introduces vulnerabilities to unauthorized access or manipulation, emphasizing the need for stringent access controls, encryption, and continuous monitoring to safeguard data integrity (refer to Table 4.5). Employing blockchain technology or cryptographic techniques for data authentication and ensuring secure data transmission protocols can fortify the integrity of data across IoT ecosystems, mitigating the risks associated with data manipulation and preserving the reliability of information within interconnected systems [8].

4.2.5 Cloud and backend attack surfaces

Cloud and backend attack surfaces refer to the vulnerable points, interfaces, or entryways within the cloud infrastructure and backend systems that support IoT devices, exposing potential weaknesses and exploitable areas susceptible to

Table 4.5 Overview of data vulnerabilities in IoT systems and mitigation strategies

Data vulnerability	Description	Examples	Potential impact	Mitigation strategies
Weak data encryption	Insufficient or weak encryption methods applied to sensitive data stored or transmitted by IoT devices	Inadequate encryption protocols Weak encryption keys	Unauthorized data access Data breaches Data interception	Implementation of robust encryption algorithms Regular encryption key updates
Insecure data storage	Vulnerabilities arising from insecure storage practices leading to unauthorized access or data leakage	Unprotected data storage Weak access controls	Unauthorized data access Data leaks Data corruption	Implementation of secure data storage protocols Encryption of stored data
Lack of data validation	Absence or insufficient validation of input data, allowing for potential exploitation or manipulation	Insufficient input validation checks Failure to sanitize data inputs	Injection attacks (e.g., SQL injection) Data manipulation Data corruption	Implementation of rigorous input validation checks Use of secure coding practices
Vulnerable data transmission	Weaknesses in data transmission protocols, susceptible to interception or manipulation during transit	Unencrypted data transmission Weak network protocols	Data interception MitM attacks Data tampering	Adoption of secure communication protocols Encryption for data transmission
Inadequate authentication	Weak or absent authentication mechanisms, leading to unauthorized access to sensitive data	Weak passwords or authentication methods Lack of multi-factor authentication	Unauthorized data access Data breaches Identity theft	Implementation of strong authentication methods Enforcement of multi-factor authentication

Poor API security	Vulnerabilities within APIs used for data communication, potentially exposing data to unauthorized access or manipulation	Insecure API endpoints Insufficient access controls	Unauthorized data access Data breaches Data manipulation	Implementation of secure API frameworks Regular security assessments of API endpoints
Data correlation risks	Risks associated with correlating data from multiple sources, potentially revealing sensitive information	Cross-referencing data from various IoT devices Data aggregation	Privacy breaches Identity exposure Data profiling	Implementation of differential privacy techniques Anonymization of aggregated data
Lack of data access controls	Absence of stringent controls over data access, leading to unauthorized handling or manipulation of sensitive information	Insufficient access restrictions Failure to segregate data	Unauthorized data access Data manipulation Data theft	Implementation of strict access controls Segregation of sensitive data
Unvalidated data sources	Risks associated with integrating data from unverified or unreliable sources into IoT systems	Use of data from untrusted sources Failure to validate incoming data	Compromise of data integrity Introduction of erroneous data Data manipulation	Validation of data sources Thorough vetting of incoming data
Lack of data lifecycle security	Insufficient security measures are implemented throughout the data lifecycle, leaving data vulnerable at various stages	Weak data disposal practices Lack of encryption in transit and storage	Data leaks at various stages Data breaches Privacy violations	Implementation of E2EE Adoption of secure data disposal practices

unauthorized access, data breaches, or service disruptions. These attack surfaces encompass vulnerabilities inherent in the cloud environment, including misconfigurations, weak access controls, or inadequate encryption protocols, which may compromise the confidentiality, integrity, or availability of data stored or processed within cloud services. Additionally, backend systems supporting IoT devices, such as authentication mechanisms, APIs, or data handling processes, form part of these attack surfaces, as weaknesses in these systems could lead to unauthorized access, data manipulation, or service interruptions, thereby impacting the functionality and security of interconnected IoT devices. Understanding and fortifying these attack surfaces are crucial to mitigating risks and ensuring the resilience and security of the cloud and backend systems that underpin IoT infrastructures [9].

Cloud infrastructure vulnerabilities: Vulnerabilities within cloud services hosting IoT data and applications encompass risks such as misconfigurations, weak access controls, or inadequate security measures. Misconfigured cloud settings might inadvertently expose sensitive data to unauthorized access or breaches. Weak access controls within cloud platforms could lead to unauthorized users gaining entry to critical resources. Insufficient encryption or weak encryption keys might compromise data confidentiality. Additionally, shared cloud environments might introduce the risk of multi-tenancy vulnerabilities, where a security breach in one instance could impact other IoT deployments, emphasizing the need for robust isolation measures and stringent access controls.

Backend system vulnerabilities: The backend systems supporting IoT devices are susceptible to various vulnerabilities, including authentication flaws, insecure APIs, or insufficient data validation mechanisms. Weaknesses in authentication mechanisms might lead to unauthorized access to backend systems or IoT device controls. Insecure APIs might expose critical functionalities to exploitation or manipulation, potentially compromising the integrity or confidentiality of data. Insufficient data validation can lead to injection attacks or data corruption, impacting the reliability and security of backend systems, which form a crucial component of IoT infrastructure.

Third-party service vulnerabilities: Integration of third-party services or APIs within cloud or backend systems introduces additional risks. These services might have vulnerabilities themselves or could introduce weaknesses into the IoT ecosystem, leading to potential security breaches or data leaks. Insecurely integrated third-party services might allow attackers to exploit vulnerabilities in these services to gain unauthorized access or manipulate data within IoT environments, underscoring the importance of rigorous vetting, security assessments, and ongoing monitoring of third-party integrations.

Data transmission security: Ensuring secure data transmission between IoT devices and cloud/backend systems is paramount. Weaknesses in communication protocols or lack of encryption during data transmission could expose data to interception, manipulation, or unauthorized access. Insecure communication channels might enable adversaries to eavesdrop on sensitive data or alter information being transmitted, compromising the confidentiality and integrity of data exchanged between IoT devices and backend/cloud systems.

DDoS and service disruption: Distributed denial of service (DDoS) attacks targeting cloud or backend infrastructure can disrupt services supporting IoT devices. These attacks flood systems with excessive traffic, rendering them inaccessible or causing service outages. Such disruptions could impact the functionality of IoT devices, interrupting critical services and affecting the reliability of interconnected systems. Implementing robust DDoS mitigation strategies, such as traffic filtering mechanisms, load balancing, or employing cloud-based DDoS protection services, is crucial to safeguard against service disruptions and maintain the availability of IoT services (refer to Table 4.6). Addressing these vulnerabilities requires a multi-layered approach involving robust security measures, including stringent access controls, encryption for data transmission, regular security assessments, secure API integrations, and DDoS mitigation strategies. These efforts are critical to fortifying cloud and backend systems supporting IoT, ensuring their resilience against potential threats and maintaining the security and functionality of interconnected devices and services [10].

Table 4.6 Summary of cloud and backend vulnerabilities in IoT systems and mitigation strategies

Vulnerability	Description	Examples	Potential impact	Mitigation strategies
Misconfigured cloud settings	Errors in cloud configuration leading to exposed sensitive data	Improperly configured access controls Inadequate encryption settings	Unauthorized access to sensitive data Data breaches	Regular security audits Configuration validation procedures
Weak backend authentication	Flaws in authentication mechanisms within backend systems	Weak or default credentials Lack of multi-factor authentication	Unauthorized access to backend systems Data manipulation	Strong authentication mechanisms Implementation of multi-factor authentication
Insecure APIs	Vulnerabilities in APIs used for communication within IoT systems	Inadequate access controls Insufficient data validation	Unauthorized access to critical functions Data manipulation	Secure API frameworks Rigorous data validation processes
Data transmission weaknesses	Vulnerabilities in data transmission channels between IoT devices and backend/cloud systems	Unencrypted data transmission Weak encryption protocols	Data interception Unauthorized access Data manipulation	Implementation of strong encryption protocols Secure communication standards

(Continues)

Table 4.6 Summary of cloud and backend vulnerabilities in IoT systems and mitigation strategies (Continued)

Vulnerability	Description	Examples	Potential impact	Mitigation strategies
Shared cloud tenancy risks	Risks associated with shared cloud environments in multi-tenant setups	Shared resources with weak isolation Inter-tenant data leakage	Cross-tenant data exposure Security breaches	Robust isolation mechanisms Strict access controls for multi-tenant environments
Insufficient data backup	Lack of robust data backup and recovery mechanisms in backend systems	Failure to regularly backup data Unreliable backup procedures	Data loss in case of system failures Inability to recover from incidents	Regular data backups Testing backup and recovery procedures
Vulnerable third-party services	Risks associated with integrating insecure third-party services or APIs	Using vulnerable third-party services Insufficiently secured integrations	Potential compromise of IoT ecosystem Data leaks	Comprehensive security assessments of third-party services Ongoing monitoring of integrations
Backend data validation issues	Lack of proper validation of data processed by backend systems	Insufficient input validation checks Failure to sanitize user inputs	Injection attacks Data corruption System compromise	Rigorous data validation procedures Adherence to secure coding practices
DDoS vulnerabilities	Weaknesses that render backend/cloud systems susceptible to DDoS attacks	Lack of DDoS protection measures Single points of failure	Service disruptions IoT device unavailability	Implementing DDoS mitigation strategies Redundancy and load balancing
Inadequate access controls	Weak or inadequate access controls within backend systems or cloud services	Improperly configured permissions Failure to enforce access policies	Unauthorized access to sensitive data Data breaches	Strengthening access control mechanisms Regular access audits and reviews

4.3 Common IoT attacks

In the realm of interconnected devices, the landscape of the IoT brings forth not just unparalleled connectivity but also a myriad of security challenges. Understanding the spectrum of threats that pervade this ecosystem is paramount. This section delves into the diverse array of common IoT attacks, encompassing various strategies utilized by malicious entities to compromise the integrity, confidentiality, or availability of IoT systems. These attacks range from exploiting vulnerabilities in devices and networks to manipulating data flows and targeting cloud or backend infrastructures. This section aims to dissect and elucidate these attack vectors, offering insights into their methodologies, potential impacts, and mitigation strategies. By comprehending the multifaceted nature of common IoT attacks, stakeholders can proactively fortify their systems, enhance resilience, and engender a more secure IoT environment for the interconnected digital world.

4.3.1 Physical security attacks

Physical security attacks pose a significant threat within the realm of IoT, targeting the tangible aspects of interconnected devices and infrastructure. These attacks, unlike purely digital threats, encompass a spectrum of methodologies aiming to exploit vulnerabilities in the physical domain, including tampering with devices, unauthorized access to hardware components, supply chain compromises, and theft. This section explores the realm of physical security attacks in the context of IoT, shedding light on the diverse strategies employed by threat actors to breach the physical barriers safeguarding these systems. By understanding these attack vectors and their potential ramifications, stakeholders can implement robust measures to fortify physical security, prevent unauthorized access, and mitigate risks inherent in the physical realm of IoT infrastructure. In the following, a popular set of physical security attacks is stated and defined.

Delayering: This attack involves removing layers of protective coatings or materials from integrated circuits to access and compromise sensitive components or data stored within the device.

Microprobing: A sophisticated attack where fine needles or probes are used to directly access and manipulate circuits or extract sensitive information from within the device without damaging the outer casing.

Circuit tracing: Adversaries attempt to trace or map the circuitry of IoT devices to understand their functionality, potentially allowing them to identify vulnerabilities or intercept data flows.

Node capture: Attackers physically capture or steal IoT nodes or devices to analyze, tamper with, or extract sensitive data or intellectual property stored within the device.

Node outage: Deliberate physical disruption or sabotage targeting IoT nodes or devices to cause service interruptions or render them temporarily or permanently non-functional.

Node malfunction: Manipulation of physical components or settings in IoT nodes to induce malfunctions, disrupt operations, or alter device functionalities, potentially leading to erroneous outputs or data corruption.

Node destruction: Physical attacks intending to completely destroy or render IoT devices inoperable, causing irreparable damage or disrupting the overall network.

Node replication: Unauthorized replication or cloning of IoT nodes, creating counterfeit devices that may infiltrate the network, leading to unauthorized access or data compromise.
Side-channel attacks: Exploiting unintended information leakage, such as power consumption or electromagnetic emissions, to gather insights or extract sensitive information from IoT devices.
Physical tampering: Any physical interference with IoT devices, including opening device casings, inserting malicious hardware, or altering device configurations to compromise their security or functionality.
Supply chain attacks: Manipulation or compromise of IoT devices or components during the manufacturing, assembly, or distribution process, introducing vulnerabilities or backdoors into the devices before they reach end-users.
Signal interception: Intercepting wireless signals or communication channels used by IoT devices to eavesdrop, manipulate, or disrupt data transmission between devices or to the cloud.
Device relocation: Physical theft or unauthorized movement of IoT devices from their designated locations to gain unauthorized access or manipulate device functionalities.
Environmental attacks: Exposing IoT devices to extreme conditions, such as temperature variations, humidity, or physical impact, causing damage or malfunction in the devices.
Physical access exploitation: Exploiting inadequate physical access controls to gain unauthorized entry or manipulate IoT devices, leading to data breaches, device compromise, or unauthorized actions.

4.3.2 Data privacy attacks

Data privacy attacks in the realm of IoT present a critical concern, aiming to compromise the confidentiality, integrity, or availability of sensitive information transmitted or stored within interconnected devices. This section explores the landscape of data privacy attacks within IoT ecosystems, encompassing a diverse range of strategies employed by malicious actors to exploit vulnerabilities and breach the privacy of user data. These attacks include unauthorized data access, data breaches, identity theft, and manipulation of sensitive information. Understanding these threats is pivotal in devising robust measures to safeguard user privacy, implement stringent data encryption, access controls, and secure data handling practices, ensuring the preservation of user confidentiality and trust within the interconnected IoT environment [11]. In the following, a popular set of data privacy attacks is stated and defined.

- **Data breaches:** Unauthorized access or exposure of sensitive information stored within IoT systems, potentially compromising user privacy and confidentiality.
- **MitM attacks:** Interception of communication between IoT devices, enabling attackers to eavesdrop on or manipulate data exchanged between devices or with backend systems.

- **Eavesdropping/sniffing:** Unauthorized interception of data transmission between IoT devices or within networks, allowing attackers to gather sensitive information.
- **Data interception:** Capturing or diverting data traffic between IoT devices or networks, potentially compromising the confidentiality of transmitted information.
- **SQL injection:** Exploiting vulnerabilities in IoT applications or databases to inject malicious SQL queries, allowing unauthorized access to sensitive data.
- **Cross-site scripting (XSS):** Injecting malicious scripts into IoT web interfaces or applications to manipulate user interactions or steal sensitive information.
- **Credential attacks:** Attempting to gain unauthorized access to IoT devices or systems by exploiting weak or default credentials, leading to data breaches or unauthorized control.
- **Phishing:** Deceptive techniques aimed at tricking users or administrators into disclosing sensitive information, credentials, or granting access to IoT systems.
- **Social engineering:** Manipulating individuals to divulge confidential information or access credentials through psychological manipulation or deception.
- **Data manipulation:** Unauthorized alteration or modification of data stored or transmitted by IoT devices, compromising data integrity or leading to false information.
- **Cryptojacking:** Exploiting IoT devices' computational resources to mine cryptocurrency without users' consent, impacting device performance and data privacy.
- **Firmware attacks:** Targeting vulnerabilities in IoT device firmware to gain unauthorized access, manipulate device functionalities, or extract sensitive data.
- **Privacy policy violations:** Breaching data privacy regulations or user agreements by mishandling or misusing collected data from IoT devices without user consent.
- **Location tracking exploitation:** Unauthorized tracking or disclosure of users' location data collected by IoT devices, compromising individual privacy.
- **IoT botnets:** Compromising multiple IoT devices to form botnets, enabling large-scale attacks that compromise data privacy, disrupt services, or facilitate other malicious activities.

4.3.3 Network attacks

Network security attacks within the IoT ecosystem present a formidable challenge, targeting the interconnected web of devices and the communication infrastructure that binds them. This section delves into the domain of network security attacks within IoT, exploring a multitude of strategies employed by malicious entities to compromise the integrity, availability, and confidentiality of data transmitted across these networks. These attacks encompass a broad spectrum, including but not limited to DDoS assaults, MitM attacks, and network sniffing, among others. Understanding these attack vectors is pivotal in fortifying the network infrastructure, implementing robust encryption, access controls, and IDSs, and fostering a resilient IoT environment capable of withstanding the diverse array of network-centric threats [12]. In the following, a popular set of network attacks is stated and defined.

DDoS attacks: Overwhelming IoT networks with an excessive volume of traffic, rendering them inaccessible and disrupting services.
MitM attacks: Intercepting communication between IoT devices or between devices and backend systems to eavesdrop or manipulate data exchanges.
Network sniffing: Monitoring and capturing data packets transmitted across IoT networks, potentially revealing sensitive information.
Spoofing attacks: Falsifying network identities or addresses to impersonate legitimate IoT devices or systems, allowing unauthorized access or data manipulation.
DNS spoofing/poisoning: Manipulating DNS records within IoT networks to redirect traffic or mislead devices, leading to unauthorized access or information theft.
ARP spoofing: Manipulating address resolution protocol (ARP) tables to associate fake MAC addresses with legitimate IP addresses, intercepting or redirecting traffic.
IoT botnets: Compromising multiple IoT devices to form botnets used for coordinated attacks, data breaches, or distributed attacks against other systems.
Rogue devices: Unauthorized or malicious devices are introduced into IoT networks, potentially causing disruptions or compromising network security.
Brute force attacks: Attempting to gain unauthorized access to IoT devices or networks by repeatedly trying various combinations of credentials or keys.
Eavesdropping: Unauthorized interception of network communications between IoT devices, potentially compromising sensitive data.
IoT worms: Self-replicating malware targeting vulnerabilities in IoT devices to spread across networks, compromising multiple devices.
Network traffic analysis: Analyzing patterns or volume of network traffic to gather insights or identify vulnerabilities within IoT networks.
Session hijacking: Unauthorized takeover of an ongoing communication session between IoT devices or between devices and backend systems.
Network layer attacks: Targeting vulnerabilities within network protocols or layers (e.g., TCP/IP vulnerabilities) to disrupt communication or compromise devices.
Zero-day exploits: Exploiting previously unknown vulnerabilities in IoT devices or network infrastructure before they are patched, potentially causing severe disruptions or breaches.

Each of these network security attacks poses significant risks to IoT environments, highlighting the importance of implementing robust security measures, such as encryption, IDSs, regular patching, network segmentation, and access controls, to mitigate these threats and safeguard interconnected IoT networks (refer to Table 4.7).

Table 4.7 Comparison between various firmware and software security attacks

Attack	Attack surface	Attack vectors	Potential impact	Mitigation strategies
Malware infections	Software	Phishing, infected downloads	Data breaches, device compromise, unauthorized control	Regular security updates, antivirus/antimalware solutions
Code injection	Software	Input manipulation	Unauthorized access, data manipulation, system compromise	Input validation, code audits, secure coding practices
Buffer overflow	Software	Input manipulation	Code execution, system crashes, unauthorized access	Code reviews, memory protection mechanisms
SQL injection	Software	Input manipulation	Data manipulation, unauthorized access, database compromise	Prepared statements, input sanitization
Cross-site scripting	Software	Input manipulation	Data theft, unauthorized actions, user session hijacking	Input validation, output encoding
Firmware tampering	Firmware	Physical access, remote	Device compromise, unauthorized access, device malfunction	Firmware integrity checks, secure boot mechanisms
Rootkit installation	Firmware	Exploiting vulnerabilities	Persistent unauthorized access, evasion of detection, remote control	Secure boot, regular firmware verification
Backdoor exploitation	Firmware	Vulnerabilities exploitation	Unauthorized access, data theft, remote control	Regular security audits, firmware signing
Zero-day exploits	Firmware/software	Unknown vulnerabilities	System compromise, data breaches, unauthorized access	Prompt patching, security monitoring
Bricking attacks	Firmware	Manipulating firmware	Device inoperability, service disruption, rendering devices irreparable	Secure boot mechanisms, regular updates
API vulnerabilities	Software	Weaknesses in API design	Unauthorized access, data manipulation, service disruptions	API security best practices, regular security assessments
Privilege escalation	Software	Vulnerability exploitation	Unauthorized access to critical functions, data breaches, system compromise	Principle of least privilege, access control mechanisms
Denial-of-service (DoS)	Software/network	Flooding, exploiting flaws	Service disruption, device unavailability, system crashes	Traffic filtering, load balancing, DoS protection services
Cryptojacking	Software	Exploiting system resources	Device slowdown, unauthorized resource usage, impact on device performance	Network segmentation, regular security scans
Firmware rollback	Firmware	Exploiting roll-back flaws	Reintroducing vulnerabilities, compromising device security, access to known weaknesses	Firmware validation mechanisms, timely updates

4.3.4 Firmware and software attacks

Software attacks within the IoT ecosystem represent a formidable threat, targeting the vulnerabilities inherent in the software systems, applications, and firmware that power interconnected devices. This section navigates through the realm of software attacks within IoT, encapsulating a myriad of tactics employed by malicious actors to exploit weaknesses, compromise security, and disrupt the functionalities of IoT devices and networks. These attacks encompass a diverse spectrum, including but not limited to malware infections, code exploits, firmware vulnerabilities, and application-level attacks. Understanding these software attack vectors is crucial in fortifying software defenses, adopting secure coding practices, implementing robust patch management, and fostering resilient IoT environments capable of withstanding the multifaceted software-centric threats prevalent in today's interconnected landscape.

4.4 Defensive strategies and mitigation techniques

In the constantly evolving landscape of the IoT, securing interconnected devices and networks stands as a paramount challenge. This section delves into a comprehensive array of defensive strategies and mitigation techniques designed to fortify IoT systems against a multitude of threats discussed in previous sections. This segment explores a broad spectrum of proactive measures, encompassing robust security protocols, encryption standards, access controls, IDSs, and resilient architectures. By navigating through these defense strategies, stakeholders gain insights into establishing secure development practices, implementing effective incident response mechanisms, fostering a culture of cybersecurity awareness, and complying with regulatory standards. Understanding and implementing these defensive approaches are instrumental in mitigating risks, ensuring data confidentiality, integrity, and availability, and fostering resilient IoT ecosystems capable of withstanding the complex and evolving threat landscape [13].

4.4.1 Secure design principles

In the realm of the IoT, secure design principles form the cornerstone of building resilient and robust systems. This section navigates through the foundational concepts of secure design principles, emphasizing the importance of integrating security measures at every stage of the IoT ecosystem's development lifecycle. These principles encompass a holistic approach, from device inception to deployment, focusing on aspects such as secure-by-design architectures, encryption protocols, data integrity mechanisms, access controls, and regular security updates. By exploring these principles, stakeholders can grasp the significance of embedding security as an integral component of IoT design, thereby fortifying devices and networks against potential vulnerabilities and ensuring a proactive defense against evolving cyber threats.

Least privilege and separation of duties: Implementing the principle of least privilege involves granting users or systems only the minimum levels of access or permissions required to perform their tasks. This principle minimizes the potential impact of a security breach or unauthorized access by restricting privileges to essential functions and preventing users or processes from accessing unnecessary or sensitive resources. Moreover, the concept of separation of duties complements this principle by assigning specific responsibilities across different entities or individuals. By delineating tasks and ensuring that no single user or entity has complete control over critical functions, separation of duties mitigates the risk of internal threats and unauthorized actions. Together, these principles bolster the overall security posture of IoT systems, reducing the attack surface and limiting the potential impact of compromised credentials or insider threats. Embracing the least privilege and separation of duties as foundational principles in IoT design not only enhances security but also fosters a more resilient and accountable ecosystem, ensuring appropriate access controls and minimizing the impact of potential security breaches.

Secure boot and hardware-based security: Secure boot and hardware-based security mechanisms play pivotal roles in safeguarding IoT devices from unauthorized access and tampering. Secure boot ensures the integrity of device firmware and software by validating the authenticity and integrity of the code during the boot process. This process involves cryptographically verifying the firmware's digital signature, ensuring it has not been altered or compromised. Additionally, leveraging hardware-based security features such as trusted platform modules (TPMs) or secure elements (SEs) provides a secure environment for storing sensitive information and executing critical operations within a tamper-resistant hardware enclave. By utilizing these hardware-based security measures, IoT devices can establish secure, isolated environments, safeguarding cryptographic keys, sensitive data, and authentication processes from external attacks or unauthorized access attempts. Integrating secure boot processes and hardware-based security measures into IoT designs significantly enhances the overall resilience and trustworthiness of devices, bolstering their defenses against a wide range of threats in the interconnected landscape.

4.4.2 Encryption and authentication

Encryption and authentication stand as fundamental pillars in fortifying the security landscape of the IoT. This section navigates through the crucial realms of encryption and authentication, emphasizing their pivotal roles in ensuring data confidentiality, integrity, and secure access control within interconnected systems. These core principles encompass the deployment of robust encryption algorithms to protect sensitive data during storage and transmission, as well as the implementation of rigorous authentication protocols to verify the identities of devices and users accessing IoT networks. By exploring these critical facets, stakeholders gain insights into the significance of adopting strong encryption methodologies, such as symmetric and asymmetric encryption, alongside multifactor authentication mechanisms. Understanding and deploying these encryption and authentication techniques effectively fortify the defense mechanisms of IoT ecosystems, mitigating potential risks, and ensuring a secure and trustworthy interconnected environment.

E2EE: Implementing E2EE serves as a cornerstone in safeguarding data confidentiality across IoT ecosystems. This robust encryption methodology ensures that data remains encrypted throughout its entire journey, from its origin to its destination, effectively shielding it from unauthorized access or interception by malicious entities. By encrypting data at its source and decrypting it only upon reaching its intended recipient, E2EE prevents unauthorized parties, including service providers and intermediaries, from deciphering or tampering with the transmitted information. This approach not only mitigates the risk of data breaches during transit but also upholds the confidentiality and privacy of sensitive data, thereby instilling trust among users and stakeholders in the secure exchange of information within interconnected IoT environments.

Two-factor authentication (2FA): 2FA stands as an indispensable layer of defense against unauthorized access to IoT devices and networks. This authentication method reinforces traditional username–password credentials with an additional layer of verification, typically involving a second factor such as a unique code sent to a user's mobile device, biometric authentication, or hardware tokens. By requiring users to provide two distinct forms of authentication, 2FA significantly enhances security, reducing the susceptibility to credential theft or brute-force attacks [14]. Even if malicious actors obtain login credentials, they are thwarted from accessing the system without the secondary authentication factor. Implementing 2FA as a standard practice across IoT devices and platforms bolsters access control measures, significantly reducing the risk of unauthorized access and fortifying the overall security posture of interconnected systems.

4.4.3 Network security measures

Network security measures serve as the cornerstone in fortifying the integrity, confidentiality, and availability of interconnected devices within the IoT. This section navigates through a comprehensive array of network security measures, focusing on diverse strategies aimed at safeguarding IoT infrastructures from an extensive range of threats. These measures encompass robust encryption protocols, IDSs, network segmentation, firewalls, and secure communication protocols.

Firewalls and IDSs: Integrating robust firewalls and IDSs stands as a fundamental strategy in bolstering the security posture of IoT networks. Firewalls act as gatekeepers, monitoring and controlling incoming and outgoing network traffic based on predefined security rules. By filtering potentially malicious traffic, firewalls prevent unauthorized access, limit exposure to known threats, and thwart malicious attempts to compromise IoT devices or networks. Complementing firewalls, IDSs continuously monitor network traffic, analyzing patterns and behaviors to identify suspicious activities or potential security breaches. IDSs provide real-time alerts or responses upon detecting anomalous behavior, enabling swift mitigation of threats before they escalate. Deploying a combination of firewalls and IDSs fortifies the defense mechanisms of IoT networks, enhancing resilience against various cyber threats and ensuring proactive threat detection and response capabilities.

Network segmentation and isolation: Employing network segmentation and isolation techniques plays a crucial role in enhancing the security and resilience of IoT environments. Network segmentation involves dividing an IoT network into smaller, isolated segments or subnetworks, each with restricted access and unique security controls. By isolating devices or systems based on functionalities, user roles, or criticality, network segmentation limits the lateral movement of threats, preventing a security breach in one segment from affecting the entire network. Moreover, network isolation involves creating separate network zones, physically or logically, to isolate critical assets or sensitive information from less secure areas. This approach significantly reduces the attack surface, mitigates the propagation of malware or unauthorized access, and enhances overall network security. Implementing robust network segmentation and isolation strategies in IoT environments ensures granular control over network resources, improves visibility into network activities, and strengthens the overall security posture against potential threats or cyber-attacks.

4.4.4 Secure software development

Secure software development serves as the bedrock of building resilient and trustworthy systems within the IoT. This section delves into the critical realm of secure software development, emphasizing the significance of integrating robust security practices throughout the software development lifecycle. This segment navigates through the principles of secure coding, rigorous testing methodologies, vulnerability assessments, and secure deployment strategies. By exploring these fundamental facets, stakeholders gain insights into embedding security as an inherent part of the software development process, ensuring the creation of robust, resilient, and secure IoT applications and systems. Understanding and implementing secure software development practices not only fortifies the defense mechanisms of IoT ecosystems but also fosters a proactive approach to mitigating potential vulnerabilities and threats inherent in the interconnected landscape [15].

Regular updates and patches: Instituting a regimen of regular updates and patches stands as a critical practice in maintaining the security and resilience of IoT software and systems. Timely updates entail the deployment of security patches, bug fixes, and software updates to address known vulnerabilities or weaknesses identified within the system. By promptly applying updates, IoT devices and applications can mitigate the risk of exploitation by cyber threats leveraging known vulnerabilities. Regular updates not only enhance the system's security posture but also ensure that the software remains up-to-date, incorporating the latest security enhancements and bug fixes. Additionally, establishing an efficient update mechanism that automates or streamlines the update process can significantly reduce the window of exposure to potential security threats, contributing to the overall robustness of IoT systems.

Code reviews and static analysis: Integrating comprehensive code reviews and static analysis methodologies forms an integral part of fostering secure software development practices within IoT ecosystems. Code reviews involve systematic examinations of software code by developers or security experts to identify vulnerabilities, coding errors, or insecure practices. Static analysis tools complement these reviews by analyzing codebases for potential security flaws, such as buffer overflows, input validation issues, or inadequate encryption practices. By leveraging these methodologies, developers can proactively identify and remediate security weaknesses in the early stages of software development. Incorporating robust code review processes and static analysis tools not only ensures the identification and mitigation of potential security vulnerabilities but also cultivates a security-oriented mindset among development teams, contributing to the creation of more secure and resilient IoT applications and systems.

4.4.5 Monitoring and incident response

Monitoring and incident response form the cornerstone of proactive defense strategies within the IoT landscape. This section navigates through the crucial domains of monitoring and incident response, emphasizing their pivotal roles in detecting, mitigating, and responding to potential threats and security breaches across interconnected IoT systems. This segment explores the deployment of robust monitoring tools, security information and event management (SIEM) systems, and incident response frameworks. By exploring these critical facets, stakeholders gain insights into establishing continuous monitoring practices, rapid incident detection, effective response protocols, and post-incident analysis. Understanding and implementing these monitoring and incident response strategies not only fortify the defense mechanisms of IoT ecosystems but also foster a proactive approach to identifying, containing, and recovering from security incidents within the interconnected IoT environment.

Anomaly detection: Implementing effective anomaly detection mechanisms plays a crucial role in fortifying the security posture of IoT environments. Anomaly detection involves continuously monitoring system behavior, network traffic, or device activities to identify deviations from established patterns or norms. Utilizing machine-learning algorithms, statistical analysis, or behavioral analytics, anomaly detection systems can detect irregularities that may indicate potential security threats or suspicious activities. By proactively identifying deviations from expected behavior, these systems enable early detection of security incidents, such as unauthorized access attempts, data breaches, or abnormal device behavior. This early detection empowers security teams to swiftly investigate potential threats, thereby mitigating risks and minimizing the impact of security breaches on IoT systems. Incorporating robust anomaly detection mechanisms forms a crucial part of a proactive defense strategy, enhancing the ability to detect and respond to emerging security threats within interconnected IoT ecosystems [16].

Rapid response to security incidents: Establishing a swift and well-defined response plan for security incidents is paramount in mitigating the impact of potential breaches within IoT environments. Rapid response protocols involve predefined steps, clear escalation procedures, and well-trained incident response teams ready to execute timely actions when security incidents occur. These protocols encompass steps such as isolating affected devices or networks, conducting forensic analysis, containing the breach, and restoring services to normal operations. Additionally, defining communication channels and responsibilities ensures swift decision-making and coordinated actions during incidents. A well-structured and rehearsed response plan enables organizations to minimize the duration and extent of security incidents, reducing potential damage, data loss, or service disruptions within the IoT ecosystem. Emphasizing rapid response capabilities as part of incident handling strategies bolsters the resilience of IoT environments, facilitating effective containment and recovery from security incidents.

4.5 Regulatory frameworks and standards

In the dynamic landscape of the IoT, regulatory frameworks and standards serve as guiding beacons, shaping the landscape of security, privacy, and operational practices. This section delves into the pivotal domain of regulatory frameworks and standards within IoT ecosystems, emphasizing their critical roles in defining compliance requirements, establishing best practices, and fostering a culture of security and accountability. This segment navigates through diverse regulatory frameworks such as GDPR (General Data Protection Regulation), NIST (National Institute of Standards and Technology), ISO (International Organization for Standardization), and industry-specific regulations. Understanding and aligning with these regulatory frameworks and standards not only fortifies the security posture of IoT ecosystems but also fosters trust among users, promotes innovation, and ensures ethical and responsible deployment of IoT technologies in compliance with legal and industry requirements.

- **GDPR:** GDPR is a comprehensive EU privacy law that mandates strict guidelines on the collection, processing, and storage of personal data of EU residents. It applies to any organization handling EU citizens' data, irrespective of the organization's location. For IoT, GDPR compliance is crucial, as IoT devices often collect sensitive personal information, necessitating clear user consent, data minimization, encryption, and robust security measures.
- **Health Insurance Portability and Accountability Act (HIPAA):** HIPAA is a US regulation governing healthcare data protection and privacy. It sets standards for safeguarding sensitive patient health information (PHI). IoT applications in healthcare, including medical devices and wearables, fall under HIPAA regulations. Compliance involves stringent measures to protect PHI, ensuring data confidentiality, integrity, and availability.
- **California Consumer Privacy Act (CCPA):** CCPA is a California-based regulation designed to enhance consumer privacy rights and control over personal information. It imposes obligations on businesses handling Californian consumers' data, providing rights such as access, deletion, and opt-out options. Compliance involves ensuring transparency in data collection, providing clear privacy notices, and granting consumers control over their data.

- **NIST cybersecurity framework:** Developed by the NIST, this framework offers guidelines for improving cybersecurity posture. While not a regulation, it provides a risk-based approach to cybersecurity, aiding organizations in managing and mitigating cyber risks in IoT deployments through identification, protection, detection, response, and recovery strategies.
- **ISO/IEC 27001:** This international standard specifies requirements for an information security management system (ISMS). It provides a systematic approach to managing sensitive information, including IoT-related data. Compliance ensures a structured framework for risk assessment, security controls, and continuous improvement in information security practices.

Adherence to these regulations and standards is essential for organizations deploying IoT solutions, as they define specific requirements related to data protection, privacy, security controls, risk management, and compliance obligations. Embracing these frameworks not only ensures legal compliance but also fosters a culture of responsible data handling and security within IoT ecosystems [17].

In the intricate web of interconnected devices and systems that constitute the IoT, ensuring robust security measures stands as an imperative. IoT security standards play a pivotal role in guiding the development, deployment, and management of these sprawling ecosystems, offering a structured framework and best practices to fortify against evolving cyber threats. These standards are a testament to the collective efforts of industry experts, regulatory bodies, and standard-setting organizations striving to establish a cohesive set of guidelines, protocols, and benchmarks aimed at enhancing the security posture of IoT deployments. The following discussion examines the diverse landscape of IoT security standards, shedding light on key frameworks such as NIST, ISO/IEC, IEC, and industry-specific standards.

- **NIST Special Publication 800-183:** NIST's publication offers guidance on IoT cybersecurity, outlining key considerations, challenges, and recommendations for securing IoT devices and systems. It addresses aspects such as device identification, secure software and firmware updates, and cybersecurity capabilities for IoT devices.
- **ISO/IEC 27000 Series:** The ISO/IEC 27000 series comprises standards related to information security management systems (ISMS). Standards such as ISO/IEC 27001, ISO/IEC 27002, and others offer guidelines for managing information security risks, including those associated with IoT devices and networks. They provide frameworks for risk assessment, security controls, and continuous improvement in information security practices within IoT ecosystems.
- **IEC 62443 Series:** The IEC 62443 series focuses on industrial automation and control systems (IACS) security. It provides guidelines and standards for securing industrial IoT (IIoT) environments, emphasizing principles, policies, and procedures for safeguarding critical infrastructure against cyber threats.
- **IoT Security Foundation (IoTSF):** Although not a standard-setting body, IoTSF provides best practices, guidelines, and frameworks for IoT security. It

offers resources, including security compliance frameworks, checklists, and guidance documents, promoting a security-first approach in IoT design, development, and deployment.

- **Trusted Computing Group (TCG):** TCG develops open standards for trusted computing and security technologies. Its specifications, such as Trusted Platform Module (TPM) and Device Identity Composition Engine (DICE), offer guidelines and protocols for hardware-based security, secure device booting, and cryptographic authentication in IoT devices.
- **Open Connectivity Foundation (OCF):** OCF develops interoperability standards for secure IoT connectivity. Its specifications define secure communication protocols and device interoperability frameworks, ensuring standardized communication and security for IoT devices across different platforms.

In navigating the increasingly complex and interconnected landscape of modern technology, industry best practices and guidelines stand as guiding principles shaping the foundation of robust and secure systems. These practices encapsulate a wealth of knowledge, insights, and methodologies gathered from diverse sectors and expert contributions, serving as a compass for organizations seeking to fortify their technological endeavors [18–20]. In the following, we embark on an exploration of industry best practices and guidelines, illuminating a collective wisdom forged through experience, research, and continuous evolution.

- **Defense-in-depth strategy:** Employing a defense-in-depth approach involves implementing multiple layers of security controls across IoT environments. This strategy encompasses diverse security measures, including encryption, access controls, network segmentation, regular updates, and monitoring, reducing the impact of a single security breach and providing comprehensive protection against various threats.
- **Secure development lifecycle (SDL):** Adopting secure development practices involves integrating security considerations at every stage of the software development lifecycle. This includes threat modeling, secure coding practices, rigorous testing, vulnerability assessments, and secure deployment, ensuring that IoT applications and devices are developed with security in mind.
- **Principle of least privilege:** Implementing the principle of least privilege restricts user or system access to only the minimum permissions necessary to perform specific tasks. Limiting access rights mitigates the risk of unauthorized actions or data breaches within IoT systems, reducing the attack surface and potential impact of security incidents.
- **Device authentication and identity management:** Implementing robust device authentication mechanisms ensures that only authorized and authenticated devices can access the IoT network. Incorporating unique device identifiers, cryptographic keys, and secure authentication protocols prevents unauthorized access and ensures the integrity of device identities within the ecosystem.
- **Regular updates and patch management:** Establishing processes for timely updates and patch management is crucial to address vulnerabilities and

mitigate security risks in IoT devices and applications. Regularly applying security patches, firmware updates, and software upgrades helps address known vulnerabilities and enhances the security posture of IoT deployments.

- **Continuous monitoring and incident response:** Implementing continuous monitoring enables real-time detection of anomalous behavior or security incidents within IoT networks. Coupled with a well-defined incident response plan, organizations can swiftly identify, contain, and respond to security breaches, minimizing the impact and mitigating risks associated with potential threats.
- **Vendor risk assessment and supply chain security:** Conducting thorough vendor risk assessments and ensuring supply chain security are essential components of IoT security. Organizations should evaluate the security practices of vendors, assess the risks associated with third-party components, and establish protocols for secure procurement, ensuring the integrity and trustworthiness of IoT devices and components. Adhering to these industry best practices and guidelines enables organizations to establish robust security postures, mitigate risks, and build resilient IoT ecosystems capable of withstanding the evolving threat landscape while ensuring the confidentiality, integrity, and availability of data within interconnected environments [21].

References

[1] Rizvi, S., Orr, R. J., Cox, A., Ashokkumar, P., and Rizvi, M. R. (2020). Identifying the attack surface for IoT network. *Internet of Things*, *9*, 100162.

[2] Barnawi, A., Gaba, S., Alphy, A., Jabbari, A., Budhiraja, I., Kumar, V., and Kumar, N. (2023). A systematic analysis of deep learning methods and potential attacks in internet-of-things surfaces. *Neural Computing and Applications*, *35*(25), 18293–18308.

[3] Zhang, M., Wang, L., Jajodia, S., and Singhal, A. (2018). Network attack surface: Lifting the concept of attack surface to the network level for evaluating networks' resilience against zero-day attacks. *IEEE Transactions on Dependable and Secure Computing*, *18*(1), 310–324.

[4] Khanji, S., Iqbal, F., and Hung, P. (2019, June). ZigBee security vulnerabilities: Exploration and evaluating. *In 2019 10th International Conference on Information and Communication Systems (ICICS)* (pp. 52–57). IEEE.

[5] Vaezi, M., Azari, A., Khosravirad, S. R., Shirvanimoghaddam, M., Azari, M. M., Chasaki, D., and Popovski, P. (2022). Cellular, wide-area, and non-terrestrial IoT: A survey on 5G advances and the road toward 6G. *IEEE Communications Surveys & Tutorials*, 24(2), 1117–1174.

[6] Stellios, I., Kotzanikolaou, P., and Grigoriadis, C. (2021). Assessing IoT enabled cyber-physical attack paths against critical systems. *Computers & Security*, *107*, 102316.

[7] Anand, P., Singh, Y., Selwal, A., Singh, P. K., Felseghi, R. A., and Raboaca, M. S. (2020). IOVT: Internet of vulnerable things? Threat architecture, attack

surfaces, and vulnerabilities in internet of things and its applications towards smart grids. *Energies*, *13*(18), 4813.

[8] Guo, H., and Yu, X. (2022). A survey on blockchain technology and its security. *Blockchain: Research and Applications*, *3*(2), 100067.

[9] Vaz, D., Matos, D., Pardal, M., and Correia, M. (2020, December). MIRES: Recovering mobile applications based on backend-as-a-service from cyber attacks. In *MobiQuitous 2020 - 17th EAI International Conference on Mobile and Ubiquitous Systems: Computing, Networking and Services* (pp. 262-272).

[10] Tushir, B., Dalal, Y., Dezfouli, B., and Liu, Y. (2020). A quantitative study of DDoS and E-DDoS attacks on WiFi smart home devices. *IEEE Internet of Things Journal*, *8*(8), 6282–6292.

[11] Rigaki, M., and Garcia, S. (2023). A survey of privacy attacks in machine learning. *ACM Computing Surveys*, *56*(4), 1–34.

[12] Khanam, S., Ahmedy, I. B., Idris, M. Y. I., Jaward, M. H., and Sabri, A. Q. B. M. (2020). A survey of security challenges, attacks taxonomy and advanced countermeasures in the internet of things. *IEEE Access*, *8*, 219709–219743.

[13] Al-Hamadi, H., Chen, R., Wang, D. C., and Almashan, M. (2020). Attack and defense strategies for intrusion detection in autonomous distributed IoT systems. *IEEE Access*, *8*, 168994–169009.

[14] Zhang, J., Tan, X., Wang, X., Yan, A., and Qin, Z. (2018). T2FA: Transparent two-factor authentication. *IEEE Access*, *6*, 32677–32686.

[15] Babiker Mohamed, M., Matthew Alofe, O., Ajmal Azad, M., Singh Lallie, H., Fatema, K., and Sharif, T. (2022). A comprehensive survey on secure software-defined network for the Internet of Things. *Transactions on Emerging Telecommunications Technologies*, *33*(1), e4391.

[16] Bernal, A. E., Monterrubio, S. M. M., Fuente, J. P., Crespo, R. G., and Verdu, E. (2021). Methodology for computer security incident response teams into IoT strategy. *KSII Transactions on Internet and Information Systems (TIIS)*, *15*(5), 1909–1928.

[17] Karie, N. M., Sahri, N. M., Yang, W., Valli, C., and Kebande, V. R. (2021). A review of security standards and frameworks for IoT-based smart environments. *IEEE Access*, *9*, 121975–121995.

[18] Saleem, J., Hammoudeh, M., Raza, U., Adebisi, B., and Ande, R. (2018, June). IoT standardisation: Challenges, perspectives and solution. *In Proceedings of the 2nd International Conference on Future Networks and Distributed Systems* (pp. 1–9).

[19] Conde Gallego, B., and Drexl, J. (2019). IoT connectivity standards: how adaptive is the current SEP regulatory framework?. *IIC-International Review of Intellectual Property and Competition Law*, *50*, 135–156.

[20] Kiela, K., Barzdenas, V., Jurgo, M., *et al.* (2020). Review of V2X–IoT standards and frameworks for ITS applications. *Applied Sciences*, *10*(12), 4314.

[21] Rao, V. V., Marshal, R., and Gobinath, K. (2021, October). The IoT Supply Chain Attack Trends-Vulnerabilities and Preventive Measures. *In 2021 4th International Conference on Security and Privacy (ISEA-ISAP)* (pp. 1–4). IEEE.

Chapter 5
Black-box machine learning for IoT security

5.1 Introduction

The proliferation of interconnected devices within the IoT has heralded an era of unparalleled convenience and efficiency, reshaping the way we interact with technology and the physical world. However, this rapid expansion has concurrently amplified the complexity and susceptibility of these interconnected ecosystems to multifaceted cyber threats. With a diverse array of devices ranging from smart home appliances and wearable gadgets to critical industrial machinery and healthcare systems, the intricate web of interconnectivity has become a breeding ground for security vulnerabilities. The inherent diversity in device types, operating systems, communication protocols, and the often insufficiently fortified security measures has exposed IoT environments to a myriad of cyber risks, including data breaches, unauthorized access, and potential manipulation of crucial systems. As these devices pervade nearly every aspect of modern life, the imperative to secure these systems has transcended beyond mere protection of data to safeguarding critical infrastructure, personal privacy, and even public safety, underscoring the urgent need for robust and innovative security measures capable of mitigating the evolving landscape of cyber threats within the expansive realm of IoT.

In securing IoT ecosystems, traditional security measures often fall short of keeping pace with the dynamic and sophisticated nature of emerging cyber threats. Black-box machine learning (ML) is a paradigm that offers a promising avenue for fortifying the security infrastructure of interconnected devices. Black-box ML encompasses a spectrum of algorithms and models designed to analyze and predict patterns within data without explicit insight into their internal workings or decision-making processes. While challenging from an interpretability standpoint, this opacity becomes a formidable asset when combating the multifaceted threats that plague IoT environments. By leveraging vast amounts of heterogeneous data generated by myriad IoT devices, black-box ML algorithms excel in detecting anomalies, identifying patterns indicative of cyber attacks, and swiftly adapting to evolving threats in real time. Their capability to discern subtle deviations from normal behavior and detect previously unknown attack vectors makes them indispensable in fortifying the security posture of IoT ecosystems. While the lack of interpretability would raise concerns, the potential of these models to autonomously learn and adapt to emerging threats presents a compelling solution to the intricate security challenges prevalent in the ever-expanding landscape of interconnected devices [1].

This chapter navigates through a comprehensive exploration of black-box ML techniques explicitly tailored to fortify the security landscape of IoT ecosystems. Beginning with an insightful journey into the fundamentals of black-box ML, it elucidates the core concepts and terminologies crucial for understanding the subsequent discussions. Delving deeper, the chapter meticulously dissects classical ML models, spanning linear models, tree-based approaches, instance-based methods, and ensemble learning, showcasing their applicability in bolstering IoT security measures. Further, an extensive exposition into deep learning unfolds, elucidating the diverse array of neural network architectures and their potential implications for securing IoT environments. As the narrative unfolds, it extends its purview into cutting-edge domains such as reinforcement learning (RL) and meta-learning, unveiling their transformative capabilities in adapting to dynamic threats and fostering resilience within IoT systems. Each section serves as a compass, guiding readers through the intricate landscape of black-box ML techniques and offering insights, applications, and potential avenues for fortifying the security posture of interconnected IoT devices [2].

5.2 Fundamentals of black-box ML

This section focuses on defining and discussing black-box ML techniques, highlighting their inherent opacity and the challenge of providing human-interpretable explanations. Black-box ML refers to a class of algorithms that excel in learning patterns and making predictions without revealing the underlying logic or decision-making process in a human-understandable manner. These techniques are characterized by their complexity and the inability to elucidate how they arrive at specific outcomes, posing a significant challenge in comprehending their inner workings. The opacity of black-box models stems from their intricate architectures, which often involve multiple layers, non-linear transformations, and complex interactions between numerous parameters. These models prioritize predictive accuracy by learning intricate patterns and relationships within data, yet they need more transparency to elucidate the reasoning behind their predictions. Unlike more transparent models like linear regression or decision trees, where the decision-making process is explicit, black-box models operate as complex computational "black boxes," making it arduous for humans to interpret and understand the reasoning behind their outputs [3].

While the lack of interpretability might raise concerns regarding trust and accountability, these black-box techniques possess significant strengths, particularly in scenarios focusing on predictive performance rather than interpretability. They excel in handling high-dimensional and heterogeneous data, making them invaluable in various domains, including IoT security, where detecting anomalies or emerging threats requires the ability to discern intricate patterns across diverse data streams. However, their inherent opacity necessitates the development of complementary methods to enhance their interpretability or the implementation of additional mechanisms to ensure transparency and ethical use in critical applications.

In the realm of ML, the concept of opacity stands as a defining characteristic of black-box models. These intricate models possess an inherent complexity that shrouds their inner workings in mystery, rendering their decision-making processes elusive to human comprehension. In contrast to transparent models where the reasoning behind predictions is explicit, black-box models—comprising deep neural networks (DNNs), ensemble methods, and other sophisticated architectures—operate as enigmatic computational "black boxes." Their predictive prowess often comes at the expense of interpretability, presenting a unique challenge in deciphering the rationale behind their outputs. This opacity poses a significant obstacle in understanding how these models arrive at specific decisions, marking a pivotal distinction from more interpretable models where the decision-making process is straightforward and intelligible.

In addition, the concept of bias embodies a crucial facet in evaluating and understanding predictive models. Bias represents the persistent discrepancy between predicted outcomes and the actual true values, showcasing a model's inclination toward consistent inaccuracies. It serves as a metric to gauge the systematic error within the model, elucidating how consistently the model's predictions deviate from reality. Models with high bias tend to exhibit consistent inaccuracies in their predictions, leaning toward consistently overestimating or underestimating the target variable. Understanding and mitigating bias within models are imperative in ensuring their reliability and accuracy, especially when aiming for an optimal balance between the model's ability to capture underlying patterns and its capability to generalize to unseen data [4].

Variance is a critical metric in ML, representing the model's sensitivity to fluctuations in the training data. Models exhibiting high variance often display substantial changes in predictions when trained on different datasets, signifying their susceptibility to minor variations in the training data. Such models tend to capture noise and specific patterns unique to the training set, leading to overfitting and compromised generalization to new, unseen data. Variance is a crucial parameter to evaluate a model's robustness, where an ideal balance is sought between capturing relevant patterns in the data without being overly sensitive to minor fluctuations in the data without being overly sensitive to minor fluctuations present in the training samples.

Overfitting emerges as a persistent challenge in ML in the pursuit of modeling complex relationships within data. Overfitting occurs when a model learns too intricately from the training data, capturing noise and irrelevant patterns specific to that dataset. Consequently, the model excels in reproducing the training set but falters in generalizing to new, unseen data due to an excessive focus on memorizing idiosyncrasies within the training samples. Recognizing and mitigating overfitting is paramount in ensuring a model's capacity to discern underlying patterns without being misled by noise or specific quirks present in the training data. Besides, underfitting, a pertinent concern in ML, signifies a model's inadequacy in capturing the underlying patterns inherent in the data. This scenario arises when a model is too simplistic or cannot grasp the complexities embedded within the dataset.

Consequently, an underfitted model performs poorly not only on the training data but also on new, unseen data due to its oversimplified representations. Underfitting reflects a failure to capture relevant relationships or nuances within the data, underscoring the necessity of developing more sophisticated models or refining existing ones to adequately capture the inherent complexities [5]. Moreover, generalization is a hallmark of a well-performing model, representing its ability to accurately apply learned patterns to new, unseen data. A model demonstrating good generalization successfully captures underlying patterns and relationships from the training data without merely memorizing specific details or noise. This capacity allows the model to make accurate predictions or inferences on previously unseen data, indicating its robustness and capability to discern essential patterns that extend beyond the training samples. The concept of feature importance plays a pivotal role in understanding a model's decision-making process by delineating the significance of each input feature in influencing predictions. Identifying which features hold more weight in determining outcomes aids in comprehending the factors that impact the model's decisions. By discerning feature importance, practitioners can focus on leveraging the most influential features while optimizing model performance and interpretability.

Regularization techniques serve as pivotal tools in mitigating overfitting within ML models. They introduce constraints or penalties to the model's parameters during training, thereby preventing excessive complexity and the tendency to fit noise present in the training data. By imposing such constraints, regularization methods like L1 (Lasso) and L2 (Ridge) regularization encourage the model to learn simpler patterns, thus enhancing generalization to new, unseen data while maintaining predictive performance. Furthermore, the balance between model interpretability and performance embodies a critical trade-off in ML. While models striving for high predictive performance might embrace complexity and opacity, achieving superior accuracy may come at the expense of interpretability. Balancing the need for a model to provide understandable insights (interpretability) against the quest for optimal predictive accuracy is crucial, particularly in domains where comprehensible explanations of model decisions are as significant as high-performance outcomes [6].

5.3 Classical ML

- Introduction to classical ML models: Overview of traditional ML algorithms applicable to IoT security.

1. Tree-based models

Gradient-boosting machines (GBM) are also considered to be black-box ML algorithms, albeit they often exhibit a slightly more interpretable nature compared to some other complex models like DNNs. The following are a few reasons why GBMs might be considered black-box models:

- Ensemble learning: GBMs are ensemble models that combine multiple weak learners (often decision trees) to create a robust predictive model. The interaction and combination of these individual learners can make understanding the collective decision-making process complex.

- Complexity: The final model is highly intricate and non-linear because GBMs are built by sequentially adding decision trees. The interactions and combinations of features across multiple trees might be challenging to interpret directly.
- Non-linearity: GBMs are adept at capturing non-linear relationships within the data. While this is beneficial for predictive performance, explaining how specific features contribute to predictions or understanding the exact decision boundaries can make it more challenging.
- Hyperparameters: GBMs have various hyperparameters that affect the model's behavior, such as the number of trees, tree depth, learning rate, and regularization parameters. Understanding the interactions between these hyperparameters and their impact on the model's predictions can be complex.

While GBMs are complex models, they have some advantages in terms of interpretability compared to other black-box models. For instance, feature importance scores can be extracted from GBMs, providing insights into which features contribute more to the model's predictions. Additionally, visualization techniques can sometimes offer a basic understanding of the decision-making process in GBMs, especially when dealing with a smaller number of trees or simpler configurations [7].

```
from sklearn.ensemble import GradientBoostingRegressor
from sklearn.model_selection import train_test_split
from sklearn.metrics import mean_squared_error

# Load dataset (example using Attack dataset)using your custom function
data = load_data()
X, y = data.data, data.target

# Split the data into training and testing sets
X_train, X_test, y_train, y_test = train_test_split(X, y, test_size=0.2,
random_state=42)

# Initialize the Gradient Boosting Regressor
gbm = GradientBoostingRegressor(n_estimators=100, learning_rate=0.1,
max_depth=3, random_state=42)

# Train the model
gbm.fit(X_train, y_train)

# Make predictions
predictions = gbm.predict(X_test)
```

AdaBoost (adaptive boosting) is a popular ensemble learning algorithm renowned for improving weak learners' performance by sequentially training models and emphasizing misclassified instances. While AdaBoost demonstrates remarkable predictive power and robustness, it also exhibits characteristics that align with the traits of black-box ML algorithms. One of the primary reasons AdaBoost is regarded as a black-box model is its inherent complexity arising from the ensemble nature of its methodology. AdaBoost operates by iteratively fitting weak learners, often decision trees, to different weighted versions of the dataset. These models are combined to form a strong learner that makes predictions based on their collective decisions. The intricate combination and weighting of these individual models within AdaBoost's ensemble can render the collective decision-making process challenging to interpret or explain directly [8].

Moreover, AdaBoost's ability to focus on misclassified instances in subsequent iterations and assign higher weights to them for subsequent models' training contributes to its opaqueness. While this adaptability enhances its predictive accuracy, it complicates the understanding of how the final model arrives at specific predictions. The complex interplay between instances, features, and weights across multiple iterations results in a highly sophisticated model, making it challenging to decipher the exact reasoning behind its decisions. Additionally, AdaBoost's reliance on tuning hyperparameters, such as the number of iterations (or estimators) and the learning rate, to achieve optimal performance contributes to its black-box nature. While crucial for model optimization, these hyperparameters influence the ensemble's complexity and behavior, requiring careful tuning without necessarily providing straightforward insights into the inner workings of the model [11].

Although AdaBoost may exhibit characteristics of a black-box algorithm due to its complexity and intricate ensemble methodology, it offers advantages such as high predictive accuracy and resilience against overfitting. Nevertheless, its opacity and the complex interactions among weighted weak learners contribute to the classification of AdaBoost within the realm of black-box ML algorithms. While AdaBoost is a powerful and widely used ensemble learning technique, its intricate ensemble structure, adaptability, and reliance on weighted models contribute to its classification as a black-box ML algorithm, presenting challenges in direct interpretability and understanding of its decision-making process.

```
from sklearn.ensemble import AdaBoostClassifier
from sklearn.model_selection import train_test_split
from sklearn.metrics import accuracy_score

# Load dataset (consider using IoT traffic dataset) using your custom
function
data = load_data()
X, y = data.data, data.target

# Split the data into training and testing sets
X_train, X_test, y_train, y_test = train_test_split(X, y, test_size=0.2,
random_state=42)

# Initialize the AdaBoost classifier
adaboost = AdaBoostClassifier(n_estimators=50, random_state=42)

# Train the model
adaboost.fit(X_train, y_train)

# Make predictions
predictions = adaboost.predict(X_test)
```

LightGBM, a gradient-boosting framework developed by Microsoft, has gained significant traction in the ML community owing to its efficiency, speed, and superior performance. While LightGBM demonstrates exceptional predictive capabilities and efficiency, its underlying characteristics align with those commonly attributed to black-box ML algorithms. The primary rationale behind categorizing LightGBM as a black-box model lies in its inherent complexity and the intricacies of its gradient-boosting methodology. LightGBM operates by constructing an ensemble of decision trees, utilizing a gradient-based boosting technique to iteratively train and improve predictive accuracy. This ensemble model, comprising numerous trees that interact to make collective predictions, can pose challenges in understanding the holistic decision-making process. One of the core features contributing to LightGBM's black-box nature is its utilization of leaf-wise growth, a strategy aimed at growing trees by selecting the leaf node that minimizes the loss function. This approach results in deeper trees and more complex interactions between features, making it challenging to interpret the specific contributions of individual features toward the model's predictions. Consequently, the resultant ensemble often comprises a large number of intricate trees, forming a highly non-linear and intricate decision surface, further obfuscating the model's interpretability [12]. Moreover, LightGBM's reliance on various hyperparameters, such as tree depth, learning rate, and the number of leaves, for optimization and fine-tuning, contributes to its classification as a black-box model. While these hyperparameters significantly influence the model's performance, their intricate interactions may not provide straightforward insights into the decision-making process, thereby impacting the model's interpretability. Despite its opacity, LightGBM offers exceptional predictive power, scalability, and efficiency, making it an attractive choice for various ML tasks. Nonetheless, its complex ensemble structure, leaf-wise growth strategy, and reliance on numerous hyperparameters reinforce its classification within the class of black-box ML algorithms, thus warranting careful consideration in scenarios where interpretability is crucial [13].

```
import lightgbm as lgb
from sklearn.model_selection import train_test_split
from sklearn.metrics import accuracy_score

# Load dataset (consider using IoT traffic dataset) using your custom
function
data = load_data()
X, y = data.data, data.target

# Split the data into training and testing sets
X_train, X_test, y_train, y_test = train_test_split(X, y, test_size=0.2,
random_state=42)

# Convert the dataset to LightGBM format
train_data = lgb.Dataset(X_train, label=y_train)
test_data = lgb.Dataset(X_test, label=y_test)
```

```
# Set hyperparameters
params = {
    'boosting_type': 'gbdt',
    'objective': 'multiclass',
    'num_class': 3,  # Number of classes in the target variable
    'metric': 'multi_error',
    'num_leaves': 31,
    'learning_rate': 0.05,
    'feature_fraction': 0.9,
    'bagging_fraction': 0.8,
    'bagging_freq': 5,
    'verbose': -1
}

# Train the model
num_round = 100
bst = lgb.train(params, train_data, num_round, valid_sets=[test_data],
early_stopping_rounds=10)

# Make predictions
predictions = bst.predict(X_test)
predicted_labels = [list(x).index(max(x)) for x in predictions]
```

CatBoost, a gradient-boosting algorithm developed by Yandex, has garnered attention in the ML community due to its robustness, ability to handle categorical variables, and competitive performance in various domains. Despite its impressive predictive power, CatBoost exhibits characteristics aligning with those commonly attributed to black-box ML algorithms. The primary reason for categorizing CatBoost as a black-box model lies in its complex ensemble structure and the intricacies of its gradient-boosting methodology. CatBoost employs an ensemble of decision trees trained sequentially, leveraging gradient boosting to optimize predictive accuracy. While decision trees individually are interpretable, the collective interaction and combination of numerous trees within the ensemble can render the overall model's decision-making process challenging to interpret holistically [14]. One of CatBoost's distinctive features that contribute to its black-box nature is its handling of categorical variables using an innovative approach called ordered boosting. This technique encodes categorical features without relying on preprocessing techniques, introducing a level of complexity that might hinder direct interpretation of the relationships between categorical variables and predictions. Moreover, CatBoost, like other gradient-boosting algorithms, involves tuning various hyperparameters such as the learning rate, tree depth, and the number of iterations. These hyperparameters significantly influence the model's behavior and performance; however, their intricate interactions and effects on the ensemble's decision-making process may not offer straightforward interpretability. Despite its black-box characteristics, CatBoost's strengths lie in its high predictive accuracy, efficient handling of categorical data, and robustness against overfitting. Nevertheless, its ensemble structure, ordered boosting methodology, and reliance on multiple hyperparameters align with the traits of black-box ML algorithms, warranting caution and careful consideration in scenarios where model interpretability is essential [15].

```
from catboost import CatBoostClassifier, Pool
from sklearn.model_selection import train_test_split
from sklearn.metrics import accuracy_score

# Load dataset (consider using IoT traffic dataset) using your custom
function
data = load_data()
X, y = data.data, data.target

# Split the data into training and testing sets
X_train, X_test, y_train, y_test = train_test_split(X, y, test_size=0.2,
random_state=42)

# Create a CatBoost Pool object
train_pool = Pool(X_train, label=y_train)
test_pool = Pool(X_test, label=y_test)

# Set hyperparameters
params = {
    'iterations': 100,
    'learning_rate': 0.1,
    'depth': 6,
    'loss_function': 'MultiClass',
    'eval_metric': 'Accuracy',
    'random_seed': 42,
    'verbose': False
}

# Initialize CatBoost classifier
catboost = CatBoostClassifier(**params)

# Train the model
catboost.fit(train_pool, eval_set=test_pool, early_stopping_rounds=10,
verbose_eval=False)

# Make predictions
predictions = catboost.predict(X_test)
```

2. Kernel methods

Support vector machines (SVMs) are indeed a popular example of kernel-based ML methods. Kernel-based methods, including SVMs, are renowned for their efficacy in solving non-linear classification and regression problems by operating in higher-dimensional spaces without explicitly transforming the input features.

SVMs are widely recognized for their effectiveness in solving classification and regression problems. Despite their successful application in various domains, SVMs exhibit characteristics that align with those commonly attributed to black-box ML algorithms. One of the primary reasons SVMs are regarded as black-box models is their inherent complexity when operating in higher dimensions. SVMs aim to find an optimal separating different feature space classes. In scenarios where a linear boundary is inadequate, SVMs use kernel methods to transform data into higher-dimensional spaces, allowing for non-linear separation. Kernel functions play a pivotal role in SVMs by implicitly projecting data into a higher-dimensional feature space. These functions, such as the linear, polynomial, radial basis function (RBF), or sigmoid kernels, define the similarity or distance measure between data points in the transformed space. This transformation can significantly complicate the interpretability of the decision boundary, especially when using complex kernel functions. Moreover, the decision function in SVMs relies on support vectors, a subset of training data points that contribute to defining the decision boundary. While this approach provides computational efficiency, understanding how each support vector influences the final decision can be challenging, particularly in scenarios with high-dimensional or transformed feature spaces. Additionally, SVMs' reliance on tuning hyperparameters, such as the choice of kernel function and regularization parameter (C), for optimal performance contributes to their classification as a black-box model. The influence and interaction of these hyperparameters on the resulting decision boundary might not offer straightforward insights into the model's decision-making process. Despite being categorized as a black-box model, SVMs exhibit advantages such as high accuracy, versatility in handling non-linear data, and resilience against overfitting in high-dimensional spaces. Nevertheless, their complex decision boundary, reliance on kernel functions, and hyperparameter tuning align with the characteristics of black-box ML algorithms, warranting consideration of the trade-offs between performance and interpretability in practical applications [16].

```
from sklearn.model_selection import train_test_split
from sklearn.svm import SVC
from sklearn.metrics import accuracy_score

# Load dataset (consider using IoT traffic dataset) using your custom
function
data = load_data()

# Split the data into training and testing sets
X_train, X_test, y_train, y_test = train_test_split(X, y, test_size=0.2,
random_state=42)

# Initialize SVM with RBF kernel
svm = SVC(kernel='rbf', gamma='scale', C=1.0, random_state=42)

# Train the model
svm.fit(X_train, y_train)

# Make predictions
predictions = svm.predict(X_test)

# Evaluate the model
accuracy = accuracy_score(y_test, predictions)
print(f"Accuracy: {accuracy}")
```

5.4 Deep learning

Deep learning, a subset of ML, has emerged as a transformative technology offering unparalleled capabilities in various domains, including IoT security. Among the most influential architectures within deep learning are DNNs, lauded for their ability to extract intricate patterns and representations from complex data. In the context of IoT security, DNNs have been instrumental in threat detection, anomaly identification, and data analysis. However, their exceptional performance often comes with a trade-off—the inherent black-box nature of these networks.

DNNs, comprised of numerous layers of interconnected neurons, demonstrate remarkable prowess in learning intricate data representations. This architecture excels in automatically extracting hierarchical features from raw data, enabling the identification of subtle patterns often imperceptible to traditional algorithms. In IoT security applications, DNNs have been leveraged to detect abnormal behavior, classify malicious activities, and secure interconnected devices within complex networks. Despite their effectiveness, DNNs exhibit a black-box nature, encapsulating complexities in their decision-making processes. The internal workings of these networks, particularly in deeper architectures, are characterized by intricate transformations of data through multiple layers. While this allows DNNs to model highly complex relationships, it often obscures the interpretability of their decisions [17]. Understanding the specific reasoning behind DNN predictions or identifying the features crucial to a decision remains challenging due to the complex and abstract representations learned by these networks. This lack of interpretability raises significant concerns in security-critical applications where transparency and explainability of decisions are imperative. In scenarios where DNNs are employed for IoT security, their black-box nature poses challenges in understanding the rationale behind identified threats or potential vulnerabilities, hindering the ability to comprehend and mitigate security risks effectively. In the subsequent sections, we delve deeper into different types of DNNs in the context of IoT security.

5.4.1 Feedforward neural networks

Feedforward neural networks (FNNs), also known as multilayer perceptrons, represent a fundamental architecture in deep learning. FNNs are structured as a collection of interconnected neurons organized into layers, with information flowing in a unidirectional manner from the input layer through hidden layers to the output layer. This architecture enables FNNs to model complex relationships in data by learning hierarchical representations.

The architecture of an FNN consists of multiple layers:

- **Input layer:** The input layer receives the raw data features and transmits them to the subsequent layers.
- **Hidden layers:** Intermediate layers between the input and output layers constitute the hidden layers. These layers are composed of neurons that apply non-linear transformations to the input data, enabling the network to extract complex and abstract features.

- **Output layer:** The output layer generates the final predictions or classifications based on the processed information from the hidden layers. The number of neurons in this layer depends on the type of problem (e.g., regression and binary/multiclass classification).

Despite their effectiveness in learning intricate patterns from data, FNNs are often characterized by their black-box nature. This characteristic arises due to several reasons: first, FNNs perform complex and non-linear transformations on input data as it propagates through multiple hidden layers. The learned representations become increasingly abstract, making it challenging to interpret how specific features influence the final predictions. Second, in deeper networks with numerous neurons and layers, understanding the specific contributions of individual neurons or features becomes increasingly challenging. This high-dimensional space makes it difficult to intuitively grasp the decision-making process of the network. Third, the inherent complexity of FNNs obscures the rationale behind their predictions. While they can achieve high accuracy, understanding the precise reasons for a particular prediction or classification can be elusive, especially in scenarios involving high-dimensional data. The black-box nature of FNNs poses challenges in interpretability and transparency, particularly in security-critical or sensitive applications. Despite their exceptional ability to learn complex patterns, the lack of explainability might limit their adoption in domains where interpretability is essential.

```
import torch.nn as nn

class FeedForwardNN(nn.Module):
    def __init__(self, input_size, hidden_sizes, output_size):
        super(FeedForwardNN, self).__init__()
        layer_sizes = [input_size] + hidden_sizes + [output_size]
        self.layers = nn.ModuleList([nn.Linear(layer_sizes[i],
layer_sizes[i + 1]) for i in range(len(layer_sizes) - 1)])
        self.activations = nn.ModuleList([nn.ReLU() for _ in
range(len(hidden_sizes))])
        self.softmax = nn.Softmax(dim=1)

    def forward(self, x):
        for layer, activation in zip(self.layers[:-1], self.activations):
            x = activation(layer(x))
        x = self.layers[-1](x)
        x = self.softmax(x)
        return x
```

5.4.2 Recurrent neural networks

Recurrent neural networks (RNNs) stand as a pivotal architecture within the realm of deep learning, specifically designed to process sequential data by preserving information or memory of past inputs. Unlike traditional feedforward neural networks that process each input independently, RNNs possess a form of internal memory, allowing them to retain and utilize past information while considering current inputs. This capability makes RNNs particularly suitable for handling sequences such as time series, natural language, and audio.

RNNs are uniquely structured to handle sequences of data by iteratively processing each element while retaining a memory state or context vector. Each element in the sequence interacts with the RNN's internal state, influencing subsequent predictions or outputs. One of the defining characteristics of RNNs is their recurrent connections, enabling them to maintain a memory of previous inputs. This feedback loop mechanism allows information to persist across time steps, ensuring that the network can consider the sequential context when making predictions or generating outputs. RNNs utilize shared parameters across different time steps, enabling them to process sequences of varying lengths. This parameter-sharing characteristic allows the network to learn from and generalize to sequences of different lengths, a feature crucial in handling dynamic or variable-length data. RNNs theoretically have the capability to capture long-range dependencies within sequences. However, traditional RNNs may struggle to retain information over many time steps, leading to difficulties in learning and preserving long-term dependencies. This limitation has led to the development of more sophisticated RNN variants like long short-term memory networks (LSTMs) and gated recurrent units (GRUs), which address the vanishing or exploding gradient problems commonly encountered in traditional RNNs.

```
import torch
import torch.nn as nn

class VanillaRNN(nn.Module):
    def __init__(self, input_size, hidden_size, output_size):
        super(VanillaRNN, self).__init__()
        self.hidden_size = hidden_size
        self.rnn = nn.RNN(input_size, hidden_size, batch_first=True)
        self.fc = nn.Linear(hidden_size, output_size)
        self.softmax = nn.Softmax(dim=1)

    def forward(self, x, h=None):
        if h is None:
            h = torch.zeros(1, x.size(0), self.hidden_size).to(x.device)

        out, h = self.rnn(x, h)
        out = self.fc(out[:, -1, :])  # Taking the output of the last time step
        out = self.softmax(out)
        return out, h
```

LSTMs represent a sophisticated variant of RNNs specifically designed to mitigate the limitations of traditional RNNs in capturing and preserving long-range dependencies within sequential data. LSTMs excel in learning and retaining information over extended time steps, making them pivotal in various applications such as natural language processing, time series analysis, and speech recognition.

The architecture of an LSTM network comprises several interconnected units or cells, each equipped with specialized mechanisms to regulate the flow of information. Key components of an LSTM include:

- Cell state: LSTMs maintain a cell state that runs linearly through the entire chain of units. This continuous flow of information allows LSTMs to retain and pass along information across multiple time steps without degradation.
- Three gates: LSTMs employ three gates to control the flow of information:
  - Forget gate: Determines what information to discard or forget from the cell state.
  - Input gate: Modulates the information to be updated and added to the cell state.
  - Output gate: Regulates the information to be output based on the updated cell state.
- Activation functions: LSTMs utilize activation functions, such as the sigmoid and hyperbolic tangent (tanh), to control the flow of information through the gates and cell state.

The architecture of LSTMs allows them to capture both short-term patterns and long-term dependencies within sequential data. The design of the gates and the cell state enables LSTMs to learn and retain information over extended time horizons, addressing the vanishing and exploding gradient problems commonly encountered in traditional RNNs [18–20].

Despite their effectiveness in capturing long-range dependencies, LSTMs exhibit a black-box nature due to several causes. For instance, the intricate interactions between gates, cell states, and activation functions result in a complex flow of information within LSTMs. Understanding how specific inputs or features influence the gating mechanisms and subsequent predictions can be challenging. LSTMs generate highly abstract representations of sequential data, making it difficult to interpret the learned representations or comprehend the precise rationale behind specific predictions or classifications. As with FNN, the non-linear transformations applied within the gates and cell states contribute to the opaque decision-making process of LSTMs. The complexity of these transformations hampers the direct interpretability of the network's decisions. While LSTMs excel in modeling temporal dependencies and preserving information over long sequences, their black-box nature poses challenges in interpretability, making it challenging to explain the decision-making process explicitly.

```
import torch
import torch.nn as nn

class VanillaLSTM(nn.Module):
    def __init__(self, input_size, hidden_size, output_size):
        super(VanillaLSTM, self).__init__()
        self.hidden_size = hidden_size
        self.lstm = nn.LSTM(input_size, hidden_size, batch_first=True)
        self.fc = nn.Linear(hidden_size, output_size)
        self.softmax = nn.Softmax(dim=1)
```

```
    def forward(self, x, h=None):
        if h is None:
            h = (torch.zeros(1, x.size(0), self.hidden_size).to(x.device),
                 torch.zeros(1, x.size(0), self.hidden_size).to(x.device))

        out, h = self.lstm(x, h)
        out = self.fc(out[:, -1, :])  # Taking the output of the last time
step
        out = self.softmax(out)
        return out, h
```

GRUs represent a type of gated recurrent neural network architecture designed to address some of the limitations found in traditional RNNs while maintaining efficiency in learning and processing sequential data. GRUs, akin to LSTMs, are structured to capture and retain long-term dependencies in sequential data, making them valuable in various applications such as natural language processing, time series prediction, and speech recognition.

Architecture of GRUs:

The architecture of a GRU consists of units or cells that integrate specialized gating mechanisms to regulate the flow of information within the network. Key components of GRUs include:

- **Update gate:** GRUs incorporate an update gate that decides how much of past information to retain and how much of new information to incorporate into the current state. This gate regulates the flow of information, facilitating the selective update of the memory.
- **Reset gate:** GRUs feature a reset gate that determines which parts of the past state should be forgotten or reset, allowing the network to adaptively control the relevance of past information.
- **Hidden state:** GRUs maintain a hidden state that evolves as they process sequential data. The hidden state, akin to the cell state in LSTMs, serves as the memory component that stores and propagates information across time steps.

Compared to LSTMs, GRUs have a more streamlined architecture, integrating fewer parameters and computations. This simplicity often leads to faster training times and is less prone to overfitting, making GRUs computationally more efficient. However, while GRUs provide efficiency in capturing long-range dependencies and adaptability to varying sequence lengths, their black-box nature poses challenges in interpretability, making it challenging to explain the decision-making process explicitly.

```
import torch
import torch.nn as nn

class VanillaGRU(nn.Module):
    def __init__(self, input_size, hidden_size, output_size):
        super(VanillaGRU, self).__init__()
        self.hidden_size = hidden_size
        self.gru = nn.GRU(input_size, hidden_size, batch_first=True)
        self.fc = nn.Linear(hidden_size, output_size)
        self.softmax = nn.Softmax(dim=1)

    def forward(self, x, h=None):
        if h is None:
            h = torch.zeros(1, x.size(0), self.hidden_size).to(x.device)

        out, h = self.gru(x, h)
        out = self.fc(out[:, -1, :])  # Taking the output of the last time
step
        out = self.softmax(out)
        return out, h
```

5.4.3 Convolutional neural networks

Convolutional neural networks (CNNs) represent a specialized class of neural networks particularly suited for processing and analyzing visual data, such as images and videos. CNNs have revolutionized computer vision tasks by efficiently learning hierarchical representations of images, enabling tasks like object recognition, image classification, object detection, and more.

CNNs employ convolutional layers that systematically apply filters (also known as kernels) across the input image. These filters perform convolutions, extracting local features by computing element-wise multiplications and summations at different spatial positions. This process allows the network to capture various patterns such as edges, textures, or shapes present in the image.

CNNs often incorporate pooling layers that reduce the dimensionality of the feature maps generated by the convolutional layers. Pooling operations (e.g., max pooling and average pooling) downsample the features, retaining essential information while enhancing computational efficiency and reducing overfitting. CNNs learn hierarchical representations of the input data. Lower layers in the network typically capture low-level features like edges and textures, while deeper layers progressively learn more complex and abstract features, combining these low-level features to identify higher-level patterns and objects. CNNs leverage parameter sharing, allowing the same set of weights (kernel) to be applied across different regions of the input image. This shared weight scheme significantly reduces the number of parameters to learn and enables the network to generalize better to variations in the input. Similar to other neural networks, CNNs incorporate activation functions (e.g., ReLU and sigmoid) to introduce non-linearities. At the end of the convolutional layers, fully connected layers combine the learned features to perform classification or regression tasks, producing the final output. CNNs have significantly advanced the field of computer vision by effectively learning and extracting hierarchical features from visual data. Their ability to automatically learn relevant features from raw pixels, along with their hierarchical architecture, makes them powerful tools for various image-related tasks.

In the early development of CNN, LeNet, developed by Yann LeCun, was one of the pioneering CNN architectures for handwritten digit recognition. Its architecture comprises convolutional layers followed by subsampling (pooling) layers, leading to hierarchical feature learning. LeNet introduced the concept of using convolutions and pooling layers alternately to capture spatial hierarchies in images.

```
import torch
import torch.nn as nn

class LeNet(nn.Module):
    def __init__(self, num_classes):
        super(LeNet, self).__init__()
        self.conv1 = nn.Conv2d(in_channels=1, out_channels=6,
kernel_size=5, stride=1)
        self.conv2 = nn.Conv2d(in_channels=6, out_channels=16,
kernel_size=5, stride=1)
        self.fc1 = nn.Linear(16 * 5 * 5, 120)
        self.fc2 = nn.Linear(120, 84)
        self.fc3 = nn.Linear(84, num_classes)
        self.relu = nn.ReLU()
        self.maxpool = nn.MaxPool2d(kernel_size=2, stride=2)
        self.flatten = nn.Flatten()

    def forward(self, x):
        x = self.maxpool(self.relu(self.conv1(x)))
        x = self.maxpool(self.relu(self.conv2(x)))
        x = self.flatten(x)
        x = self.relu(self.fc1(x))
        x = self.relu(self.fc2(x))
        x = self.fc3(x)
        return x
```

AlexNet, a breakthrough in the ImageNet large-scale visual recognition challenge (ILSVRC) in 2012, demonstrated the effectiveness of deep CNNs. It featured multiple convolutional layers, pooling layers, and fully connected layers, significantly deeper than previous architectures.

The architecture employed ReLU activation functions, dropout for regularization, and data augmentation techniques, enhancing performance.

In addition, VGG networks emphasized depth in CNNs, consisting of up to 19 layers, with small receptive fields (3×3 convolutional filters) and max-pooling layers. The uniform architecture with stacked convolutional layers and pooling layers made VGG networks easy to understand and implement. ResNet introduced residual connections or skip connections to tackle the vanishing gradient problem in very deep networks. These skip connections allow information to flow directly across layers, enabling the network to learn residual mappings, making it easier to train extremely deep networks (e.g., hundreds of layers). Moreover, Inception Networks, notably GoogLeNet, employed inception modules comprising multiple parallel convolutional operations of varying filter sizes. These modules aimed to capture features at different scales and achieve competitive accuracy while reducing computational complexity.

CNNs, including the aforementioned architectures, exhibit a black-box nature due to their inherent complexity and layered structures. The high number of parameters, intricate transformations within layers, and hierarchical feature learning make it difficult to interpret the decision-making process of these networks. While these CNN architectures have demonstrated exceptional performance in various image-related tasks, their black-box nature presents challenges in explaining the precise reasoning behind their predictions or feature representations. Interpreting the learned features and understanding the internal workings of these networks remains a significant area of research and concern, especially in critical applications where transparency and interpretability are crucial.

5.4.4 Generative adversarial networks

Generative adversarial networks (GANs) represent a powerful class of ML architectures introduced by Ian Goodfellow and his colleagues in 2014. GANs are composed of two neural networks, a generator and a discriminator, engaged in a competitive learning process. This innovative framework aims to generate new data samples that resemble the training data distribution, enabling the creation of synthetic data with high fidelity.

The generator network within a GAN creates synthetic data samples by mapping random noise or a latent space to the data space. It learns to generate increasingly realistic samples by transforming random input into data that resembles the distribution of the training data. Conversely, the discriminator, a separate neural network, acts as a binary classifier that evaluates whether a given sample is real (from the actual training data) or fake (generated by the generator). It aims to distinguish between real and generated samples, providing feedback to the generator on how well it produces realistic data.

The core principle of GANs involves a competitive training process between the generator and the discriminator. The generator strives to produce data that fools the discriminator into accepting the generated samples as real, while the discriminator aims to become more accurate in distinguishing real from fake samples. GANs leverage a minimax game framework where the generator and discriminator play a game against each other. The generator aims to minimize the probability of the discriminator correctly classifying generated samples as fake, while the discriminator aims to maximize its accuracy in distinguishing real from fake samples. During training, both networks are optimized iteratively. The generator improves its ability to generate realistic data as it receives feedback from the discriminator, while the discriminator enhances its discriminative ability by accurately classifying real and generated samples [21].

GANs have gained immense popularity due to their ability to generate high-quality synthetic data samples, which can be utilized for various creative and practical purposes. However, training GANs can be challenging and requires careful balancing between the generator and the discriminator to achieve convergence and generate realistic output.

```
import torch
import torch.nn as nn

class Generator(nn.Module):
    def __init__(self, input_size, hidden_size, output_size):
        super(Generator, self).__init__()
        self.fc1 = nn.Linear(input_size, hidden_size)
        self.relu = nn.ReLU()
        self.fc2 = nn.Linear(hidden_size, output_size)
        self.tanh = nn.Tanh()

    def forward(self, x):
        x = self.relu(self.fc1(x))
        x = self.tanh(self.fc2(x))
        return x

class Discriminator(nn.Module):
    def __init__(self, input_size, hidden_size, output_size):
        super(Discriminator, self).__init__()
        self.fc1 = nn.Linear(input_size, hidden_size)
        self.relu = nn.ReLU()
        self.fc2 = nn.Linear(hidden_size, output_size)
        self.sigmoid = nn.Sigmoid()

    def forward(self, x):
        x = self.relu(self.fc1(x))
        x = self.sigmoid(self.fc2(x))
        return x
```

Deep convolutional GAN (DCGAN) was proposed to employ convolutional networks in the generator and discriminator, enabling them to generate high-resolution and high-quality images. They leverage convolutional and transpose convolutional layers for generating images from random noise. DCGANs exhibit a black-box nature due to their complex hierarchical structures and the intricate relationships between convolutional layers. Understanding the learned representations and how individual features contribute to the generated output can be challenging.

Conditional GAN (cGAN) was proposed to extend the GAN framework by conditioning both the generator and the discriminator on additional information, such as class labels or auxiliary data. This conditioning allows for the controlled generation of data belonging to specific classes or categories. The conditional aspect in cGANs adds complexity, making it challenging to interpret how conditional information influences the generation process, contributing to their black-box nature. Moreover, least squares GAN (LSGAN) was proposed to modify the loss function by using least squares loss instead of the binary cross-entropy loss employed in traditional GANs. They aim to stabilize the training process by penalizing the generator for generating samples that the discriminator scores with low confidence. While LSGANs improve stability in training, the underlying mechanisms of how the least squares loss affects the learning dynamics and the decision boundaries of the discriminator and generator remain challenging to interpret.

InfoGAN was designed to learn interpretable and disentangled representations by introducing additional constraints on the learned latent space. It aims to encourage learning structured and meaningful representations from latent variables. While InfoGAN strives for disentangled representations, the complexity of the learned latent space and the interdependencies between different learned factors make it difficult to explicitly interpret or control individual factors in the representation. Wasserstein GAN (WGAN) utilized the Wasserstein distance as a measure of discrepancy between the generated and real data distributions. They aim to overcome mode collapse and instability issues by introducing a more stable training objective. WGANs improve training stability, yet the intricacies of how the Wasserstein distance affects the learning process and how the discriminator evaluates the discrepancy between distributions contribute to their black-box nature. Furthermore, CycleGAN is specialized for image-to-image translation tasks without paired training data. It learns to translate images from one domain to another while maintaining semantic consistency between domains. CycleGAN's ability to perform unsupervised image translation, manipulating image style and content, presents challenges in interpreting the learned representations and how the translation process occurs across domains.

5.4.5 Autoencoders

Autoencoders represent a class of neural networks employed for unsupervised learning and dimensionality reduction tasks. They aim to learn efficient representations of input data by training the network to reconstruct its input. This architecture consists of an encoder and a decoder, working together to encode the input into a latent space representation and then reconstruct the original input from this representation.

- Encoder: The encoder component of an autoencoder receives the input data and maps it to a compressed representation in a lower-dimensional space (latent space).
- It consists of a series of layers, typically neural network layers, that perform transformations to reduce the input data into a compressed representation.
- Latent space representation: The compressed representation obtained by the encoder is called the latent space representation or encoding. This latent space contains a compressed and more informative representation of the input data, capturing its essential features.
- Decoder: The decoder takes the compressed representation from the latent space and reconstructs the original input data. It mirrors the encoder's architecture but in reverse, aiming to generate an output that closely resembles the input data.
- Objective function (loss function): Autoencoders are trained by minimizing a loss function that measures the difference between the input and the reconstructed output. Common loss functions used in training autoencoders include mean squared error (MSE) or binary cross-entropy, depending on the nature of the input data.

Autoencoders can serve as effective tools for dimensionality reduction by learning a compact representation of high-dimensional data in the latent space. They aim to reconstruct the input faithfully, encouraging the network to learn meaningful representations that capture important features of the data. The evolution of ML many variants of autoencoders, such as variational autoencoders (VAEs), denoising autoencoders, and sparse autoencoders, and introduces additional constraints or modifications to enhance their capabilities, such as generating new data samples, denoising corrupted data, or learning sparse representations.

VAEs aim to generate new data samples by learning a probabilistic distribution of the input data in a latent space. They consist of an encoder that maps input data to a distribution in the latent space and a decoder that generates data from sampled points in the latent space. VAEs exhibit a black-box nature due to the probabilistic modeling of the latent space, making it challenging to interpret individual latent variables and the generation process. The complexity lies in understanding the learned distribution and how individual latent variables influence the generated output.

Denoising autoencoders are trained to reconstruct clean data from noisy or corrupted input. They learn to remove noise and extract meaningful features from the input by minimizing the reconstruction error between the clean input and the denoised output. Understanding the precise representations learned by denoising autoencoders and how they handle various types and levels of noise can be challenging. The inherent complexity lies in deciphering the learned representations' robustness to different noise patterns.

Sparse autoencoders introduce sparsity constraints during training to encourage the network to learn sparse representations of the input data. They aim to learn compact and informative representations by activating only a few neurons in the hidden layers for a given input. Deciphering the specific patterns or features encoded sparsely by the network and understanding the mechanisms that lead to sparsity can be intricate due to the complex interactions and learned representations within the network.

Contractive autoencoders are designed to learn robust representations by penalizing changes in the latent space with respect to small variations in the input. They aim to capture invariant or essential features by minimizing the sensitivity of the latent space to slight changes in input data. Understanding the exact criteria by which contractive autoencoders learn invariant representations and their robustness against input variations presents challenges in interpretation due to the learned manifold and latent space transformations. Adversarial autoencoders combine the traditional autoencoder structure with elements of GANs to generate realistic data samples. They consist of an encoder-decoder structure trained along with a discriminator to produce high-quality data samples. The interplay between the encoder, the decoder, and the discriminator in adversarial autoencoders introduces complexities in understanding how each component contributes to the generation process and the learned representations.

```
import torch
import torch.nn as nn

class Autoencoder(nn.Module):
    def __init__(self, input_size, hidden_size):
        super(Autoencoder, self).__init__()
        self.encoder = nn.Sequential(
            nn.Linear(input_size, hidden_size * 2),
            nn.ReLU(True),
            nn.Linear(hidden_size * 2, hidden_size),
            nn.ReLU(True),
            nn.Linear(hidden_size, hidden_size // 2)
        )
        self.decoder = nn.Sequential(
            nn.Linear(hidden_size // 2, hidden_size),
            nn.ReLU(True),
            nn.Linear(hidden_size, hidden_size * 2),
            nn.ReLU(True),
            nn.Linear(hidden_size * 2, input_size),
            nn.Sigmoid()
        )

    def forward(self, x):
        x = self.encoder(x)
        x = self.decoder(x)
        return x
```

5.4.6 Transformers

Transformer networks are a type of neural network architecture introduced by Vaswani *et al.* in 2017, primarily designed for sequence-to-sequence tasks in natural language processing (NLP). Unlike recurrent or convolutional architectures, transformers rely on self-attention mechanisms to capture dependencies between different elements within a sequence, making them highly effective for processing sequential data.

The core principle of transformers is the self-attention mechanism, enabling the network to weigh the significance of different elements (words and tokens) within a sequence concerning each other. Self-attention allows the model to compute contextual representations by assigning weights to each element in the sequence based on its relevance to other elements, capturing long-range dependencies efficiently. In addition, transformers typically consist of an encoder and a decoder component. The encoder processes the input sequence, generating contextual representations for each element by applying self-attention mechanisms and feed-forward neural network layers. In sequence-to-sequence tasks, the decoder uses the encoded representations to generate an output sequence while incorporating its own self-attention and additional attention over the encoder's outputs. Moreover, transformers employ multi-head attention mechanisms, where multiple self-attention heads run in parallel, allowing the network to attend to different parts of the input sequence. Each attention head learns different representations, which are later concatenated and linearly transformed to capture diverse aspects of the sequence relationships.

Transformers do not inherently possess positional information as recurrent architectures do. Therefore, positional encodings (e.g., sinusoidal embeddings) are added to the input embeddings to convey the sequence's positional information. Positional encodings enable the model to differentiate the positions of elements within the sequence. Transformers incorporate FFNN layers after the self-attention modules in both the encoder and the decoder. These FFNN layers provide additional transformation and non-linearities to the learned representations. Transformers utilize masking techniques, particularly in the decoder, to ensure that the model attends only to positions preceding the current position in an autoregressive manner. This prevents information leakage from future tokens during training for sequence generation tasks. Transformers have demonstrated exceptional performance in various NLP tasks such as machine translation, text generation, summarization, and question answering. Their ability to capture long-range dependencies and their parallelizable nature make them highly efficient for processing sequential data.

The evolution of different types of transformers, including bidirectional encoder representations from transformers (BERT), generative pre-trained transformer (GPT), and others. BERT is a transformer-based model introduced by Google in 2018, designed for various NLP tasks, including sentence classification, named entity recognition, and question answering. It leverages bidirectional attention mechanisms to capture contextual information from both the left and right contexts of a word in a sentence. BERT's large-scale architecture and bidirectional context modeling result in a black-box nature, making it challenging to interpret the specific learned representations and the reasoning behind its predictions. Understanding how each token's context contributes to the final output remains intricate. Additionally, GPT, was developed by OpenAI, is a series of autoregressive language models that predict the next token in a sequence, introduced in different versions (GPT-1, GPT-2, GPT-3, and GPT-4). GPT models are trained to generate coherent and contextually relevant text, demonstrating capabilities in text completion, language generation, and more. GPT's autoregressive nature and reliance on context and history for generating text contribute to its black-box nature. The complex interdependencies between tokens and the learned representations across multiple layers pose challenges in interpreting its decision-making process.

T5 (text-to-text transfer transformer), also from Google, adopts a text-to-text framework where various NLP tasks are framed as converting one text sequence to another. It follows a unified architecture that operates on input-output text pairs, aiming for simplicity and generalizability across different tasks. T5's unified text-to-text framework poses challenges in understanding the transformations and learned representations across diverse tasks. Deciphering the reasoning behind its text transformations and how it captures task-specific nuances remain complex. Besides, RoBERTa (robustly optimized BERT pretraining approach) is an optimized version of BERT that incorporates modifications in pretraining objectives, hyperparameters, and training data, aiming to improve performance on various downstream NLP tasks. It achieves state-of-the-art results in tasks like natural language understanding, text classification, and language modeling. Moreover,

XLNet is a model that incorporates ideas from autoregressive models (like GPT) and autoencoding models (like BERT) to improve learning capabilities. It uses a permutation-based training objective to capture bidirectional context and learn representations effectively.

Each of these transformer-based models showcases a black-box nature due to their large-scale architectures, intricate attention mechanisms, and complex interactions between tokens, making it challenging to interpret the learned representations and the decision-making processes involved in various NLP tasks.

```
import torch
import torch.nn as nn
import torch.nn.functional as F

class Transformer(nn.Module):
    def __init__(self, input_size, num_heads, hidden_size, num_layers):
        super(Transformer, self).__init__()
        self.embedding = nn.Embedding(input_size, hidden_size)
        self.positional_encoding = PositionalEncoding(hidden_size)
        self.transformer_layers = nn.ModuleList([
            TransformerEncoderLayer(hidden_size, num_heads) for _ in
range(num_layers)
        ])
        self.fc = nn.Linear(hidden_size, input_size)

    def forward(self, x):
        x = self.embedding(x)
        x = self.positional_encoding(x)
        for layer in self.transformer_layers:
            x = layer(x)
        x = self.fc(x)
        return F.log_softmax(x, dim=-1)

class PositionalEncoding(nn.Module):
    def __init__(self, hidden_size, max_len=1000):
        super(PositionalEncoding, self).__init__()
        pe = torch.zeros(max_len, hidden_size)
        position = torch.arange(0, max_len,
dtype=torch.float).unsqueeze(1)
        div_term = torch.exp(torch.arange(0, hidden_size, 2).float() * (-
math.log(10000.0) / hidden_size))
        pe[:, 0::2] = torch.sin(position * div_term)
        pe[:, 1::2] = torch.cos(position * div_term)
        pe = pe.unsqueeze(0)
        self.register_buffer('pe', pe)

    def forward(self, x):
        return x + self.pe[:, :x.size(1)]

class TransformerEncoderLayer(nn.Module):
    def __init__(self, hidden_size, num_heads):
        super(TransformerEncoderLayer, self).__init__()
        self.self_attn = nn.MultiheadAttention(hidden_size, num_heads)
        self.linear1 = nn.Linear(hidden_size, hidden_size)
        self.linear2 = nn.Linear(hidden_size, hidden_size)
        self.norm1 = nn.LayerNorm(hidden_size)
        self.norm2 = nn.LayerNorm(hidden_size)
        self.dropout1 = nn.Dropout(0.1)
        self.dropout2 = nn.Dropout(0.1)
```

```
    def forward(self, x):
        x = x.permute(1, 0, 2)  # Transformer expects (sequence length,
batch size, hidden size)
        attn_output, _ = self.self_attn(x, x, x)
        x = x + self.dropout1(attn_output)
        x = self.norm1(x)
        linear_output = self.linear2(F.relu(self.linear1(x)))
        x = x + self.dropout2(linear_output)
        x = self.norm2(x)
        return x.permute(1, 0, 2)  # Restore to original shape
```

5.4.7 Memory-augmented neural networks

Memory-augmented neural networks (MANNs) are a specialized class of neural networks designed to efficiently incorporate external memory components. They merge neural network architectures with external memory access, enabling the model to store and retrieve information dynamically during computation. MANNs aim to address tasks that demand complex reasoning and require the retention of information over extended MANNs consisting of a neural network coupled with an external memory component that the model can read from and write to during computations. The architecture enables the model to learn how to interact with the memory, deciding what information to read, write, and retain for subsequent processing steps.

MANNs allow dynamic allocation and usage of memory, where the model learns to access and modify memory content based on the current task or context. The network learns to attend to specific memory locations and selectively update or retrieve information as needed for improved performance. MANNs employ mechanisms such as attention mechanisms or content-based addressing to interact with the external memory. Attention mechanisms enable the model to focus on specific memory locations, attending to relevant information based on the current context or query. MANNs excel at tasks requiring information retention over extended periods, allowing the model to remember and access information from past experiences or training instances. This capability enables MANNs to perform tasks involving reasoning, logical inference, language understanding, and sequential decision-making.

Different variants of memory-augmented architectures exist, including neural turing machines (NTMs), differentiable neural computers (DNCs), etc. NTMs are a type of MANNs proposed by Graves *et al.*, integrating a neural network with an external memory matrix. They use attention mechanisms for read and write operations to access the external memory, enabling flexible memory interaction. NTMs exhibit a black-box nature due to their complex architecture involving neural networks and external memory. Deciphering the mechanisms by which the model interacts with memory, learns addressing, and stores information poses challenges in interpretation.

DNCs, an advancement over NTMs, introduced by DeepMind, feature a controller network interacting with an external memory matrix via read and write heads. They use content-based addressing and dynamic memory allocation, enhancing memory interaction and learning capabilities. DNCs inherit the black-box nature of NTMs, given their similar architecture with improved memory interaction mechanisms. Understanding the interactions between the controller, memory, and the learned representations remains challenging. Memory networks are a class of MANNs focused on explicitly storing information in a memory component as key-value pairs. They consist of modules for reading, writing, and addressing the memory, allowing for efficient retrieval based on content or addressing mechanisms. Memory networks, while explicit in memory management, possess a black-box nature due to the complexity of information retrieval processes. Understanding how the model accesses and utilizes stored key-value pairs for inference remains intricate.

5.4.8 Spiking neural networks

Spiking neural networks (SNNs) are a type of artificial neural network inspired by the biological behavior of neurons and their communication through spikes or action potentials. SNNs model neural computation based on the timing and spiking activity of neurons, enabling them to represent and process information in a manner similar to the brain's neural networks. SNNs are composed of spiking neurons that communicate through discrete, asynchronous pulses called spikes or action potentials. Neurons integrate input signals over time and generate spikes when their membrane potential surpasses a certain threshold.

SNNs utilize the timing of spikes to encode information, employing a temporal code where the timing of spikes carries information about input stimuli. Information is processed based on the precise timing of spikes, enabling SNNs to encode temporal patterns and process time-varying signals effectively. Unlike conventional neural networks that operate based on continuous computation, SNNs function in an event-driven manner, processing information only when spikes occur. This event-driven processing results in energy-efficient computation, as the network primarily activates when required by input spikes. SNNs model synaptic connections between neurons, incorporating principles of synaptic plasticity that enable the modification of synaptic strengths based on learning rules. Synaptic plasticity allows SNNs to adapt their connections, strengthening or weakening synapses based on learning experiences. SNNs can comprise multiple spiking layers interconnected through synaptic connections, resembling the layered architecture of traditional neural networks.

Layers of neurons communicate through spike transmissions, and each layer processes information through spike-based computations.

The research provides different types of SNNs, including spike timing-dependent plasticity (STDP), liquid state machines (LSM), and neuromorphic computing models. STDP is a learning rule observed in biological synapses, where the efficacy of synaptic connections between neurons is adjusted based on the timing of pre- and postsynaptic spikes. It reinforces synapses when presynaptic spikes occur before postsynaptic spikes and weakens them when the order is reversed. In addition, LSMs are a type of SNN that emphasizes the dynamics of recurrently connected spiking neurons in a large-scale network, mimicking the dynamics of a “liquid” neural substrate. They leverage the complex dynamics of recurrent connections in a spiking network to process and represent temporal information. Neuromorphic computing models encompass a broader range of SNN-based computational paradigms that aim to emulate brain-like functionalities using hardware or software implementations. These models focus on mimicking the brain’s neural structure, dynamics, and computation principles, aiming for energy-efficient and parallel information processing. SNNs are implemented in specialized hardware architectures called neuromorphic chips or neuromorphic computing systems. These hardware systems are designed to efficiently simulate the behavior of SNNs and leverage the parallelism and low-power consumption inherent in spiking neuron models.

```
import torch
import torch.nn as nn

class SpikingNeuron(nn.Module):
    def __init__(self, num_inputs, threshold=1.0, decay=0.9):
        super(SpikingNeuron, self).__init__()
        self.num_inputs = num_inputs
        self.threshold = threshold
        self.decay = decay
        self.membrane_potential = nn.Parameter(torch.zeros(num_inputs),
requires_grad=False)
        self.spike = nn.Parameter(torch.zeros(num_inputs),
requires_grad=False)

    def forward(self, inputs):
        self.membrane_potential = self.membrane_potential * self.decay +
inputs
        self.spike = torch.zeros_like(self.membrane_potential)
        self.spike[self.membrane_potential >= self.threshold] = 1.0
        self.membrane_potential[self.membrane_potential >= self.threshold]
= 0.0
        return self.spike

class SpikingNeuralNetwork(nn.Module):
    def __init__(self, input_size, hidden_size, output_size):
        super(SpikingNeuralNetwork, self).__init__()
        self.input_layer = SpikingNeuron(input_size)
        self.hidden_layer = SpikingNeuron(hidden_size)
        self.output_layer = SpikingNeuron(output_size)
        self.fc1 = nn.Linear(input_size, hidden_size)
        self.fc2 = nn.Linear(hidden_size, output_size)

    def forward(self, x):
        spike_input = self.input_layer(x)
        spike_hidden = self.hidden_layer(self.fc1(spike_input))
        spike_output = self.output_layer(self.fc2(spike_hidden))
        return spike_output
```

5.4.9 Graph neural networks

Introduce the general concept of GNN, architecture, and its main principles.

Discuss different types of GNNs:

Graph convolutional networks (GCN), graph attention networks (GAN), GraphSAGE, etc.

```
from torch.nn import Linear
import torch.nn.functional as F
from torch_geometric.nn import GCNConv
from torch_geometric.nn import global_mean_pool

class GCN(torch.nn.Module):
    def __init__(self, hidden_channels):
        super(GCN, self).__init__()
        torch.manual_seed(12345)
        self.conv1 = GCNConv(dataset.num_node_features, hidden_channels)
        self.conv2 = GCNConv(hidden_channels, hidden_channels)
        self.conv3 = GCNConv(hidden_channels, hidden_channels)
        self.lin = Linear(hidden_channels, dataset.num_classes)

    def forward(self, x, edge_index, batch):
        # 1. Obtain node embeddings
        x = self.conv1(x, edge_index)
        x = x.relu()
        x = self.conv2(x, edge_index)
        x = x.relu()
        x = self.conv3(x, edge_index)

        # 2. Readout layer
        x = global_mean_pool(x, batch)  # [batch_size, hidden_channels]

        # 3. Apply a final classifier
        x = F.dropout(x, p=0.5, training=self.training)
        x = self.lin(x)

        return x
```

5.5 Reinforcement learning

Reinforcement learning (RL) is a subfield of ML concerned with agents learning to make sequences of decisions by interacting with an environment. Unlike supervised learning, RL does not rely on labeled datasets but instead learns from trial and error, aiming to maximize cumulative reward signals received from the environment through its actions. RL models adopt a sequential decision-making framework, where an agent learns to navigate an environment by taking actions and receiving feedback in the form of rewards or penalties, enabling it to discover optimal strategies or policies.

In the context of IoT security, RL presents significant implications and relevance. IoT ecosystems consist of numerous interconnected devices operating in complex and dynamic environments. The security of these systems is paramount due to their susceptibility to cyber threats, privacy breaches, and potential consequences of compromised data integrity. Traditional security measures often fall short of addressing the evolving and intricate nature of IoT vulnerabilities. RL offers a promising approach by providing adaptive, learning-based solutions that can dynamically respond to emerging threats and evolving attack vectors within IoT networks. The application of RL in IoT security encompasses various domains. RL algorithms can be employed to develop intelligent intrusion detection systems capable of learning and adapting to new attack patterns, anomalous behaviors, or zero-day vulnerabilities within IoT networks. Furthermore, RL facilitates the design of robust access control mechanisms that dynamically adjust authorization policies based on contextual changes or user behaviors, mitigating potential security breaches [9,10].

Moreover, RL techniques contribute to enhancing the resilience of IoT systems against adversarial attacks. By continuously learning from interactions with the environment, RL-based agents can anticipate potential threats, proactively identify vulnerabilities, and autonomously reconfigure security protocols or response strategies to prevent or mitigate security breaches. Additionally, RL-driven anomaly detection mechanisms assist in identifying deviations from normal behavior patterns in IoT devices, signaling potential security threats or malfunctions.

Q-learning is a fundamental model-free RL algorithm used for solving sequential decision-making problems in an environment without prior knowledge. Developed by Watkins in 1989, Q-learning operates by learning an action-value function that represents the expected cumulative reward an agent can attain by taking a specific action in a particular state and following a policy to maximize long-term rewards. The core principle underlying Q-learning revolves around the notion of *Q*-values. The *Q*-value (also known as the action–value function) of a state–action pair in an RL context represents the expected cumulative reward an agent receives by performing a specific action in a given state and following a particular policy thereafter. Formally, the *Q*-value is represented as $Q(s, a)$, denoting the value of taking action $'a'$ in state $'s'$.

The Q-learning algorithm iteratively updates *Q*-values based on the agent's experiences in the environment. Through exploration and exploitation, the agent interacts with the environment, observes states, takes actions, receives rewards, and updates *Q*-values accordingly to refine its decision-making process. The *Q*-value update rule is based on the Bellman equation, allowing the agent to iteratively improve its action–value estimates:

$$Q(s,a) \leftarrow (1-\alpha)\cdot Q(s,a) + \alpha \cdot \left[r + \gamma \cdot \max_{a'} Q\left(s',a'\right)\right] \tag{5.1}$$

where $Q(s, a)$ represents the *Q*-value of taking action "*a*" in state "*s*." r is the immediate reward received after taking action "*a*" in state "*s*." α is the learning rate determining the extent to which newly acquired information overrides previous *Q*-value estimates. γ is the discount factor balancing immediate and future rewards. "*s*" represents the resulting state after taking action "*a*" in state "*s*." a' is the subsequent action chosen based on the maximum *Q*-value in state "*s*."

The Q-learning algorithm operates through an iterative process of exploration and exploitation, gradually updating Q-values for each state-action pair until convergence to an optimal policy that maximizes long-term rewards. Despite its simplicity and efficiency in learning optimal policies in discrete state and action spaces, Q-learning faces challenges in scalability to larger, continuous state spaces and high-dimensional environments due to its requirement of storing Q-values for each state-action pair.

Deep Q-networks (DQNs) represent a significant advancement in RL, combining DNNs with Q-learning to address complex, high-dimensional state spaces in environments. Developed by researchers at DeepMind, DQN extends Q-learning by leveraging DNNs to approximate the action–value function, enabling effective learning in environments with large state spaces, such as high-resolution images or continuous sensor data.

The primary principle underlying DQN involves approximating the action–value function $Q(s, a)$ using a DNN, commonly referred to as the Q-network. This network takes a state s as input and outputs Q-values for each possible action a in that state. The Q-network aims to estimate action values similar to Q-learning, where the agent selects actions that maximize long-term rewards. The key innovation of DQN lies in its ability to handle high-dimensional input spaces by utilizing DNNs as function approximators. The Q-network is trained to minimize the difference between predicted Q-values and target Q-values obtained from the Bellman equation. DQN utilizes experience replay and target networks to stabilize and enhance learning efficiency.

1. **Experience replay:** DQN incorporates experience replay, storing experiences (consisting of state, action, reward, and next state) in a replay buffer. During training, random mini-batches of experiences are sampled from the replay buffer, breaking correlations in sequential experiences and improving data efficiency and learning stability.
2. **Target network:** DQN employs a separate target network to stabilize training, periodically updating its parameters with the Q-network's weights. This target network is used to generate target Q-values during training, reducing the potential for overestimation bias and providing more stable training.

The DQN algorithm aims to iteratively update the parameters of the Q-network by minimizing a loss function that quantifies the discrepancy between predicted Q-values and target Q-values. The loss function is typically calculated using a mean squared error between the predicted Q-values and the target values derived from the Bellman equation:

$$loss = \frac{1}{N}\sum_{i=1}^{N}\left(Q(s_i, a_i) - \left(r_i + \gamma \cdot \max_{a'} Q\left(s_i' a' ; \theta\right)\right)\right)^2 \tag{5.2}$$

where $Q(s, a)$ represents predicted Q-values by the Q-network. r_i is the immediate reward received after taking action a_i in state s_i. s_i' represents the resulting state after taking action a_i in state s_i. a' is the subsequent action chosen based on the

Q-value from the target network. γ is the discount factor balancing immediate and future rewards. θ denotes the parameters of the target network.

Actor–critic models represent a class of RL algorithms that combine elements from policy-based methods (the actor) and value-based methods (the critic) to address sequential decision-making tasks. These models aim to learn both a policy, specifying the agent's action selection, and an estimated value function, determining the quality of the chosen actions, to optimize long-term rewards.

The primary distinction in actor-critic models lies in their dual components:

- **The actor:** This component learns a policy, which is essentially a mapping from states to actions. It explores the environment and suggests actions that maximize expected cumulative rewards. The policy is typically represented as a neural network, outputting action probabilities or deterministic actions based on states.
- **The critic:** The critic estimates the value function, evaluating the goodness of actions taken by the actor. It provides a feedback signal to the actor by assessing the value of the state–action pairs. The critic estimates expected future rewards, guiding the actor towards actions that lead to higher cumulative rewards. This component often takes the form of a value network approximated by a neural network.

The actor-critic architecture combines the advantages of policy-based and value-based approaches. It enables efficient exploration of the action space while simultaneously learning a value function to estimate the long-term consequences of actions. This integration leads to more stable learning compared to purely value-based or policy-based methods.

The learning process in actor-critic models involves updating both the actor's policy and the critic's value estimation network based on the experiences gained from interactions with the environment. The critic evaluates actions chosen by the actor, providing feedback on their quality. The actor updates its policy by maximizing the estimated values provided by the critic. Simultaneously, the critic learns to estimate values by minimizing the temporal difference error between predicted and actual returns.

The update equations in actor-critic models often involve minimizing the temporal difference (TD) error to update both the policy and value function:

$$\text{TD error} = r + \gamma \cdot V\left(s'\right) - V(s) \tag{5.3}$$

$$\text{Policy update}: \theta_{\text{Actor}} \leftarrow \theta_{\text{Actor}} + \alpha_{\text{Actor}} \cdot \nabla_{\theta_{\text{Actor}}} \log \pi(a|s).\text{TD error} \tag{5.4}$$

$$\text{Value update}: \theta_{\text{Critic}} \leftarrow \theta_{\text{Critic}} + \alpha_{\text{Critic}} \cdot \text{TD error} \cdot \nabla_{\theta_{\text{Criti}}} V(s) \tag{5.5}$$

where r is the immediate reward received after taking an action. γ is the discount factor. $V(s)$ represents the estimated value of state s. $\pi(a|s)$ denotes the policy (probability of taking action a given state s). α_{Actor} and α_{Critic} are learning rates for the actor and critic, respectively. θ_{Actor} and θ_{Critic} denote the parameters of the actor and critic networks. Actor–critic models offer a versatile approach in RL by combining policy-based and value-based methods. Through the collaboration between the actor responsible for action selection and the critic, responsible for value estimation, these models achieve efficient and stable learning in complex environments.

5.6 Meta-learning

Meta-learning, often referred to as "learning to learn," represents a paradigm within ML that focuses on developing models or algorithms capable of acquiring knowledge or learning strategies from multiple tasks or domains. Unlike traditional ML, where models are trained on specific datasets to perform well on a particular task, meta-learning aims to generalize learning across various tasks, enabling models to adapt and learn more efficiently from new tasks with limited data. The core concept of meta-learning revolves around the idea of learning higher-level knowledge or representations that facilitate faster learning or adaptation to new tasks. Meta-learning algorithms strive to extract common patterns, structures, or priors from a diverse set of tasks, allowing models to acquire transferable knowledge that aids in rapid adaptation, generalization, and improved performance on unseen tasks [21].

Meta-learning focuses on learning strategies or representations that enable models to adapt to new tasks efficiently. This adaptation often involves updating model parameters based on a small amount of data or experiences from multiple related tasks. Meta-learning commonly involves two distinct phases: meta-training and meta-testing. During meta-training, the model learns from a diverse set of tasks or datasets, acquiring knowledge that facilitates rapid adaptation. In meta-testing, the model's adaptability is evaluated on new, unseen tasks, where it learns to perform with minimal data or experiences. Meta-learning algorithms often emphasize the extraction of useful representations or features that capture task-agnostic information. These representations help models generalize across tasks by capturing underlying patterns or inductive biases present in the training data. Many meta-learning approaches employ gradient-based optimization methods, such as gradient descent or higher-order derivatives, to update model parameters for efficient adaptation to new tasks. These methods aim to find optimal initialization or update rules that facilitate quick adaptation.

Model-agnostic meta-learning (MAML) is a prominent meta-learning framework that seeks to learn an initialization of model parameters that can be quickly fine-tuned or adapted to new tasks. It optimizes model parameters to enable fast adaptation with minimal task-specific data. The ultimate goal of meta-learning is to develop models or algorithms capable of leveraging prior knowledge gained from diverse tasks or domains to enhance learning efficiency and generalize effectively to new, unseen tasks. By capturing underlying patterns or inductive biases, meta-learning approaches enable models to acquire transferable knowledge, thereby reducing the need for extensive training data and accelerating learning on novel tasks. This academic-style introduction provides an overview of meta-learning, highlighting its fundamental concepts, principles, and objectives. It emphasizes the focus on learning adaptable strategies, extracting transferable representations, and optimizing models to facilitate rapid adaptation and improved generalization across diverse tasks or domains in ML.

Siamese networks represent a class of neural network architectures designed for learning similarity metrics or embeddings between pairs of inputs. These networks have garnered attention for their effectiveness in tasks involving similarity-based comparisons, such as image similarity, signature verification, and sentence similarity. The fundamental principle underlying Siamese networks involves learning a shared representation or embedding space for input pairs by using identical subnetworks (or twin networks) with shared weights. The term "Siamese" originates from the resemblance to the conjoined twins, reflecting the symmetric nature of the network architecture. Siamese networks have two identical neural network branches (or arms) that share the same architecture and weights. These branches process each input separately, extracting meaningful representations or embeddings. The primary objective of Siamese networks is to compare pairs of inputs and determine their similarity or dissimilarity in the learned embedding space. This comparison is often achieved by measuring the distance or similarity metric between the embeddings of the input pairs. Siamese networks aim to learn a suitable distance metric or similarity function in the embedding space, enabling the network to distinguish between similar and dissimilar pairs effectively. Common distance metrics include Euclidean distance, cosine similarity, or contrastive/hinge loss functions. Siamese networks are trained using pairs of input samples along with their corresponding similarity labels (or dissimilarity labels). The network learns to minimize the distance or dissimilarity between similar pairs while maximizing the distance between dissimilar pairs, typically through contrastive loss or triplet loss functions. Siamese networks excel in scenarios with limited labeled training data for similarity-based tasks. By learning a meaningful embedding space and similarity metrics, these networks facilitate effective comparison and decision-making between input pairs, enabling tasks that require discerning similarity or dissimilarity.

Reptile is a meta-learning algorithm that facilitates rapid adaptation or few-shot learning in ML models. Developed by OpenAI, Reptile focuses on meta-learning, where models aim to quickly adapt to new tasks or environments with limited training samples.

The fundamental principle underlying Reptile involves a form of meta-learning that optimizes model parameters to enable faster adaptation across tasks. Unlike traditional gradient-based optimization methods, Reptile employs a simple and efficient approach emphasizing task adaptation through iterative updates to the model parameters. Reptile operates in two distinct phases: meta-training and meta-testing. During meta-training, the model learns across a diverse set of tasks, updating its parameters to facilitate rapid adaptation. In meta-testing, the model's ability to adapt is evaluated on new tasks, where it is required to learn from a few examples or shots. Reptile adopts a straightforward approach by initializing model parameters and iteratively updating them toward the parameters of task-specific models. It involves multiple gradient steps on a new task, making updates to the model's parameters in the direction that facilitates faster convergence on that task. The objective of Reptile is to enable models to generalize from a small number of training examples or shots. By performing multiple updates on the model parameters using data from new tasks, Reptile aims to enhance the model's ability to adapt and generalize quickly.

Reptile's optimization strategy focuses on achieving fast adaptation to new tasks by adjusting model parameters to perform well on those tasks. It aims to learn a set of parameters that exhibit good generalization across a variety of related tasks. The update process in Reptile resembles gradient descent, where the model parameters are updated in the direction that reduces the loss or error on the new task. However, instead of converging to the optimal parameters for a single task, Reptile aims to find parameters that adapt well across multiple tasks. Reptile demonstrates its effectiveness in scenarios where models need to quickly adapt to new tasks or domains with limited training data. Its simplicity in updating model parameters based on a few examples from new tasks enables efficient adaptation and generalization, making it suitable for few-shot learning and rapid task acquisition.

Meta-SGD (Meta-stochastic gradient descent) represents a meta-learning approach aimed at enhancing the optimization process in ML models across diverse tasks or datasets. Developed by researchers, Meta-SGD focuses on learning an optimizer that adapts to different tasks, aiming to improve the efficiency and convergence speed of gradient-based optimization methods. The fundamental principle underlying Meta-SGD involves learning an adaptive optimization algorithm, or meta-optimizer, that adjusts the learning rate or updates the rule of the underlying stochastic gradient descent optimizer. This meta-optimizer learns to generalize across tasks and dynamically adjusts optimization hyperparameters to facilitate faster convergence and better generalization. Meta-SGD operates through a meta-training phase, where the meta-optimizer is trained across a range of tasks or datasets. During meta-testing, the learned meta-optimizer is evaluated on new tasks to assess its adaptability and efficiency in optimizing model parameters. The objective of Meta-SGD is to learn parameters or update rules for the optimization algorithm that can adapt to diverse tasks. Instead of using a fixed learning rate or optimization strategy, Meta-SGD aims to dynamically adjust these parameters based on task-specific characteristics. Meta-SGD utilizes gradient-based optimization to update the meta-parameters of the optimizer. It leverages gradients derived from the performance of the optimizer on different tasks to improve its adaptability and effectiveness. Meta-SGD focuses on learning an adaptive learning rate or update rule that optimally adjusts the step size of the optimization process. This adaptive learning rate aims to enhance convergence speed and generalization across tasks by effectively utilizing the available information.

Despite the considerable promise of meta-learning methods in reshaping security intelligence within the IoT landscape, a persistent challenge lies in their inherent black-box nature, contributing to limited interpretability. These meta-learning techniques, while offering impressive capabilities in rapid adaptation, few-shot learning, and task generalization across diverse IoT security scenarios, often operate as opaque systems, making it challenging to decipher the reasoning behind their decisions or understand their internal mechanisms comprehensively. This lack of interpretability poses significant hurdles, especially in security-critical contexts, where transparent and explainable models are essential for identifying vulnerabilities, understanding attack patterns, and making informed decisions. Consequently, the deployment of meta-learning approaches in IoT security applications encounters obstacles related to trust, transparency, and regulatory compliance. As such, efforts to enhance the interpretability of these black-box meta-learning models within IoT security remain crucial to foster greater trust, facilitate human understanding of their decision-making processes, and ultimately augment their effectiveness in safeguarding IoT ecosystems.

References

[1] Adadi, A. and Berrada, M. (2018). Peeking inside the black-box: A survey on explainable artificial intelligence (XAI). *IEEE Access*, 6, 52138–52160.

[2] Chennam, K. K., Mudrakola, S., Maheswari, V. U., Aluvalu, R., and Rao, K. G. (2022). Black box models for eXplainable artificial intelligence. In *Explainable AI: Foundations, Methodologies and Applications* (pp. 1–24). Cham: Springer International Publishing.

[3] Guidotti, R., Monreale, A., Ruggieri, S., Turini, F., Giannotti, F., and Pedreschi, D. (2018). A survey of methods for explaining black box models. *ACM Computing Surveys (CSUR)*, *51*(5), 1–42.

[4] Mehrabi, N., Morstatter, F., Saxena, N., Lerman, K., and Galstyan, A. (2021). A survey on bias and fairness in machine learning. *ACM Computing Surveys (CSUR)*, *54*(6), 1–35.

[5] Cunningham, P., and Delany, S. J. (2021). Underestimation bias and underfitting in machine learning. In *Trustworthy AI-Integrating Learning, Optimization and Reasoning: First International Workshop, TAILOR 2020, Virtual Event, September 4–5, 2020, Revised Selected Papers 1* (pp. 20–31). Springer International Publishing.

[6] Soumare, H., Benkahla, A., and Gmati, N. (2021). Deep learning regularization techniques to genomics data. *Array*, *11*, 100068.

[7] Konstantinov, A. V., and Utkin, L. V. (2021). Interpretable machine learning with an ensemble of gradient boosting machines. *Knowledge-Based Systems*, *222*, 106993.

[8] Kiran, R. J., Sanil, J., and Asharaf, S. (2024). A novel approach for model interpretability and Domain Aware Fine-Tuning in AdaBoost. *Human-Centric Intelligent Systems*, 1–23.

[9] Chen, W., Qiu, X., Cai, T., Dai, H. N., Zheng, Z., and Zhang, Y. (2021). Deep reinforcement learning for Internet of Things: A comprehensive survey. *IEEE Communications Surveys & Tutorials*, *23*(3), 1659–1692.

[10] Uprety, A., and Rawat, D. B. (2020). Reinforcement learning for IOT security: A comprehensive survey. *IEEE Internet of Things Journal*, *8*(11), 8693–8706.

[11] Huang, X., Li, Z., Jin, Y., & Zhang, W. (2022). Fair-AdaBoost: Extending AdaBoost method to achieve fair classification. *Expert Systems with Applications*, *202*, 117240.

[12] Ke, G., Meng, Q., Finley, T., *et al.* (2017). Lightgbm: A highly efficient gradient boosting decision tree. *Advances in Neural Information Processing Systems*, *30*.

[13] Karampinis, I., Bantilas, K. E., Kavvadias, I. E., Iliadis, L., and Elenas, A. (2024). Seismic response prediction of rigid rocking structures using explainable LightGBM models. *Mathematics*, *12*(14), 2280.

[14] Prokhorenkova, L., Gusev, G., Vorobev, A., Dorogush, A. V., and Gulin, A. (2018). CatBoost: unbiased boosting with categorical features. *Advances in Neural Information Processing Systems*, *31*.

[15] Maouche, I., Terrissa, L. S., Benmohammed, K., and Zerhouni, N. (2023). An explainable AI approach for breast cancer metastasis prediction based on clinicopathological data. *IEEE Transactions on Biomedical Engineering*, *70* (12), 3321–3329.

[16] Vishwanathan, S. V. M., and Murty, M. N. (2002, May). SSVM: A simple SVM algorithm. In *Proceedings of the 2002 International Joint Conference on Neural Networks. IJCNN'02 (Cat. No. 02CH37290)* (Vol. 3, pp. 2393-2398). IEEE.

[17] LeCun, Y., Bengio, Y., and Hinton, G. (2015). Deep learning. *Nature*, *521* (7553),

[18] Medsker, L., and Jain, L. C. (eds.). (1999). *Recurrent Neural Networks: Design and Applications*. CRC press.

[19] Lalapura, V. S., Amudha, J., and Satheesh, H. S. (2021). Recurrent neural networks for edge intelligence: a survey. *ACM Computing Surveys (CSUR)*, *54*(4), 1–38.

[20] Cossu, A., Carta, A., Lomonaco, V., and Bacciu, D. (2021). Continual learning for recurrent neural networks: an empirical evaluation. *Neural Networks*, *143*, 607–627.

[21] Saxena, D., & Cao, J. (2021). Generative adversarial networks (GANs) challenges, solutions, and future directions. *ACM Computing Surveys (CSUR)*, *54*(3), 1–42.

Chapter 6
Explainable artificial intelligence for safeguarding IoT

6.1 Introduction

Explainable artificial intelligence (XAI) represents a crucial advancement within the field of artificial intelligence (AI), aimed at unraveling the complex decision-making processes inherent in AI algorithms. In the multifaceted landscape of the IoT, where a multitude of interconnected devices autonomously interact, XAI assumes paramount significance in ensuring the transparency and interpretability of AI-driven functionalities. As IoT ecosystems continue to expand across diverse domains, encompassing healthcare, smart homes, transportation, and industrial sectors, the integration of AI introduces unparalleled complexities and challenges. XAI emerges as an indispensable framework that not only elucidates the functioning of intricate machine-learning models but also empowers stakeholders to comprehend the rationale behind AI-generated decisions. Within this context, the significance of XAI in IoT security lies not solely in demystifying opaque algorithms but also in fortifying the trustworthiness and reliability of AI systems operating within these interconnected environments [1].

The intricate nature of IoT networks, characterized by diverse data streams and interconnected nodes, underscores the criticality of transparent AI systems in safeguarding against potential threats and vulnerabilities. XAI assumes a pivotal role in enhancing the explainability and accountability of AI models deployed within IoT architectures. By providing insights into the decision-making process of AI algorithms, XAI enables stakeholders, including developers, end-users, and regulatory bodies, to scrutinize, validate, and comprehend the outcomes produced by AI-driven devices. Moreover, the amalgamation of XAI principles into IoT security strategies serves as a catalyst for advancing ethical considerations and addressing inherent biases within AI systems. Consequently, this augments the security posture of IoT ecosystems and fosters a conducive environment for responsible and trustworthy AI deployment, thus affirming the integrity and resilience of interconnected devices in an increasingly digital landscape.

Transparency and interpretability are foundational pillars in ensuring the efficacy and trustworthiness of AI systems within the intricate framework of the IoT security. In the context of interconnected devices and AI-driven decision-making, transparency pertains to the comprehensibility and visibility of the underlying mechanisms guiding AI algorithms. It encompasses the accessibility of information regarding the functioning, rationale, and decision-making processes of these complex systems. Interpretability, on the other hand, delves deeper into the capacity to understand and explain the logic behind AI-generated outcomes. Within IoT environments, characterized by diverse data streams and intricate network interactions, the integration of transparent and interpretable AI systems becomes imperative in discerning the cause-effect relationships and ensuring the accountability of automated decisions. This integration not only fosters comprehension among stakeholders but also enables the identification and mitigation of potential biases or erroneous functionalities, thus fortifying the overall security fabric of interconnected IoT ecosystems [2].

The significance of transparency and interpretability in AI systems for IoT security extends beyond mere comprehension; it resonates profoundly with ethical considerations and regulatory imperatives. Ensuring transparent and interpretable AI models within IoT environments aligns with ethical frameworks, fostering responsible AI deployment. It empowers users, developers, and regulatory bodies to scrutinize the functioning of AI algorithms, ascertain the fairness and accountability of decisions, and detect any inherent biases or discriminatory patterns. This proactive approach bolsters the trust and acceptance of AI-driven systems and contributes to compliance with regulatory standards and guidelines. Moreover, by promoting transparency and interpretability, AI systems operating within IoT ecosystems become more resilient to adversarial attacks or manipulations, reinforcing interconnected devices' security posture. Thus, the emphasis on transparency and interpretability in AI systems for IoT security serves as a cornerstone for comprehensibility and underpins the ethical, regulatory, and security dimensions essential for the integrity of AI-enabled IoT environments [3].

6.2 Principles of XAI

Understanding the key principles and characteristics of XAI models is pivotal in unraveling the intricacies of complex machine-learning algorithms while ensuring transparency and comprehensibility. One fundamental principle of XAI involves the concept of transparency, which revolves around elucidating the inner workings of AI systems to make their decision-making processes accessible and understandable. Transparency allows stakeholders to discern the rationale behind AI-driven outputs, enabling scrutiny and validation of the models' decisions. Another essential principle lies in interpretability, which emphasizes the ability to comprehend and explain the outcomes generated by AI models. Interpretability aims to provide human-understandable explanations, allowing users to grasp how inputs are transformed into predictions or decisions, fostering trust and confidence in the AI's functionality within various domains, including healthcare, finance, and autonomous systems [4].

Furthermore, XAI models exhibit characteristics such as causability, which delineates the capacity to identify causal relationships between input variables and model predictions. By discerning causality, XAI models move beyond correlations, enabling a deeper understanding of why certain predictions or decisions are made, which is vital for critical domains like healthcare diagnostics or autonomous vehicles. Trustworthiness is another pivotal characteristic, emphasizing the reliability and consistency of AI systems. Trustworthiness ensures that AI models provide consistent and reliable outcomes across diverse scenarios, instilling confidence in stakeholders regarding the reliability of AI-driven decisions. Additionally, the concept of fairness underscores the importance of mitigating biases and ensuring equitable outcomes from AI models, aligning with ethical imperatives. These key principles and characteristics collectively form the foundation of XAI models, driving the quest for transparent, interpretable, and ethical AI systems, thereby enhancing their applicability and acceptance in various real-world contexts [5].

In our endeavor to delve into the intricate realm of XAI, it becomes imperative to establish a foundational understanding by elucidating a set of pivotal concepts and terminologies intricately associated with this burgeoning field. As we navigate the landscape of XAI, comprehending the intricacies of its principles and vocabulary stands as a fundamental precursor to unraveling the complexities inherent in AI systems' interpretability and transparency. By defining and elucidating these key concepts, ranging from transparency, interpretability, and accountability to fairness, causability, and trustworthiness, we aim to construct a comprehensive framework that lays the groundwork for a nuanced comprehension of XAI's principles and their implications within the dynamic context of contemporary AI. This groundwork not only fosters a deeper understanding of the theoretical underpinnings but also provides a scaffold upon which practical applications and advancements in XAI can be comprehensively understood and explored within various domains and applications [6].

Definition 1: *Transparency is defined as the quality or state of an AI system being open, understandable, and clear in its operations, decisions, and processes. Transparent AI systems allow stakeholders to comprehend and inspect the inner workings, logic, and algorithms, facilitating an understanding of how and why specific decisions or predictions are made.*

Definition 2: *Opaqueness is defined as the characteristic of an AI system that lacks clarity, accessibility, or visibility in its decision-making processes or inner mechanisms. Opaque AI systems conceal their operations, making it challenging for stakeholders to discern or comprehend the logic behind their outputs, leading to limited or no understanding of how decisions are arrived at.*

In the domain of IoT security, transparency is pivotal to ensuring that interconnected devices' operations and decision-making processes remain comprehensible and open to scrutiny. Transparent AI models embedded within IoT systems allow stakeholders, including users and developers, to inspect and comprehend the underlying mechanisms, logic, and algorithms guiding the behavior of these devices. This transparency facilitates an understanding of how specific security decisions or protocols are implemented, aiding in identifying potential vulnerabilities or threats. Conversely, opaqueness represents a significant challenge, characterized by a lack of clarity or visibility into the decision-making processes of interconnected devices. Opaque AI systems operating within IoT frameworks hinder stakeholders' ability to discern or comprehend the logic behind security-related decisions or functionalities. In such scenarios, the concealment of operations makes it arduous to evaluate the security posture or identify potential vulnerabilities within IoT systems. This lack of transparency could lead to limitations in understanding the root causes of security breaches or failures, hindering the ability to implement effective security protocols or address emerging threats promptly. Opaque AI systems in IoT security may compromise trust among users and impede efforts to ensure a robust and resilient security framework within interconnected devices [7].

Definition 3: *Explainability is defined as the capacity of an AI system to provide understandable and clear explanations regarding its decision-making processes, enabling users to comprehend the rationale behind its outputs.*

Definition 4: *Interpretability is defined as the attribute of an AI model that allows human users to understand and explain its predictions or decisions by providing insights into how the model arrives at specific outcomes based on input data.*

Explainability holds immense importance in IoT security, defining the capability of AI systems embedded within interconnected devices to offer clear and understandable explanations for their security-related decision-making processes. Within IoT ecosystems, XAI enables stakeholders to comprehend the rationale behind the implementation of specific security measures or protocols. Additionally, explainability facilitates the identification of potential security gaps or flaws, allowing for timely interventions and enhancements to fortify the IoT security infrastructure. On the other hand, interpretable AI systems enable users and developers to gain an understanding of how and why certain security-related outcomes are derived from the input data. In the context of interconnected devices, interpretability ensures that the logic behind security-related decisions is transparent and understandable, allowing for the validation and refinement of security protocols. This attribute fosters user confidence and facilitates the detection and resolution of security issues, contributing to the overall robustness and reliability of IoT security frameworks.

Definition 5: *Causability refers to the AI system's ability to identify and explain cause-effect relationships between input variables and the resulting predictions or decisions, enabling an understanding of why certain outcomes are produced.*

In IoT security, causability embodies the AI system's capability to elucidate the cause-and-effect relationships between input variables and the resulting security-related decisions or predictions. Understanding causality within interconnected devices aids in comprehending why specific security measures or responses are triggered based on various environmental or input factors. This attribute empowers stakeholders to discern the underlying reasons behind security incidents or vulnerabilities, facilitating proactive measures to address root causes and strengthen the overall security posture of IoT ecosystems [8].

Definition 6: *Trustworthiness is defined as the degree of reliability, consistency, and predictability exhibited by an AI system, instilling confidence in stakeholders regarding its performance and ethical usage across diverse scenarios.*

Trustworthiness is paramount in ensuring the reliability, consistency, and ethical usage of AI-driven security mechanisms within IoT environments. Trustworthy AI systems operating within interconnected devices instill confidence in stakeholders regarding the reliability and predictability of security-related decisions and actions. Trustworthy IoT security frameworks foster user trust, ensuring that security measures are consistently applied across diverse scenarios, contributing to the stability and resilience of interconnected device networks.

Definition 7: *Interactive AI is defined as the AI systems designed to engage in bidirectional communication or collaboration with users, allowing for feedback, adaptation, and refinement based on user interactions or inputs.*

Interactive AI systems within IoT security enable bidirectional communication and collaboration between users and devices. This attribute allows stakeholders to provide feedback, adapt security protocols, and refine decision-making processes based on user interactions or changing environmental conditions. In IoT ecosystems, interactive AI fosters dynamic security responses, allowing for real-time adjustments to mitigate evolving threats and vulnerabilities, thereby enhancing the adaptability and responsiveness of security measures.

Definition 8: *Robust AI is defined as the capability of an AI model or system to maintain performance and accuracy even in the presence of noisy, incomplete, or adversarial data, ensuring resilience against potential disruptions or perturbations.*

Robustness in IoT security denotes the capability of AI models or systems to maintain optimal performance and accuracy even in challenging scenarios such as noisy data, incomplete information, or deliberate adversarial attacks. Within interconnected device networks, robust AI ensures that security measures remain effective and reliable despite potential disruptions or attempts to compromise the system's integrity. Robust AI-driven security mechanisms bolster the resilience of IoT ecosystems, minimizing the impact of anomalies or adversities, and sustaining consistent protection against potential threats.

Definition 9: *Reproducible AI is defined as the quality of an AI model or experiment to be replicated or reproduced by others, ensuring consistency in results and methodologies when applied in similar contexts or scenarios.*

Reproducibility within IoT security refers to the quality of AI models or experiments being replicable or reproducible by others in similar contexts or scenarios. This attribute ensures consistency in the results and methodologies employed in implementing security measures across different IoT environments. Reproducible AI in IoT security allows for the validation and verification of security protocols, contributing to the reliability and standardization of effective security practices within interconnected device networks [9].

Definition 10: *Bias is defined as the systematic and unfair favoritism or prejudice in AI systems toward certain groups or individuals, leading to discriminatory outcomes or decisions based on characteristics such as race, gender, or socio-economic status.*

Definition 11: *Fairness is defined as the principle ensuring equitable and impartial treatment of individuals or groups within AI systems, aiming to mitigate biases and ensure unbiased outcomes for all stakeholders involved.*

Bias within IoT security represents systematic favoritism or prejudices embedded within AI systems that may lead to discriminatory outcomes or decisions concerning security measures. Detecting and mitigating biases in interconnected devices are crucial to ensure equitable and unbiased security-related actions. Addressing biases helps in foster fairness, minimize discrimination, and ensure that security protocols within IoT ecosystems serve all stakeholders equally and ethically.

Definition 12: *Accountability is defined as the responsibility and obligation of individuals, organizations, or AI systems to justify and take ownership of their actions, decisions, or consequences, particularly in contexts where they impact users or society.*

Definition 13: *Responsibility is defined as the ethical obligation of AI developers, users, and stakeholders to employ AI technologies in a manner that aligns with societal values, ensuring that AI systems contribute positively and ethically to the welfare of individuals and communities.*

In IoT security, accountability ensures that the entities responsible for implementing security measures or responding to security breaches are transparent about their actions, fostering a culture of responsibility and trust among stakeholders. It also helps in identifying and addressing potential gaps in security protocols within interconnected device networks. In addition, responsible AI deployment within interconnected devices ensures that security measures contribute positively and ethically to the safety and well-being of individuals and communities. Ethical considerations guide the development, implementation, and usage of security protocols within IoT ecosystems, promoting a secure, reliable, and socially responsible environment for interconnected devices [10].

6.3 The need for explainability in IoT systems

Unexplainable AI poses profound implications within the intricate landscape of the IoT, raising substantial concerns and challenges for the reliability, transparency, and security of interconnected devices. In IoT scenarios, deploying unexplainable AI models engenders a significant lack of transparency, hindering stakeholders' ability to comprehend the decision-making processes underlying AI-driven functionalities. This opacity obstructs the visibility into how and why specific security measures or automated actions are enacted within IoT systems, impeding the validation and understanding of the security protocols adopted by interconnected devices. Moreover, unexplainable AI within IoT environments exacerbates the difficulty in identifying and rectifying potential security vulnerabilities or malfunctions. The absence of transparent explanations behind AI-generated decisions amplifies the risk of misinterpretations, limiting the capability effectively anticipate and mitigate security threats. Besides, unexplainable AI systems operating within IoT ecosystems present significant ethical and regulatory challenges. The lack of explainability within AI models deployed in interconnected devices raises ethical concerns surrounding accountability and responsibility. In scenarios where unexplainable AI dictates critical security decisions within IoT environments, the inability to provide clear and understandable rationales for these decisions impedes stakeholders' capacity to assign accountability or address potential biases or discriminatory outcomes. Additionally, regulatory compliance becomes a formidable challenge in the presence of opaque AI systems in IoT. Regulatory standards often mandate transparency, fairness, and user understanding in AI-driven systems, particularly within security-critical applications. The prevalence of unexplainable AI models in IoT security hampers compliance efforts, posing obstacles to adhering to established regulations and guidelines. Therefore, the implications of unexplainable AI in IoT scenarios extend beyond technical challenges, encompassing ethical, regulatory, and security concerns that impede the optimal functioning and trustworthiness of interconnected devices [11,12].

Integrating black-box models within IoT applications introduces a myriad of inherent risks and consequential challenges that significantly impact the reliability, transparency, and security of interconnected devices. Black-box models, characterized by their opaque nature and complex inner workings, present a substantial risk within IoT environments by obscuring the decision-making processes of AI-driven functionalities. This opacity impedes stakeholders' understanding of how these interconnected devices arrive at specific security-related decisions, obstructing the ability to comprehend and validate the security protocols or measures enacted. Moreover, the inherent lack of transparency in black-box models within IoT applications amplifies the difficulty in detecting and mitigating potential security vulnerabilities or threats. This creates a precarious situation where critical security decisions made by interconnected devices lack comprehensible explanations, heightening the potential for misinterpretations or failures to proactively address emerging security risks.

The consequences of deploying black-box models in IoT applications extend beyond technical complexities, delving into ethical and regulatory quandaries. The inherent opacity of black-box models hinders the establishment of accountability and responsibility within IoT environments. In scenarios where these models dictate crucial security decisions, the inability to elucidate the reasoning behind these decisions limits stakeholders' capacity to assign accountability or rectify potential biases or discriminatory outcomes. Furthermore, utilizing black-box models in IoT applications poses challenges regarding compliance with regulatory standards.

Ethical frameworks and regulatory guidelines often emphasize the importance of transparency, fairness, and user comprehension in AI-driven systems, particularly in security-critical applications. The prevalence of black-box models within IoT exacerbates compliance challenges, impeding efforts to adhere to established regulations and guidelines [13].

XAI is a crucial catalyst in addressing paramount concerns surrounding accountability, trust, and ethical considerations in AI applications. By providing transparent and understandable explanations for AI-driven decision-making processes, XAI fundamentally enhances accountability in AI systems. Through transparent explanations of how decisions are made, stakeholders gain insights into the rationale behind these decisions, enabling them to trace back the causes of outcomes or actions. In contexts like IoT applications, XAI fosters accountability by ensuring the reasoning behind interconnected devices' security-related decisions is accessible and comprehensible. This transparency enables stakeholders to assess the ethical implications and potential biases embedded in these decisions, facilitating the identification and rectification of biases or discriminatory patterns. Consequently, XAI's emphasis on transparency empowers stakeholders to hold AI systems accountable for their actions, fostering a culture of responsibility and ethical usage within AI-driven ecosystems. Moreover, XAI plays a pivotal role in nurturing trust among users and stakeholders by elucidating the inner workings of AI models and the reasoning behind their outputs. In IoT applications, where interconnected devices operate autonomously, transparent AI systems instill confidence and trustworthiness by providing clear and understandable explanations for security-related decisions or actions. This transparency in AI-driven functionalities allows users and developers to comprehend and verify the reliability and fairness of these decisions, thus enhancing trust in the AI systems deployed within IoT ecosystems. Furthermore, by addressing ethical concerns related to biases, fairness, and accountability, XAI contributes to building trust by ensuring that AI systems within IoT environments align with ethical standards and societal values [14] (Figure 6.1).

The multifaceted nature of XAI in the context of IoT security extends beyond mere technical intricacies, encapsulating a spectrum of motives and reasons that underscore its indispensability. These motives can be summarized as follows:

(a) Explain to justify
This motive embodies the role of XAI in providing comprehensible explanations for the decisions made by AI-driven systems within IoT security. These explanations enable stakeholders to comprehend and validate the rationale behind security-related decisions, justifying the necessity and appropriateness of specific actions taken by interconnected devices. By offering transparent insights into the decision-making processes, XAI allows users and developers to justify the security measures enacted, promoting accountability and confidence in the AI-driven functionalities embedded within IoT ecosystems.

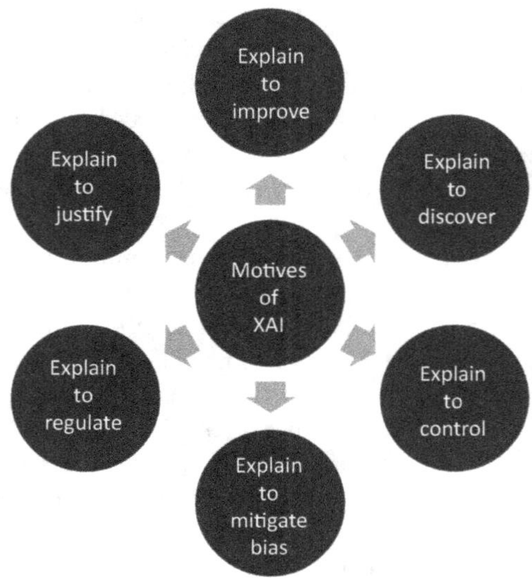

Figure 6.1 Motives for XAI in IoT security

(b) Explain to control
This motive delineates XAI's function in empowering users to have better control over the AI-driven security measures employed within IoT environments. By offering understandable explanations, XAI enables users to exert informed control over interconnected devices' security actions. This transparency facilitates effective control mechanisms where users can comprehend and manage the operations of AI systems, ensuring that the security protocols align with desired outcomes. Consequently, XAI empowers stakeholders to exert necessary controls and interventions when required, thereby enhancing the adaptability and responsiveness of IoT security systems.

(c) Explain to improve
This motive pertains to its capacity to provide transparent explanations that facilitate ongoing improvements and refinements within IoT security measures. By elucidating the reasoning behind AI-driven security decisions, XAI allows stakeholders to identify shortcomings, vulnerabilities, or inefficiencies. These insights enable continuous evaluation and enhancements, fostering an iterative improvement cycle within interconnected devices' security frameworks. The transparent explanations provided by XAI aid in identifying areas for enhancement, thereby facilitating the optimization of security measures for better efficacy and resilience against emerging threats.

(d) Explain to discover
This motive highlights XAI's function in aiding stakeholders to discover and comprehend unforeseen patterns or anomalies within IoT security systems. By offering transparent explanations for AI-driven decisions, XAI enables users to

delve deeper into the underlying patterns or trends. This attribute facilitates the discovery of unexpected behaviors or deviations, enabling stakeholders to identify potential security breaches, anomalies, or irregularities within interconnected devices' operations. Consequently, XAI supports proactive measures for discovering and addressing emerging security risks, contributing to the robustness and vigilance of IoT security frameworks.

(e) Explain to mitigate bias
This motive underscores XAI's pivotal role in mitigating biases inherent in AI systems deployed within IoT security contexts. By providing transparent explanations for decision-making processes, XAI aids in identifying and rectifying biases present in AI algorithms. This transparency enables stakeholders to comprehend and address biases or discriminatory patterns within interconnected devices' security measures. Consequently, XAI fosters fairness and equity, ensuring that security protocols remain unbiased and equitable for all users, contributing to a more ethical and reliable IoT security landscape.

(f) Explain to ensure regulatory compliance
This motive emphasizes XAI's function in enabling adherence to regulatory standards and guidelines governing IoT security. Transparent explanations offered by XAI aid in ensuring compliance with established regulations that mandate transparency, fairness, and ethical usage of AI systems within security-critical applications. By providing comprehensible insights into AI-driven security decisions, XAI supports efforts to align IoT security measures with regulatory frameworks, ensuring that interconnected devices operate ethically and in accordance with established standards. Consequently, XAI fosters a compliant and accountable IoT security environment [15].

6.4 Dimensions of explainability in IoT security

Within IoT security, the concept of explainability transcends a singular facet, encompassing multiple dimensions crucial for fostering transparency and comprehension in AI-driven systems. Recognizing the intricate interplay between AI and the security landscape within interconnected devices, our focus converges on delineating and addressing distinct dimensions of explainability. Specifically, our aim in this section is to delve into the intricate facets of algorithmic, data, contextual, and human-centric explainability tailored to the unique challenges posed by IoT security frameworks. These dimensions represent crucial pillars that intricately contribute to elucidating the decision-making processes, mechanisms, and outcomes of AI models embedded within interconnected devices [16].

6.4.1 Algorithmic explainability

Algorithmic explainability refers to the capacity of an AI system to provide understandable and transparent insights into the decision-making processes, logic, and functioning of the underlying algorithms. It involves elucidating how AI algorithms operate, enabling stakeholders to comprehend the reasoning behind the system's predictions or decisions. Algorithmic explainability focuses on making the complex inner workings of AI models accessible and interpretable, allowing users to gain insights into the relationship between input data and output decisions, thus fostering trust and confidence in the system's functionality.

Exploring the attainability of explainability within various AI algorithms deployed in IoT systems is imperative to ensure transparency and trustworthiness in their decision-making processes. Achieving algorithmic explainability involves employing techniques that demystify the operations of AI models, making them comprehensible to users and stakeholders. Decision trees, rule-based systems, and model distillation are prominent methodologies contributing to this pursuit. Decision trees offer a transparent structure, mapping out decision paths based on input data, enabling users to trace the rationale behind the system's choices. Rule-based systems utilize explicit rules and conditions, providing a clear framework for decision-making and enhancing interpretability. Additionally, model distillation simplifies complex AI models by distilling them into more understandable and interpretable versions. These techniques serve as pivotal tools in making AI algorithms within IoT systems explainable, offering insights into the decision-making processes and enabling stakeholders to comprehend and validate the logic behind AI-driven decisions. By employing these methodologies, IoT developers and practitioners can pave the way toward transparent and interpretable AI systems, fostering reliability and enhancing user trust within interconnected device networks.

6.4.2 Data explainability

Data explainability refers to the comprehensibility and transparency of the data utilized in AI systems, especially in explaining how data is collected, processed, and utilized in making decisions or predictions. It makes the data and its various attributes understandable to stakeholders, allowing them to discern the sources, quality, biases, and transformations applied to the data throughout its lifecycle within AI models. Data explainability focuses on elucidating the characteristics, origins, and modifications of data to enhance transparency and trust in the decision-making processes of AI systems [17].

Transparent data collection and preprocessing play a pivotal role in fortifying the security posture of IoT ecosystems. The significance lies in ensuring that data collection methods adhere to ethical standards and regulatory requirements and that the collected data is representative, reliable, and devoid of biases that could impact AI-driven security measures. Transparent data collection guarantees that stakeholders understand the sources and acquisition methods of data used in AI models deployed in interconnected devices, mitigating concerns about data manipulation or unauthorized access. Furthermore, preprocessing steps involving data cleaning, normalization, and feature engineering must maintain transparency and traceability to preserve data integrity. Issues surrounding data provenance, which delineate the

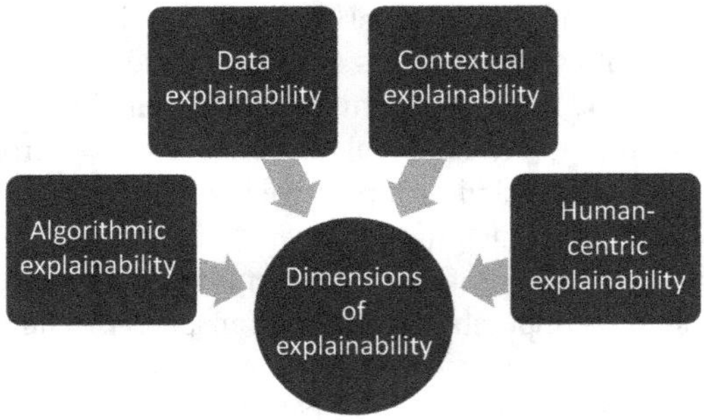

Figure 6.2 Dimensions of explainability in IoT security

origins and lineage of data, become paramount in IoT security. Similarly, data quality assurance measures and strategies to address data biases are indispensable. These facets emphasize the criticality of transparent data collection and pre-processing methodologies in fortifying the reliability, fairness, and trustworthiness of AI systems operating within IoT environments (Figure 6.2).

The assurance of data provenance, data quality, and the mitigation of biases stand as linchpins in ensuring the integrity and reliability of AI-driven security measures within IoT ecosystems. Data provenance delineates the lineage and history of data, allowing stakeholders to track the origins and transformations applied to the data, thereby verifying its authenticity and reliability. Additionally, emphasizing data quality involves ensuring that collected data meets established standards of accuracy, completeness, and consistency, which is indispensable for robust and reliable AI models in IoT security. Addressing data biases, such as those related to demographic, temporal, or sampling biases, is crucial to mitigate unfair or discriminatory outcomes in AI-driven security decisions. Detecting, acknowledging, and rectifying biases in data contribute to fostering fairness and equity within IoT security frameworks. Implementing strategies that prioritize transparent data provenance, uphold stringent data quality assurance, and mitigate biases serve as fundamental pillars in fortifying the trustworthiness and ethical underpinnings of AI systems within interconnected devices, thus enhancing the resilience and reliability of IoT security measures.

6.4.3 Contextual explainability

Contextual explainability refers to the ability of AI systems to elucidate decisions or predictions by considering and integrating contextual information specific to the environment in which they operate. This form of explainability focuses on providing insights into how AI models interpret and utilize environmental or situational factors to make decisions, thus offering a clearer understanding of the rationale behind the system's outputs. Contextual explainability encompasses the incorporation of relevant contextual information within AI models, enabling stakeholders to comprehend how environmental nuances influence the decision-making processes, fostering transparency and trust in the system's operations [18].

Understanding the contextual information inherent in IoT environments serves as a cornerstone in enhancing explainability within interconnected device networks. Incorporating contextual factors, such as environmental conditions, location-based data, temporal information, and device interactions, significantly contributes to elucidating the decision-making processes of AI-driven security measures. Contextual explainability, within IoT security frameworks, hinges upon recognizing how environmental specifics impact security-related decisions made by interconnected devices. For instance, IoT devices operating in varying environmental conditions might trigger distinct security responses based on contextual factors like weather, time of day, or user behavior patterns. Acknowledging and integrating these environmental nuances into AI models empower stakeholders to comprehend the rationale behind security-related decisions, consequently enhancing transparency and bolstering the explainability of AI systems within IoT environments.

Situational awareness and environment-specific factors are pivotal in shaping the contextual explainability of AI systems operating within IoT environments. Situational awareness encompasses the AI system's capability to understand and adapt to dynamic environmental conditions and evolving contexts. Within IoT security, situational awareness aids in interpreting and incorporating real-time information and context-specific cues into the decision-making processes of interconnected devices. Moreover, environment-specific factors such as geographical location, physical surroundings, or device interactions contribute to tailoring AI models to comprehend and respond to specific environmental nuances. By considering these factors, AI models operating within IoT ecosystems gain the capability to provide transparent insights into how situational awareness and environment-specific elements influence security-related decisions, thereby reinforcing the explainability and reliability of AI-driven security measures in interconnected device networks.

6.4.4 Human-centric explainability

Human-centric explainability refers to the attribute of AI systems to provide transparent and understandable explanations tailored for human users. This form of explainability prioritizes presenting information in a comprehensible manner that resonates with human cognition and decision-making processes. Human-centric explainability bridges the gap between complex AI-driven decisions and human comprehension, ensuring that explanations provided by AI systems within IoT environments are accessible, interpretable, and meaningful to users. It emphasizes designing and delivering explanations that cater to human cognitive capacities and preferences, thereby enhancing user trust, comprehension, and acceptance of AI-driven decisions [19].

The role of human users in comprehending AI-driven decisions within IoT security frameworks is paramount in fostering transparency and trust. Human-centric explainability emphasizes tailoring explanations generated by AI systems to suit human understanding and interpretation. Users within IoT environments need to comprehend the decisions made by interconnected devices to trust and rely on their security measures effectively. Designing interfaces and visualizations that effectively communicate the reasoning behind AI-driven security decisions becomes imperative. User-friendly interfaces and visual representations that simplify complex AI outputs into intuitive and understandable formats enable users to grasp the rationale behind security-related decisions. Utilizing visual aids, such as diagrams, graphs, or interactive displays, facilitates the effective communication of AI-driven decisions, empowering users to comprehend and trust the actions taken by interconnected devices. By prioritizing human-centric explanations and employing intuitive interfaces, IoT developers and practitioners can bridge the gap between intricate AI processes and human understanding, fostering transparency and trust within IoT security frameworks.

6.5 Regulatory landscape and standards

Several regulatory frameworks and standards have been emerging to address the deployment of XAI within IoT security, although the landscape might have evolved since then.

One notable framework is the European Union's General Data Protection Regulation (GDPR), which emphasizes transparency, fairness, and accountability in automated decision-making processes. GDPR's provisions mandate that individuals have the right to receive meaningful information about the logic behind automated decisions that affect them, a principle in line with XAI's objectives. Article 22 specifically addresses automated individual decision-making, including profiling, and grants individuals the right not to be subject to a decision based solely on automated processing, while ensuring meaningful information about the logic involved. Compliance with GDPR's requirements significantly impacted XAI implementation in IoT systems, prompting developers to incorporate explainability mechanisms into their AI models to ensure compliance with these regulations while deploying IoT systems.

Additionally, the Institute of Electrical and Electronics Engineers (IEEE) P7003 Standard for Algorithmic Bias Considerations aims to address biases in AI systems, encouraging transparency and fairness. Although not specifically focused on XAI, the principles outlined in IEEE P7003 intersect with XAI goals, particularly in ensuring fairness and transparency in AI algorithms used in IoT security. Adherence to this standard urges developers to employ XAI techniques to mitigate biases, thereby enhancing fairness and transparency in decision-making processes within IoT systems. Adherence to such standards prompts IoT developers to integrate XAI techniques to ensure fair and transparent AI-driven security measures.

Moreover, the AI Ethics Guidelines issued by various institutions and organizations, such as the OECD (Organisation for Economic Co-operation and

Development) and the World Economic Forum, emphasize the importance of explainability, accountability, and transparency in AI. These guidelines highlight ethical considerations in deploying AI systems and emphasize the need for XAI techniques in IoT security to ensure adherence to ethical norms and guidelines.

The impact of these regulatory frameworks and standards on the development and deployment of XAI in safeguarding IoT systems is substantial. They necessitate that AI-driven IoT systems be designed and implemented with transparency, interpretability, and accountability in mind. Developers are prompted to integrate XAI techniques into IoT security measures to ensure compliance with these regulations and standards, thereby enhancing the explainability of AI systems used in IoT, fostering user trust, and mitigating potential risks associated with opaque or biased AI models within interconnected devices. Additionally, these frameworks encourage ongoing research and development in XAI to meet the evolving regulatory requirements, thereby contributing to the improvement and standardization of XAI practices within IoT security frameworks [20].

6.6 Frameworks and tools for implementing explainable AI in IoT security

Given the growing significance of XAI in various domains, there arises a pressing need within the research community for the creation of tools and frameworks. These resources play a pivotal role in simplifying the implementation and replication of diverse explainability techniques for a wide spectrum of stakeholders. Therefore, this section of the document emphasizes the prevalent open-source frameworks tailored for XAI. Table 6.1 is dedicated to delineating and contrasting the distinct attributes of available open-source XAI toolkits, facilitating a comprehensive comparison among them [21].

- **AIX360:** AIX360 stands as an extensive open-source toolkit developed by IBM Research that offers a wide array of interpretable AI algorithms and explainability techniques. This comprehensive framework covers various aspects of XAI, providing tools for model explanations, fairness assessment, and bias mitigation. AIX360 serves as a versatile resource enabling users to explore and apply diverse XAI methods, enhancing transparency and interpretability in AI models.
- **Alibi:** Alibi, an open-source Python library, focuses on providing tools for model-agnostic explanations, fairness evaluation, and counterfactual generation. This framework integrates various XAI methods, including feature attributions and counterfactual generation, facilitating transparent and interpretable AI model assessments. Alibi's user-friendly interface allows users to delve into model explanations and fairness assessment, promoting transparency and trust in AI decision-making.
- **DALEX:** DALEX, a versatile R package, specializes in model explanations and interpretations across various machine-learning models. It offers

Table 6.1 Comparative review between common open-source XAI frameworks

Frame-works/tools	Language	Counter-facuals	Data types	Deploy-ment options	Explicitly interpre-table	Local	Global	Evaluation measures
Alibi [22]	Python	✓	✓	✓	×	✓	✓	×
Skater	Python	×	×	×	✓	✓	✓	×
tf-explain	Python	×	×	×	×	✓	✓	×
EthicalML-XAI [23]	Python	×	×	×	×	×	✓	×
iNNvestigate		×	×	×	×	×		×
Captum [23]	Python	×	✓	×	×	×	✓	✓
InterpretML [24]	Python	×	×	×	✓	✓	✓	×
DALEX [25]	Python, R	×	×	×	×	✓	✓	✓
XAITK	Python	✓	✓	×		✓	✓	✓
Quantus [26]		✓	✓	✓	✓	✓	✓	✓
AIX360 [27–28]	Python	✓	✓	✓	✓	✓	✓	✓

visualization tools and methods for understanding model behavior, feature contributions, and global model explanations. DALEX assists users in comprehending complex models' decisions, fostering transparency and interpretability crucial for stakeholders' trust in AI-driven systems.

- **EthicalML-XAI:** EthicalML-XAI is an open-source Python library designed to address ethical considerations in AI and XAI. This framework focuses on integrating ethical principles into XAI methodologies, emphasizing fairness, accountability, and transparency. EthicalML-XAI provides tools for bias detection, fairness evaluation, and ethical impact assessments, ensuring AI systems adhere to ethical standards.
- **H2O:** H2O stands as an open-source platform offering an extensive suite of machine learning and XAI functionalities. This framework provides tools for model explanations, feature importance analysis, and interpretability across a range of machine-learning models. H2O's intuitive interface and diverse XAI capabilities cater to users seeking transparent and interpretable AI solutions.
- **InterpretML:** InterpretML, a Python package, specializes in interpretable machine-learning techniques. It offers various XAI methods, including global and local explanations, feature attributions, and model-agnostic interpretability. InterpretML facilitates model explanations, enabling users to understand and trust AI predictions.
- **Quantus:** Quantus is an open-source framework designed for fairness assessment and explanations in machine-learning models. It focuses on evaluating model fairness, bias mitigation, and providing transparent explanations for stakeholders. Quantus aids users in ensuring fairness and transparency in AI systems, promoting ethical and accountable AI practices.

- **Skater:** Skater, an open-source Python library, focuses on model interpretation and explanation. It provides tools for model-agnostic explanations, feature attributions, and interpretability across different machine-learning models. Skater's functionalities aid users in understanding and validating AI models' decisions.
- **Tf-explain:** Tf-explain, tailored for TensorFlow users, is an open-source toolkit dedicated to providing model interpretability and explanations within TensorFlow-based models. This framework offers various visualization and interpretation methods, facilitating transparent and interpretable AI solutions.
- **The Explainable AI Toolkit (XAITK):** The XAITK serves as a comprehensive open-source toolkit designed to offer a wide range of XAI functionalities. It encompasses tools for model explanations, fairness evaluation, and interpretability techniques. XAITK caters to diverse XAI needs, enabling users to explore and apply various explainability methods across AI models.
- **What-If Tool:** The What-If Tool, developed by Google, focuses on providing an interactive and visual platform for model interpretation and explanations. It offers tools for understanding model predictions, exploring what-if scenarios, and assessing model fairness. The tool's interactive interface facilitates intuitive exploration and comprehension of AI model behaviors, promoting transparency and trust in AI-driven decisions.

References

[1] Kalakoti, R., Bahsi, H., and Nõmm, S. (2024). Improving IoT security with explainable AI: Quantitative evaluation of explainability for IoT Botnet detection. *IEEE Internet of Things Journal*, *11*(10), 18237–18254.

[2] Moustafa, N., Koroniotis, N., Keshk, M., Zomaya, A. Y., and Tari, Z. (2023). Explainable intrusion detection for cyber defences in the internet of things: Opportunities and solutions. *IEEE Communications Surveys & Tutorials*, *25* (3), 1775–1807.

[3] Jagatheesaperumal, S. K., Pham, Q. V., Ruby, R., Yang, Z., Xu, C., and Zhang, Z. (2022). Explainable AI over the Internet of Things (IoT): Overview, state-of-the-art and future directions. *IEEE Open Journal of the Communications Society*, *3*, 2106–2136.

[4] Kök, Ibrahim, Feyza Yıldırım Okay, Özgecan Muyanlı, and Suat Özdemir. (2023). Explainable artificial intelligence (xai) for internet of things: A survey. *IEEE Internet of Things Journal* 10(16), 14764–14779.

[5] Arrieta, A. B., Díaz-Rodríguez, N., Del Ser, J., *et al.* (2020). Explainable Artificial Intelligence (XAI): Concepts, taxonomies, opportunities and challenges toward responsible AI. *Information fusion*, *58*, 82–115.

[6] Ding, W., Abdel-Basset, M., Hawash, H., and Ali, A. M. (2022). Explainability of artificial intelligence methods, applications and challenges: A comprehensive survey. *Information Sciences*, *615*, 238–292.

[7] Wu, H., Han, H., Wang, X., and Sun, S. (2020). Research on artificial intelligence enhancing internet of things security: A survey. *IEEE Access*, *8*, 153826–153848.

[8] Sahoh, B., Kaewrat, C., Yeranee, K., Kittiphattanabawon, N., and Kliangkhlao, M. (2022). Causal AI-powered event interpretation: A cause-and-effect discovery for indoor thermal comfort measurements. *IEEE Internet of Things Journal*, *9*(22), 23188–23200.

[9] Raghupathi, W., Raghupathi, V., and Ren, J. (2022). Reproducibility in computing research: An empirical study. *IEEE Access*, *10*, 29207–29223.

[10] Fares, N. Y., and Jammal, M. (2023). AI-Driven IoT Systems and Corresponding Ethical Issues. *In AI and Society* (pp. 233–248). Chapman and Hall/CRC.

[11] Hoenig, A., Roy, K., Acquaah, Y., Yi, S., and Desai, S. (2024). Explainable AI for Cyber-Physical Systems: Issues and Challenges. *IEEE Access*.

[12] Zolanvari, M., Yang, Z., Khan, K., Jain, R., and Meskin, N. (2021). Trust XAI: Model-agnostic explanations for ai with a case study on iiot security. *IEEE Internet of Things Journal*, *10*(4), 2967–2978.

[13] Carotenuto, R., Merenda, M., Della Corte, F. G., and Iero, D. (2023). Online black-box modeling for the IoT digital twins through Machine Learning. *IEEE Access*, *11*, 48158–48168.

[14] Kök, I., Okay, F. Y., Muyanlı, Ö., and Özdemir, S. (2023). Explainable artificial intelligence (XAI) for internet of things: a survey. *IEEE Internet of Things Journal*, *10*(16), 14764–14779.

[15] Gerlings, J., Shollo, A., and Constantiou, I. (2020). *Reviewing the need for explainable artificial intelligence (xAI). arXiv preprint arXiv:2012.01007.*

[16] McDermid, J. A., Jia, Y., Porter, Z., and Habli, I. (2021). Artificial intelligence explainability: The technical and ethical dimensions. *Philosophical Transactions of the Royal Society A*, *379*(2207), 20200363.

[17] Dwivedi, R., Dave, D., Naik, H., *et al.* (2023). Explainable AI (XAI): Core ideas, techniques, and solutions. *ACM Computing Surveys*, *55*(9), 1–33.

[18] Al-Shedivat, M., Dubey, A., and Xing, E. (2020). Contextual explanation networks. *Journal of Machine Learning Research*, *21*(194), 1–44.

[19] Horvatić, D., and Lipic, T. (2021). Human-centric AI: The symbiosis of human and artificial intelligence. *Entropy*, *23*(3), 332.

[20] Nannini, L., Balayn, A., and Smith, A. L. (2023, June). Explainability in AI policies: A critical review of communications, reports, regulations, and standards in the EU, US, and UK. In *Proceedings of the 2023 ACM Conference on Fairness, Accountability, and Transparency* (pp. 1198-1212).

[21] Le, P. Q., Nauta, M., Van Bach Nguyen, S. P., Pathak, S., Schlötterer, J., and Seifert, C. (2023, August). Benchmarking eXplainable AI: A survey on available Toolkits and open challenges. In *IJCAI* (pp. 6665–6673).

[22] Klaise, J., Van Looveren, A., Vacanti, G., and Coca, A. (2021). Alibi explain: Algorithms for explaining machine learning models. *Journal of Machine Learning Research*, 22,1–17.

[23] N. Kokhlikyan, Miglani, V., Martin, M., *et al.* (2020). Captum: A unified and generic model interpretability libray for PyTorch. http://arxiv.org/abs/2009.07896.

[24] Nori, H., Jenkins, S., Koch, P., and Caruana, R. (2019). *InterpretML: A unified framework for machine learning interpretability*. http://arxiv.org/abs/1909.09223.

[25] Baniecki, H., Kretowicz, W., Piatyszek, P., Wisniewski, J., and Biecek, P. (2021). Dalex: Responsible machine learning with interactive explainability and fairness in python. *Journal of Machine Learning Research*, 22, 1–7.

[26] Hedström, A., Weber, L., Krakowczyk, D., *et al.* (2023). Quantus: An explainable AI toolkit for responsible evaluation of neural network explanations and beyond. *Journal of Machine Learning Research*, 24(34), 1–11.

[27] Arya, V., Bellamy, R. K. E.; Chen, P.-Y., *et al.* (2019). *One explanation does not fit all: A toolkit and taxonomy of AI explainability techniques*. http://arxiv.org/abs/1909.03012.

[28] Arya, V., Bellamy, R. K. E., Chen, P.-Y., *et al.* (2021). *AI explainability 360: Impact and design*. http://arxiv.org/abs/2109.12151.

Chapter 7

Explainability methods in explainable security intelligence: fine-grained taxonomy

7.1 Introduction

The integration of machine-learning models within the expansive landscape of Internet-of-Things (IoT) devices has ushered in unparalleled opportunities but also raised concerns regarding security, transparency, and trust in automated decision-making processes. The overarching purpose of this chapter is to delve into the pivotal role of eXplainable Artificial Intelligence (XAI) methods specifically tailored for IoT security. By elucidating the fine-grained taxonomy of XAI techniques, this chapter aims to comprehensively address the inherent complexities and challenges in understanding, interpreting, and trusting machine-learning models operating within the intricate IoT security ecosystem. In navigating the complex terrain of securing IoT environments, the delineation of a fine-grained taxonomy within explainable security intelligence assumes paramount importance. Understanding the diverse nuances and intricacies across XAI methods tailored for IoT security becomes imperative for stakeholders to ensure robustness, transparency, and compliance in the decision-making processes of machine-learning models. Thus, the development and adoption of a granular taxonomy offer a structured approach that not only enhances interpretability but also provides a systematic framework for practitioners, researchers, and policymakers to navigate the multifaceted realm of securing IoT devices through transparent and accountable AI systems [1,2].

This chapter meticulously structures its exploration by offering a comprehensive overview of various categories and dimensions within XAI methods specifically designed for IoT security. It systematically divides its focus into elucidating the dichotomy between model-agnostic versus model-specific methods, dissecting post-hoc versus intrinsic approaches, delineating global versus local interpretability paradigms, delving into the realms of human-understandable versus machine-oriented explanations, addressing domain-specific XAI techniques, exploring quantitative versus qualitative explanation strategies, and scrutinizing interpretability types. Each section delineates a distinct dimension, contributing to the holistic understanding and integration of XAI within the intricacies of securing IoT ecosystems. By traversing through this structured exploration of explainability methods in IoT security, this chapter endeavors to equip stakeholders with a nuanced understanding of the purpose, significance, and taxonomy intricacies of

XAI. As such, the chapter not only aims to elucidate the importance of a fine-grained taxonomy within explainable security intelligence but also provides a roadmap for comprehending, selecting, and integrating appropriate XAI techniques tailored for the transparent and secure deployment of machine-learning models in the IoT security landscape.

7.2 Model-agnostic versus model-specific XAI methods

Understanding the fundamental dichotomy between model-agnostic and model-specific XAI methods is crucial for elucidating the diverse approaches toward achieving transparency and interpretability within IoT security. Model-agnostic techniques encompass a spectrum of approaches that transcend individual model architectures, focusing on generating explanations applicable across various algorithms, while model-specific methods are tailored intricately to the interpretability of a particular model or class of models (refer to Table 7.1). This section endeavors to delineate the nuanced differences between these methodologies, elucidating their unique attributes, application domains, and significance in fortifying the security aspects of IoT ecosystems. Through an analytical lens, we explore the theoretical underpinnings, practical implications, and contextual relevance of both model-agnostic and model-specific XAI methodologies within the dynamic landscape of IoT security [3].

Table 7.1 Comparison of model-agnostic and model-specific XAI methods based on different aspects

Aspects	Model-agnostic XAI methods	Model-specific XAI methods
Applicability	Applicable to any machine learning model architecture	Tailored for specific types of models or architectures
Interpretability scope	Focuses on general interpretability across model types	Customized to provide detailed insights for specific models or architectures
Flexibility	Offers flexibility in usage across various model types	Might lack adaptability for models beyond the specified type
Complexity handling	Handles complex models without prior architecture knowledge	Capitalizes on inherent architecture characteristics for interpretability
Performance impact	May have higher computational overhead for explanations	Often designed to minimize performance impact on specific models
Explainability depth	Provides a broad overview of model behavior and predictions	Offers detailed insights tailored to specific model intricacies
User expertise requirement	Requires less model-specific expertise for interpretation	Often demands familiarity with the particular model type for meaningful insights
Application range	Widely applicable across diverse domains and model architectures	Often optimized for specific domains or applications with tailored models

Model-agnostic XAI methods represent a class of interpretability techniques designed to provide insights into the decision-making processes of machine-learning models, irrespective of their underlying architectures or complexities. These methods operate independently of the internal workings of the predictive models, focusing on generating explanations that are universally applicable across diverse algorithms, including neural networks, decision trees, support vector machines, and ensemble models. By dissociating from the intricate details of model architectures, model-agnostic methods aim to offer transparency and comprehensibility to black-box models, thereby enhancing their trustworthiness and facilitating human comprehension of model predictions in various domains, including but not limited to healthcare, finance, and cybersecurity [4].

An essential characteristic of model-agnostic techniques is their capacity to produce explanations that are not limited by the peculiarities inherent in particular machine-learning models. The methods mentioned earlier comprise a variety of tools, including LIME (Local Interpretable Model-agnostic Explanations), SHAP (SHapley Additive Explanations), Anchors, and others. These tools implement a wide range of strategies, including input feature perturbation, Shapley value utilization, and simplified decision rule definition. Model-agnostic methods employ these strategies to generate explanations that are comprehensible to humans and clarify the role of input features in model predictions. This empowers stakeholders to comprehend the reasoning behind the model's decisions without requiring an in-depth knowledge of the model's inner workings. Model-agnostic methods are crucial in promoting trust, facilitating regulatory compliance, and cultivating wider societal adoption of machine learning models utilized in critical decision-making situations due to their universal applicability and adaptability.

Contrary to the broad applicability of model-agnostic approaches, model-specific XAI methods are specifically engineered to unravel the intricacies and decision-making processes inherent to particular machine-learning architectures or classes of models. These techniques delve deep into the internal mechanics of a targeted model, aiming to elucidate the rationale behind its predictions while leveraging the model's specific design principles, structures, and functionalities. Model-specific methods are inherently focused on providing insights into the interpretability of a single or a set of related models, showcasing their strengths in offering detailed and tailored explanations that align with the nuances of these models. Their specificity enables a more detailed comprehension of the decision-making process, which is particularly beneficial in domains such as image recognition, natural language processing, and anomaly detection within IoT security contexts. The crux of model-specific methods lies in their tailored interpretability strategies for individual or specific classes of models. These methods encompass various techniques such as layer-wise relevance propagation (LRP) for neural networks, decision tree visualization algorithms like classification and regression trees (CART), or rule extraction methods specific to ensemble models. By capitalizing on the inherent characteristics and architectural features unique to the target models, these methods provide comprehensive and detailed insights into the decision-making processes, allowing stakeholders to grasp intricate model behaviors and decision rationales. Despite their narrower scope compared to model-agnostic techniques, model-specific methods are indispensable tools for in-depth model understanding and domain-specific interpretability enhancement within IoT security frameworks [5].

7.3 Post-hoc versus intrinsic XAI methods

The realm of XAI manifests through a dichotomy of approaches: post-hoc and intrinsic methods, both offering distinct avenues to unravel the black-box nature of machine-learning models employed within the IoT security framework. This section embarks on an exploratory journey, dissecting the differential attributes, operational mechanisms, and applicability of post-hoc and intrinsic XAI methodologies within the dynamic landscape of IoT security. Through an analytical lens, we delve into the theoretical underpinnings and pragmatic implications of both paradigms, aiming to illuminate their respective roles in fortifying the interpretability and trustworthiness of machine-learning models operating in IoT security contexts.

Post-hoc XAI methods represent a distinct category of interpretability techniques characterized by their capacity to unravel the decision-making process of machine learning models post-training, without influencing the model's learning dynamics. These techniques aim to provide insights into the opaque nature of black-box models by generating explanations retrospectively, thereby enabling stakeholders to comprehend and scrutinize model predictions without altering the underlying model architecture or learning algorithms. Notably, post-hoc methods encompass a diverse array of approaches such as saliency maps, surrogate models, and feature importance algorithms. Leveraging these techniques, stakeholders gain valuable insights into the factors influencing model predictions, fostering transparency and trust in decision-making systems operating within the purview of IoT security [6]. One of the primary advantages of post-hoc methods lies in their non-intrusive nature, allowing the application of interpretability without requiring modifications to the original model training process. Techniques such as LIME perturb input features around specific instances to discern local model behavior. In contrast, surrogate models approximate the black-box model's decision surface to provide a more understandable representation. These methods find utility across various IoT security scenarios, facilitating comprehension of model predictions and assisting domain experts in validating the credibility of machine-generated insights. However, the post-hoc nature of these techniques may pose challenges in fully explaining complex model behaviors, often offering only partial insights into the inner workings of sophisticated models, thereby necessitating a holistic understanding when interpreting model decisions within IoT security frameworks [7].

Intrinsic XAI methods represent a class of interpretability techniques meticulously woven into the fabric of machine-learning models during their training phase, aiming to imbue inherent interpretability into the models themselves. Unlike post-hoc methods applied externally after model training, intrinsic methods integrate interpretability principles within the model architecture, fostering transparency and comprehensibility without compromising predictive performance. These techniques encompass various strategies such as incorporating attention mechanisms in neural networks, using decision paths in decision trees, or designing rule-based models like decision sets. Within the realm of IoT security, these methods hold promise in facilitating not only transparent but also inherently understandable models crucial for critical decision-making scenarios. The distinctive feature of intrinsic methods lies in their proactive integration of interpretability elements during the model's development, aiming to strike a balance between accuracy and comprehensibility from the model's inception. Techniques like attention mechanisms enable neural networks to assign varying degrees of importance to different input features, thereby spotlighting crucial elements in the decision-making process. Decision tree-based models, on the other hand, inherently create interpretable structures allowing direct interpretation of decision paths. These methods cater to the specific needs of IoT security by providing interpretable and transparent models vital for ensuring robustness, trustworthiness, and compliance with regulatory requirements. However, the interpretability achieved through intrinsic methods may face constraints in handling complex relationships within high-dimensional data, necessitating a delicate trade-off between interpretability and model complexity within the IoT security landscape [8].

7.4 Global versus local interpretability

The dichotomy between global and local interpretability represents a fundamental distinction in the realm of XAI; delineating approaches focused on understanding machine-learning models at varying granularities. Global interpretability pertains to overarching insights that elucidate the model's behavior and decision-making processes holistically, providing comprehensive understanding across the entire model (refer to Table 7.2). In contrast, local interpretability zooms into specific instances or subsets of data, offering fine-grained explanations for individual predictions or localized regions within the model's decision space. This section explores these distinctive interpretability paradigms, deciphering their differential attributes, operational frameworks, and applicability within the intricate landscape of IoT security. Through an analytical lens, we delve into the theoretical foundations and pragmatic implications of global and local interpretability, aiming to illuminate their respective roles in fortifying the transparency and trustworthiness of machine learning models deployed in IoT security environments [9]. Global interpretability within the domain of XAI embodies techniques and approaches aimed at unraveling comprehensive insights into machine-learning models' behavior across their entirety. These methods strive to offer a macroscopic view of the model's decision-making process, elucidating overarching patterns, dependencies, and feature importance, thus providing stakeholders with a holistic understanding of the model's behavior. Global interpretability techniques encompass diverse

Table 7.2 Comparison of local XAI and global XAI across different aspects

Aspects	Local XAI	Global XAI
Scope	Focuses on explaining individual model predictions	Aims to provide insights into the model's overall behavior
Explanation granularity	Offers detailed insights into specific predictions	Provides an overview of general model behaviors
Interpretability	Helps understand specific model decisions	Offers an understanding of the model's broader tendencies
Computation	May involve higher computational overhead	Often computationally more efficient due to global scope
Complexity handling	Handles complexities of individual predictions	Deals with complexities associated with entire model
Transparency	Offers transparency at a local prediction level	Enhances overall transparency of the model's functioning
Use cases	Useful for understanding specific instances	Beneficial for understanding global model behaviors
Decision relevance	Useful in cases where local decisions matter	Relevant when broader insights into model are necessary
Stakeholder focus	More relevant for end-users or domain experts	Valuable for data scientists, auditors, or model developers
Examples of methods	LIME and SHAP	Global surrogate models, LRP, principal component analysis

methodologies, including feature importance ranking, model-specific visualization methods, and statistical summaries of model behavior. These techniques operate on the principle of highlighting the general trends and influential factors driving the model's predictions, thereby facilitating high-level comprehension of the model's functioning. Within IoT security, global interpretability plays a pivotal role in not only fostering transparency but also validating model reliability and identifying potential biases that might impact the overall performance of machine-learning models [10]. One of the primary advantages of global interpretability methods lies in their capacity to offer comprehensive and bird's-eye views of model behaviors, empowering stakeholders with insights into how the model utilizes input features to arrive at predictions across the entire dataset. Techniques like feature importance ranking through methods such as permutation importance or SHAP contribute by quantifying the impact of input features on model predictions on a global scale. Visualizations depicting decision boundaries or class separability also aid in comprehending model behaviors at a broader level. While global interpretability provides invaluable insights into the overall model behavior, it might oversimplify intricate relationships or nuances present in specific subsets of data, requiring a nuanced balance between global comprehensibility and local insights within the context of IoT security.

Local interpretability constitutes a facet of XAI dedicated to unraveling explanations for individual predictions or localized segments within the decision space of machine-learning models. These techniques aim to provide fine-grained insights into specific instances or subsets of data, facilitating a detailed understanding of why the model made a particular prediction. Unlike global interpretability methods that offer overarching views, local interpretability approaches zoom into the inner workings of the model, shedding light on the rationale behind specific predictions for individual data points. These methods encompass diverse strategies such as instance-level feature importance, local surrogate models, and perturbation-based explanations, enabling stakeholders to glean insights into the model's decision-making process at a micro-level. Within the domain of IoT security, local interpretability techniques serve as invaluable tools for validating model predictions, debugging individual instances, and ensuring the robustness and reliability of machine-learning systems in real-time, critical decision-making scenarios. The distinguishing feature of local interpretability lies in its capability to provide explanations tailored to individual predictions, thereby offering a more nuanced understanding of model behaviors within specific contexts. Techniques like instance-level feature importance, such as LIME, dissect the contribution of features for a particular prediction, providing localized insights. Additionally, local surrogate models attempt to approximate the black-box model's decision boundary around specific instances, facilitating comprehendible explanations. However, while local interpretability methods offer granular insights, they might lack the comprehensive overview that global interpretability provides, potentially limiting the generalizability of the obtained explanations across the entire dataset. Thus, a balanced integration of both global and local interpretability techniques becomes imperative in the context of IoT security to ensure a thorough understanding of machine-learning model behaviors [11].

7.5 Complexity and granularity

The complex terrain of XAI encompasses a significant contrast between granularity and complexity, which are critical factors that impact the extent and thoroughness of interpretability in machine-learning models utilized for IoT security. In this section, we undertake an exhaustive investigation, deconstructing the complex relationship between granularity and complexity in XAI methodologies. By employing an analytical approach, we critically examine the operational frameworks, pragmatic implications, and theoretical foundations of balancing complexity and granularity. This sheds light on how these two factors work in tandem to enhance the transparency, reliability, and usability of machine-learning models utilized for the purpose of securing IoT ecosystems.

Simple/linear methods in the realm of XAI encompass a category of techniques that prioritize simplicity and transparency, providing straightforward and understandable explanations for machine-learning model predictions within IoT security. These methods eschew complexity in favor of clear, interpretable models, often employing linear models or rule-based systems that facilitate comprehensibility without sacrificing predictive performance. By opting for simplicity, these methods aim to distill the underlying decision-making process of the model into easily understandable rules or relationships, thereby fostering trust and comprehension among stakeholders. Simple/linear methods include techniques such as linear regression, decision trees, and rule extraction algorithms, which generate easily interpretable models, albeit with potential trade-offs in predictive accuracy. Despite these trade-offs, these methods find relevance in IoT security scenarios, where transparency and comprehensibility of machine-learning models are paramount for stakeholders' trust and validation of model predictions [12].

One of the primary advantages of simple/linear methods lies in their intuitive and transparent nature, offering human-interpretable explanations that align with the decision-making process of straightforward models. Techniques like decision trees create hierarchical rules that directly link input features to model predictions, making the decision process easy to comprehend. Moreover, linear regression models provide a clear representation of feature importance and their impact on predictions through coefficients, enhancing interpretability. However, while these methods excel in simplicity and transparency, they might struggle to capture complex relationships present in high-dimensional or nonlinear data, potentially limiting their applicability in scenarios where model intricacies are pivotal. Thus, in the context of IoT security, a balanced integration of simple/linear methods alongside more complex models becomes imperative to ensure both interpretability and predictive performance.

Complex/non-linear methods in the domain of XAI encompass a spectrum of techniques designed to unravel intricate relationships and interactions within machine-learning models operating in IoT security scenarios. These methods embrace the inherent complexity of data by leveraging sophisticated models capable of capturing non-linear relationships, interactions, and dependencies among features. Unlike simple/linear methods, complex/non-linear approaches include algorithms such as neural networks, kernel methods, ensemble techniques like random forests or gradient boosting, and deep learning architectures. These methods excel in discerning intricate patterns within high-dimensional and non-linear data, offering nuanced and detailed explanations for model predictions by considering complex interactions among input features. Although these methods yield highly accurate predictions, their interpretability often poses challenges due to their inherent complexity. Nevertheless, within IoT security domains where comprehending intricate relationships in data is crucial, complex/non-linear methods play a pivotal role in providing predictive power alongside nuanced explanations.

The primary advantage of complex/non-linear methods lies in their ability to capture intricate relationships and patterns present in complex datasets, enabling detailed and accurate predictions for IoT security applications. Neural networks, for instance, can identify complex hierarchical features and relationships, providing high predictive accuracy. Ensemble methods amalgamate diverse models' predictions, considering multiple facets of the data, and thereby enhancing robustness and predictive performance. However, the interpretability of these methods remains a challenge, as their complex architectures and nonlinear transformations often obscure the direct linkage between input features and predictions. This opacity might hinder stakeholders' understanding of the underlying decision-making process. Thus, while these methods excel in predictive performance, balancing their accuracy with interpretability remains an ongoing challenge in securing IoT devices, warranting ongoing research efforts to improve their explainability without compromising their efficacy [13].

7.6 Human-understandable versus machine-oriented explanations

The contrast between human-understandable and machine-oriented explanations is a core feature of XAI, expressing different viewpoints on IoT security machine-learning model explanations. Human-understandable explanations focus on making findings palatable and coherent for human stakeholders. Machine-oriented explanations focus on model technicalities and internal representations to deliver computationally more extensive insights that may be harder to grasp. This section examines the characteristics, operational frameworks, and pragmatic implications of human-understandable and machine-oriented explanations in the complex terrain of IoT ecosystem security.

Human-understandable explanations within the realm of XAI encompass a set of techniques and approaches meticulously tailored to bridge the gap between complex machine-learning models and human comprehension. These methods prioritize the articulation of model behaviors and predictions in a format easily digestible and interpretable by individuals devoid of technical expertise in machine learning. By employing intuitive and simplified representations such as visualizations, natural language descriptions, or interactive interfaces, human-understandable explanations aim to elucidate the rationale behind model predictions in a manner accessible to diverse stakeholders, including domain experts, policymakers, and end-users within IoT security contexts. Techniques like decision trees visualization, LIME, or interactive dashboards offer transparent and intuitive representations of model predictions, enhancing trust and fostering collaboration between human stakeholders and machine-learning systems. The defining characteristic of human-understandable explanations is their capacity to condense intricate model behaviors into readily comprehensible and practicable insights, promoting transparency and confidence in machine-learning models. Visualizations, such as graphs or charts, provide clear and understandable representations of important features, decision boundaries, or model forecasts. Algorithms like as LIME generate natural language descriptions that convert model outputs into explanations that may be easily understood by non-technical audiences, hence simplifying the understanding of model conclusions. Although these strategies improve the comprehensibility of the models, they may compromise the inclusion of intricate details included in complex models in order to generate simplified representations. This could potentially restrict the depth of insights for activities that require a high level of specialization. However, in the context of IoT security frameworks, human-understandable explanations play a crucial role in guaranteeing model transparency and promoting efficient human–machine collaborations [14].

Machine-oriented explanations encompass a category of explanations specifically designed and formatted for consumption by computational systems, aiming to facilitate system-to-system communication and decision-making processes within the realm of XAI. These explanations prioritize technical specifics, internal representations, and formalized structures optimized for computational processing rather than human interpretability. Machine-oriented explanations often include detailed feature importance scores, activation patterns in neural networks, or model-specific representations that serve as inputs for subsequent machine-learning algorithms or decision-making modules. These explanations are instrumental in enhancing the interpretability and interoperability of machine-learning models, enabling seamless integration into larger systems, and aiding in automated decision-making processes within IoT security frameworks. Providing thorough and technical insights mainly for consumption by computational systems and algorithms is the defining quality of machine-oriented explanations. Complex representations, like neural network activation patterns or comprehensive model weights, are commonly used to explain these phenomena; they allow computer processes to take place without direct human interpretability. Systems are able to use and integrate the information and insights gained from complicated models to make automatic and informed decisions, thanks to these clarifications, which greatly improve the efficiency of machine learning models' interactions within IoT security ecosystems. Machine-oriented explanations are great at helping with computational processes and system-level interactions, but they may not be as easy for humans to understand, so they would not be able to be used directly in situations where humans are needed [15].

7.7 Domain-specific XAI methods

The field of XAI is constantly evolving, with a variety of methodologies developed to tackle the specific problems and requirements of interpretability in many disciplines. We will explore the unique characteristics, customized frameworks, and practical implications of domain-specific XAI approaches in different IoT security situations in this section. We want to shed light on the critical importance of domain-specific approaches in improving the interpretability, trustworthiness, and domain-specific relevance of machine-learning models used in IoT security ecosystems by providing a comprehensive analysis of these methods.

Domain-specific XAI techniques represent a specialized category of interpretability methodologies finely tuned to cater to the intricacies and exigencies of distinct domains such as healthcare, finance, or natural language processing (NLP) within the broader landscape of machine-learning applications. These techniques are meticulously crafted to address the unique challenges and requirements prevalent in each domain, offering explanations that resonate with domain experts, regulatory standards, and specific operational contexts. For instance, within healthcare, XAI techniques might focus on generating explanations that aid medical professionals in understanding diagnoses or treatment recommendations derived from complex machine learning models. Similarly, in finance, these methods could elucidate the rationale behind algorithmic trading decisions or credit scoring predictions, aligning with regulatory compliance and financial standards. Moreover, in NLP, domain-specific XAI methods could provide insights into the workings of sentiment analysis or language generation models, facilitating better human-machine interactions. By tailoring explanations to the domain's intricacies and end-users' needs, these techniques bridge the gap between complex model outputs and domain experts, fostering trust and adoption of machine learning systems within these specialized domains. The capacity to deliver customized explanations that connect with the complex needs and limitations of particular domains is the characteristic of domain-specific XAI approaches. In healthcare, for example, these methods could provide explanations in terms of biomarkers, metrics unique to diseases, or treatment protocols that doctors can understand. To ensure that model forecasts are in line with industry standards and decision-making frameworks, domain-specific explanations could cover risk considerations, market indicators, or financial rules in the finance domain. Similarly, in NLP, linguistic characteristics or semantic attributes may represent domain-tailored explanations, which help linguists and language specialists understand model outputs. While these methods are great at what they do—providing insights specific to a domain—they may not be very generalizable to other domains or may need a lot of domain knowledge to interpret. So, in the world of explainable AI, we need to find a happy medium between the two extremes [16].

7.8 Quantitative versus qualitative explanation

The distinction between quantitative and qualitative explanation methodologies constitutes a fundamental paradigm within the domain of XAI, delineating contrasting approaches in the manner and nature of explanations provided by machine-learning models. Quantitative explanation methodologies prioritize numerical, measurable, and statistical representations to explicate model decisions, emphasizing the quantitative assessment of feature importance, model behavior, and prediction mechanisms. In contrast, qualitative explanation methodologies emphasize descriptive, narrative, and non-numeric elucidations of model outputs, focusing on providing insightful narratives, visualizations, or contextual understandings of machine-learning models' decision-making processes. This section embarks on a comprehensive exploration, dissecting the distinctive attributes, operational frameworks, and pragmatic implications of quantitative and qualitative explanation paradigms within the complex landscape of securing IoT environments. Through an analytical lens, we aim to scrutinize the theoretical underpinnings and practical implications of both methodologies, elucidating their respective roles in fortifying the interpretability, transparency, and acceptance of machine learning models deployed in IoT security frameworks.

Quantitative explanations in the domain of XAI encompass a suite of methodologies aimed at delineating model decisions and behaviors through numerical, measurable, and statistical representations. These methods prioritize quantitative metrics, such as feature importance scores, model weights, or statistical measures, to elucidate the rationale behind machine-learning model predictions within IoT security contexts. Techniques like permutation importance, SHAP, or coefficient analysis in linear models quantify the impact and relevance of input features, thereby offering transparent insights into the influence of each feature on model predictions. By emphasizing numerical representations, quantitative explanations facilitate objective and quantitative assessments of model behaviors, enabling stakeholders to gauge the relative importance and contributions of features in the decision-making process. While these methods excel in providing precise and measurable insights into model mechanisms, they may sometimes lack contextual or intuitive understandings of model behaviors, necessitating a balance between quantifiable metrics and qualitative insights for comprehensive interpretability within IoT security frameworks [17]. Quantitative explanations use numbers and statistics to explain machine-learning model decision-making. SHAP values measure each feature's marginal contribution to model predictions throughout the dataset, revealing feature importance. Permutation significance quantifies the impact of input features on predictions by measuring model performance while input features are randomly shuffled. These quantitative measurements provide precision and impartiality in assessing model behaviors, but they may miss complicated feature interactions or nuanced qualitative characteristics necessary for a holistic understanding of model decisions in IoT security applications. A nuanced fusion of quantitative explanations with qualitative insights is needed to comprehend IoT security machine-learning models.

Qualitative explanations in the realm of XAI encompass a diverse array of methodologies that prioritize descriptive, narrative, and visual representations to explicate the decision-making processes of machine-learning models deployed within IoT security contexts. These methods eschew quantitative metrics in favor of descriptive narratives, visualizations, or intuitive representations aimed at conveying insights into model behaviors in a human-understandable format. Techniques such as decision boundary visualizations, saliency maps, or activation maximization generate qualitative explanations by presenting graphical or visual representations that delineate model predictions and decision boundaries. By leveraging visual aids and intuitive descriptions, qualitative explanations aim to provide stakeholders, including domain experts and end-users, with understandable narratives and visual cues that elucidate complex model behaviors. Despite their reliance on descriptive and visual representations, these methods might sometimes lack the precision and quantifiable metrics offered by quantitative explanations, necessitating a balance between qualitative insights and quantitative assessments for a holistic understanding of machine-learning models in IoT security scenarios [18].

The distinguishing feature of qualitative explanations lies in their focus on conveying model behaviors and predictions through descriptive narratives and visual representations. Visualizations, such as heatmaps highlighting important regions in images or decision boundaries depicting class separations in feature space, offer intuitive insights into model predictions. Saliency maps in convolutional neural networks, for instance, highlight crucial image regions influencing predictions, aiding in understanding model decision-making in image classification tasks. However, while these qualitative methods excel in providing intuitive and easily interpretable explanations, they might sometimes lack the precise quantification of feature importance or contributions offered by quantitative explanations, potentially limiting their applicability in scenarios where rigorous numerical assessments are required. Hence, an integrated approach that leverages both qualitative and quantitative explanations becomes essential to foster a comprehensive and interpretable understanding of machine-learning models deployed in IoT security domains.

7.9 Interpretability type

The concept of interpretability type represents a pivotal facet within the domain of XAI, delineating distinct categorizations and dimensions characterizing the nature and scope of explanations generated by machine-learning models. This section embarks on a systematic exploration of the multifaceted landscape of interpretability type, elucidating the diverse dimensions and classifications that underpin the nature of explanations offered by these models within the intricate framework of securing IoT environments.

Textual explanations in the field of XAI are a type of interpretability techniques that express model predictions and behaviors using normal language. Their purpose is to provide insights in a format that can be easily understood by humans, without the need for technical terminology. These methods employ descriptive narratives, summaries, or textual explanations to clarify the decision-making process of machine-learning models used in different situations, such as IoT security. Methods such as natural language generation models, rule-based explanations, or structured textual summaries convert intricate model outputs into easily comprehensible textual descriptions, promoting clear communication between machine-learning systems and human stakeholders. Textual explanations serve the purpose of making model predictions understandable to a wide range of people, including domain experts, policymakers, and end-users who may not have technical knowledge of machine learning. This helps in making educated decisions within IoT security frameworks. Textual explanations, although they provide insights that humans can understand, may not always be as precise and detailed as quantitative metrics or visual representations from other explanation methods. This highlights the importance of integrating a variety of interpretability techniques to achieve a comprehensive understanding of the model. The defining characteristic of textual explanations is its capacity to connect intricate model outputs with human understanding through the provision of natural language explanations for model predictions and judgments. Transformer-based architectures, which are a type of natural language generation models, produce coherent and contextually relevant textual explanations. These explanations help in comprehending the reasoning behind model predictions. Furthermore, rule-based explanations express the process of making judgments in a well-organized written manner, providing if–then rules or logical reasoning that form the basis of model decisions. Although textual methods improve interpretability by providing descriptions that can be easily understood by humans, they may face difficulties in capturing intricate interactions in data with many dimensions or lack the visual clarity provided by graphical representations. Therefore, it is important to carefully choose explanation methods that are suitable for specific contexts in the IoT security field.

In the field of XAI, visual explanations cover a range of approaches that use visuals to explain how machine-learning models make decisions and function in different contexts, such as IoT security. To simplify the presentation of complicated model behaviors and predictions, these methods emphasize the use of visual aids like heatmaps, bar charts, decision boundaries, or interactive visualizations. The most important parts of model predictions and feature contributions can be easily seen with the help of techniques like saliency maps, gradient-based visualizations, and feature significance plots. Helping domain experts, legislators, and end-users comprehend model predictions within IoT security frameworks, visual explanations provide clear and understandable graphical insights into model mechanisms. However, it is essential to note that visual explanations can improve interpretability by providing intuitive graphical representations. However, there are instances where visual explanations oversimplify complex model behaviors or miss subtle interactions between features in complex datasets. In such cases, it is necessary to combine different explanation methods to understand the model fully. Visual explanations are unique because they simplify difficult-to-understand model outputs and decision-making processes by using visually appealing representations. To better understand model decisions in image recognition tasks, heatmaps can be created using methods such as Gradient-weighted Class Activation Mapping (Grad-CAM), which identifies critical areas in images that impact predictions. To better understand how the model sorts of data, decision boundary visualizations highlight feature-space class separations or model predictions. It is important to use a combination of visual and other explanation techniques to fully understand machine-earning models used in IoT security applications, because although these visual methods can provide intuitive insights, they may not always be as precise as numerical metrics or textual descriptions [19].

Symbolic explanations, which fall under the domain of XAI, are a group of interpretability methods that simplify intricate model actions into symbolic depictions or rule-driven systems. This enables machine-learning models implemented in various contexts, such as IoT security, to be explained in a transparent and comprehensible manner. The aforementioned techniques give precedence to the retrieval of symbolic representations, logical expressions, or explicit principles that provide clarity regarding the decision-making mechanisms of machine-learning models. Symbolic representations are produced by methods such as knowledge-based systems, rule extraction algorithms, and decision trees, which convert model predictions into rule-based or symbolic formats. These representations provide comprehensible and transparent insights into the model's decision logic. Symbolic explanations are highly effective instruments for communicating clear and intuitive explanations that are accessible to stakeholders and domain experts who may not have extensive knowledge of complex machine-learning algorithms. Nevertheless, symbolic methods demonstrate exceptional proficiency in generating rules that are comprehensible to humans. However, they may encounter difficulties in capturing intricate model behaviors or oversimplifying complex relationships in high-dimensional data. As a result, IoT security frameworks must strike a delicate balance between transparency and model performance (refer to Table 7.3).

Symbolic explanations are distinguished by their ability to represent the decisions of machine-learning models in a symbolic or rule-based format that corresponds to human comprehension. As an example, decision trees provide a hierarchical arrangement of if–then conditions that establish a direct connection between input features and model predictions, thereby augmenting clarity and comprehensibility. However, it should be noted that although symbolic methods improve interpretability through the provision of transparent rules, they may not fully encompass the intricacies of nonlinear relationships found in specific datasets or fail to depict subtle interactions between features. This underscores the importance of integrating symbolic approaches with other explanation methods in order to obtain a comprehensive understanding of machine-learning models utilized in IoT security scenarios.

Interactive explanations within the realm of XAI encapsulate methodologies that facilitate user engagement and comprehension by enabling stakeholders to interactively explore, manipulate, and query machine-learning models' decisions in real-time. These methods prioritize user-centric interfaces, interactive dashboards, or query-based systems that empower users, including domain experts, policy-makers, or end-users, to probe model predictions, feature contributions, or decision-making processes within IoT security frameworks. Techniques such as interactive visualizations, adjustable parameters, or feature manipulation interfaces offer dynamic and user-controlled platforms that allow stakeholders to explore and query the model's decisions. Interactive explanations foster transparent and dynamic engagement by allowing users to test hypotheses, validate assumptions, or gain insights into model behaviors, thereby enhancing trust and facilitating informed decision-making. However, while these interactive methods offer dynamic exploration and user engagement, they might sometimes overwhelm users with excessive information or interactions, necessitating intuitive design and user-friendly interfaces for effective utilization within IoT security contexts [20]. The property of interactive explanations lies in their ability to empower users with dynamic and user-controlled interfaces, enabling them to actively explore and query model predictions or behaviors. Interactive visualizations, for example, allow users to zoom, pan, or filter specific aspects of model outputs, providing dynamic insights into feature importance or model predictions. Adjustable parameters in interactive dashboards enable users to manipulate variables and observe real-time changes in model outputs, fostering a deeper understanding of model behaviors. Nevertheless, while interactive methods promote user engagement, they might

Table 7.3 Comparison of different types of explanations in XAI for IoT security

Aspect	**Textual explanations**	**Visual explanations**	**Symbolic explanations**	**Interactive explanations**
Representation format	Natural language descriptions	Graphical and visual representations	Rule-based or symbolic representations	User-controlled interfaces, interactive visuals
Methodology	Uses text to convey model decisions	Utilizes graphs, charts, heatmaps, etc.	Derives explicit rules or logical expressions	Employs user-driven exploration of model decisions
Interpretability	Human-readable and comprehensible	Intuitive insights, may lack precision	Transparent rules, may oversimplify behaviors	Engages users in dynamic model exploration
Usability	Understandable to non-technical stakeholders	Offers intuitive insights, lacks granularity	Provides human-interpretable rules	Requires user-friendly design for effective usage
Complexity handling	May oversimplify complex model behaviors	May lack precision in representing details	Might struggle with high-dimensional data	Potential information overload for users
Application	Suitable for domain experts and end-users	Provides intuitive insights for stakeholders	Enhances transparency but may sacrifice complexity	Empowers users to interactively explore model outputs
Comprehensiveness	Provides descriptive and context-rich explanations	Offers visual cues but may lack textual clarity	Offers explicit but potentially oversimplified rules	Enables dynamic exploration but may overwhelm with information
Precision	Subjective in conveying precise model decision logic	Provides intuitive insights, may lack quantification	Offers explicit rules but may miss nuanced relationships	Allows real-time exploration but may lack granularity
Integration capability	Compatible with narrative-based analysis	Integrates well with visual analytics	Compatible with rule-based systems	Requires seamless integration with interactive tools
Domain adaptability	Adaptable to various domains and use cases	Applicable across different visualization contexts	May need adjustments for different rule-based systems	Requires customization for various interactive interfaces
Context sensitivity	Can convey nuanced contextual information	Offers intuitive visual context but may lack specifics	May capture context through rule-based expressions	Facilitates context-based exploration but may need refinement

present challenges related to information overload or interface complexity, requiring careful design considerations to ensure usability and effectiveness within IoT security environments. Therefore, integrating interactive features that balance usability and informativeness becomes pivotal for facilitating meaningful interactions and enhancing stakeholders' understanding of machine-learning models in IoT security applications.

References

[1] Nushi, B., Kamar, E., and Horvitz, E. (2018, June). Towards accountable AI: Hybrid human-machine analyses for characterizing system failure. In *Proceedings of the AAAI Conference on Human Computation and Crowdsourcing* (Vol. 6, pp. 126-135).

[2] Slota, S. C., Fleischmann, K. R., Greenberg, S., *et al.* (2023). Many hands make many fingers to point: challenges in creating accountable AI. *Ai & Society*, 1–13.

[3] Bonifazi, G., Cauteruccio, F., Corradini, E., *et al.* (2024). A model-agnostic, network theory-based framework for supporting XAI on classifiers. *Expert Systems with Applications*, *241*, 122588.

[4] Zolanvari, M., Yang, Z., Khan, K., Jain, R., and Meskin, N. (2021). Trust XAI: Model-agnostic explanations for AI with a case study on IIoT security. *IEEE Internet of Things Journal*, *10*(4), 2967–2978.

[5] Dwivedi, R., Dave, D., Naik, H., *et al.* (2023). Explainable AI (XAI): Core ideas, techniques, and solutions. *ACM Computing Surveys*, *55*(9), 1–33.

[6] Moradi, M., and Samwald, M. (2021). Post-hoc explanation of black-box classifiers using confident item sets. *Expert Systems with Applications*, *165*, 113941.

[7] Kenny, E. M., Ford, C., Quinn, M., and Keane, M. T. (2021). Explaining black-box classifiers using post-hoc explanations-by-example: The effect of explanations and error-rates in XAI user studies. *Artificial Intelligence*, *294*, 103459.

[8] Mi, J. X., Jiang, X., Luo, L., and Gao, Y. (2024). Toward explainable artificial intelligence: A survey and overview on their intrinsic properties. *Neurocomputing*, *563*, 126919.

[9] Bassan, S., Amir, G., and Katz, G. (2024). Local vs. Global Interpretability: A Computational Complexity *Perspective. arXiv preprint arXiv:2406. 02981.*

[10] Zhang, Y., Tiňo, P., Leonardis, A., and Tang, K. (2021). A survey on neural network interpretability. *IEEE Transactions on Emerging Topics in Computational Intelligence*, *5*(5), 726–742.

[11] Saleem, R., Yuan, B., Kurugollu, F., Anjum, A., and Liu, L. (2022). Explaining deep neural networks: A survey on the global interpretation methods. *Neurocomputing*, *513*, 165–180.

[12] Amparore, E., Perotti, A., and Bajardi, P. (2021). To trust or not to trust an explanation: Using LEAF to evaluate local linear XAI methods. *PeerJ Computer Science*, *7*, e479.

[13] Zhang, Y., He, H., Tan, Z., and Yuan, Y. (2023). Trade-off between efficiency and consistency for removal-based explanations. *Advances in Neural Information Processing Systems*, *36*, 25627–25661.

[14] Paleja, R., Ghuy, M., Ranawaka Arachchige, N., Jensen, R., and Gombolay, M. (2021). The utility of explainable ai in ad hoc human-machine teaming. *Advances in Neural Information Processing Systems*, *34*, 610–623.

[15] Moyle, S., Martin, A., and Allott, N. (2024). XAI human-machine collaboration applied to network security. *Frontiers in Computer Science*, *6*, 1321238.

[16] Paredes, J. N., Teze, J. C. L., Simari, G. I., and Martinez, M. V. (2021). On the importance of domain-specific explanations in AI-based cybersecurity systems (technical report). *arXiv preprint arXiv:2108.02006.*

[17] Nauta, M., Trienes, J., Pathak, S., *et al.* (2023). From anecdotal evidence to quantitative evaluation methods: A systematic review on evaluating explainable ai. *ACM Computing Surveys*, *55*(13s), 1–42.

[18] Johs, A. J., Agosto, D. E., and Weber, R. O. (2022). Explainable artificial intelligence and social science: Further insights for qualitative investigation. *Applied AI Letters*, *3*(1), e64.

[19] Muddamsetty, S. M., Jahromi, M. N., Ciontos, A. E., Fenoy, L. M., and Moeslund, T. B. (2022). Visual explanation of black-box model: Similarity difference and uniqueness (SIDU) method. *Pattern Recognition*, *127*, 108604.

[20] Bertrand, A., Viard, T., Belloum, R., Eagan, J. R., and Maxwell, W. (2023, April). On selective, mutable and dialogic XAI: A review of what users say about different types of interactive explanations. *In Proceedings of the 2023 CHI Conference on Human Factors in Computing Systems* (pp. 1–21).

Chapter 8
Intrinsically explainable security intelligence

8.1 Introduction

In today's interconnected world, the proliferation of Internet of Things (IoT) devices has ushered in unprecedented convenience and connectivity. Yet, amid this technological marvel lies a pressing concern—the security of these interconnected systems. As IoT systems become increasingly intertwined with our daily lives, ensuring their security has emerged as a critical priority. At the heart of this challenge lies the need for transparency and interpretability in the artificial intelligence (AI) systems governing these IoT networks [1].

Intrinsically explainable AI (XAI) stands as a beacon of hope in addressing the complexities of ensuring IoT security. Unlike traditional black-box AI algorithms, XAI approaches offer a fundamental shift by emphasizing interpretability and transparency as integral components of AI models. Specifically tailored for IoT systems, these intrinsically explainable methods hold the promise of not only delivering robust security measures but also empowering users and stakeholders with insights into the decision-making processes of these interconnected devices. However, the journey toward fortified IoT security via AI is marred by a series of challenges stemming from the opacity and lack of interpretability in existing AI systems. As IoT devices multiply, the reliance on AI for managing, analyzing, and securing the vast data streams they generate grows exponentially. Yet, this reliance often comes at the cost of understanding how and why AI-driven decisions are made within these systems. The black-box nature of many AI algorithms poses a substantial barrier, hindering the comprehension of crucial security-related actions, leaving users and administrators in the dark regarding potential vulnerabilities and threat patterns [2].

Moreover, within the context of IoT security, the lack of transparency amplifies concerns about accountability and trustworthiness. As these AI-driven systems autonomously make decisions that impact security protocols, the inability to decipher their decision-making rationale raises significant apprehensions. It complicates identifying and mitigating vulnerabilities and raises ethical and regulatory issues surrounding liability and responsibility in the event of security breaches or system malfunctions. In addressing these challenges, the quest for intrinsically XAI methods becomes imperative. By bridging the gap between complex AI algorithms and human comprehension, these methods offer a pathway to fortify IoT security while simultaneously ensuring a deeper understanding of the decision-making processes driving these systems. Through this lens, this chapter embarks on an exploratory journey into the realm of intrinsically XAI methods within the context of IoT security. In this chapter, the focus revolves around categorizing intrinsically XAI models into four primary categories, each representing a distinct approach toward achieving interpretability and transparency in AI systems. These categories encompass linear models, graphical models, tree-based models, and K-nearest neighbors (KNNs), each offering unique characteristics and advantages in facilitating explainability [3].

8.2 Linear models

Linear models represent a foundational approach in the realm of interpretable AI. They include regression models, linear classifiers (such as logistic regression), and linear support vector machines. These models are characterized by their simplicity and ease of interpretation due to their clear representation of the relationships between input features and output predictions. Coefficients assigned to each feature provide straightforward insights into their influence on the model's decision-making process, enhancing transparency [4].

8.2.1 Linear regression

Linear regression is a fundamental statistical method used for modeling the relationship between a dependent variable (often denoted as y) and one or more independent variables (denoted as $x_1, x_2, \ldots, x_n$). It is a linear approach to modeling the relationship between a scalar response and one or more explanatory variables. The fundamental assumption is that this relationship can be expressed as a straight-line equation:

$$y = \beta_0 + \beta_1 x_1 + \ldots + \beta_n x_n + \epsilon \tag{8.1}$$

where y represents the dependent variable being predicted. $x_1, x_2, ..., x_n$ are the independent variables (features). β_0 is the intercept term (the value of y when all value of x are zero). $\beta_1, \beta_2, ..., \beta_n$ are the coefficients (slopes) associated with each independent variable. ϵ denotes the error term or residual (the difference between the predicted and actual values not explained by the model).

The effectiveness of the linear regression model is contingent upon the fulfillment of several critical assumptions. The evaluation of whether the model accurately represents the relationships within the data relies heavily on verifying these assumptions. The primary assumptions of linear regression encompass linearity, normality, homoscedasticity, independence, fixed features, and the absence of multicollinearity.

- **Linearity:** One fundamental assumption in linear regression is the linearity of the relationship between the dependent and independent variables. This assumption implies that the change in the dependent variable associated with a unit change in the independent variable(s) is consistent across the entire range of the data. Assessing linearity involves examining scatterplots and residual plots to confirm that the relationship appears linear and that the residuals exhibit a random pattern around zero along the fitted line.

- **Normality:** Another crucial assumption pertains to the normality of residuals. It presupposes that the residuals (errors) resulting from the model should follow a normal distribution. This assumption allows for reliable statistical inference, hypothesis testing, and confidence interval estimation. Evaluating normality typically involves graphical assessments, such as histograms or *Q–Q* plots, along with statistical tests like the Shapiro–Wilk test or the Kolmogorov–Smirnov test.
- **Homoscedasticity:** Homoscedasticity implies that the variance of the residuals remains constant across all levels of the independent variables. In simpler terms, the spread of the residuals should exhibit uniformity along the range of predicted values. Detecting violations of homoscedasticity can be accomplished by examining residual plots against fitted values or through formal statistical tests like the Breusch–Pagan test or White's test.
- **Independence:** The independence assumption signifies that the residuals are independent of each other. In other words, the error terms should not exhibit any systematic patterns or correlations. Verification of independence can be assessed through residual autocorrelation plots or the Durbin–Watson statistic, which examines the presence of autocorrelation in the residuals.
- **Fixed features:** Linear regression assumes that the independent variables are measured without error or variability. This assumption implies that the independent variables are fixed and do not change in response to the outcome or other variables.
- **Absence of multicollinearity:** Multicollinearity refers to the presence of high correlations among independent variables. This assumption assumes that independent variables are not excessively correlated with each other, as high multicollinearity can lead to unstable estimates of coefficients. Variance inflation factor (VIF) analysis or correlation matrices are commonly used to detect multicollinearity.

The best-fit line is the crux of linear regression, representing the linear equation that minimizes the discrepancy between observed data points and their corresponding predicted values. The essence lies in identifying the line that best captures the relationship between the dependent and independent variables. This line is determined by adjusting the coefficients (slopes) and the intercept to minimize the sum of squared differences between the observed and predicted values. The goal is to ascertain a line that optimally represents the trend within the data, enabling accurate predictions and inferences [5].

The methodology employed to estimate the coefficients in linear regression is ordinary least squares (OLS). OLS seeks to determine the values of the coefficients that minimize the sum of squared residuals, representing the vertical distances between the observed data points and the fitted line.

$$\hat{\beta} = \arg\min_{\beta_0,\ldots,\beta_p} \sum_{i=1}^{n} \left(y^{(i)} - \left(\beta_0 + \sum_{j=1}^{p} \beta_j x_j^{(i)} \right) \right)^2 \tag{8.2}$$

By minimizing these squared differences, OLS calculates the most suitable coefficients that define the best-fit line. This methodological approach ensures that the model effectively captures the relationship between variables by providing estimates that best approximate the observed data points.

The interpretation of coefficients in linear regression is pivotal for comprehending how changes in the independent variables influence the dependent variable. Each coefficient in the linear equation signifies the effect of a unit change in the corresponding independent variable on the dependent variable, holding other variables constant. These coefficients encapsulate the slope of the line and represent the magnitude and direction of the impact of each independent variable on the dependent variable. Consequently, interpreting coefficients involves understanding how a one-unit change in an independent variable influences the predicted change in the dependent variable, facilitating insights into the relationships embedded within the model.

Regarding numerical features, a standardized framework for interpretation involves discerning how a one-unit increment in the specific feature x_k influences the predicted outcome y, while keeping all other features constant. This standardized explanation asserts that for each unitary increase in the numerical feature *xkxk*, the predicted outcome y is expected to augment by β_k units. On the other hand, when handling categorical features, the interpretation is contingent upon the transformation of the categorical feature x_k from its reference category to an alternative category. The interpretation asserts that upon switching the categorical feature from the reference category to another specified category, the predicted outcome y is anticipated to increase by β_k units, provided that all other features remain unchanged and constant [6].

Another pivotal metric in comprehending the implications of linear models is the R-squared statistic. This metric serves as an indicator of the extent to which the model accounts for the overall variability observed in the target outcome. Essentially, R-squared quantifies the proportion of variance in the dependent variable that can be elucidated by the model's independent variables. Higher values of R-squared signify a greater degree of explanatory power within the model, signifying that a larger percentage of the variability in the target outcome is captured by the model's predictors. In essence, an elevated R-squared value indicates a superior ability of the model to elucidate and account for the observed variability in the dataset. The R-squared is calculated with the following formula:

$$R^2 = 1 - \frac{SSE = \sum_{i=1}^{n} \left(y^{(i)} - \tilde{y}^{(i)}\right)^2}{SST = \sum_{i=1}^{n} \left(y^{(i)} - \bar{y}\right)^2} = \tag{8.3}$$

One aspect to consider is that R-squared tends to rise as more features are included in the model, even if some of these features lack meaningful information about the target variable. This occurrence poses a challenge as R-squared may inflate due to the addition of irrelevant or redundant features, erroneously suggesting improved model performance. Thus, employing the adjusted *R*-squared metric becomes advantageous as it adjusts for the number of features incorporated in the model. This adjustment helps mitigate the issue by penalizing models with a higher number of features, thereby offering a more accurate assessment of the model's performance in explaining the variance in the target variable. The calculation for adjusted *R*-squared involves a formula that takes into account both the *R*-squared value and the number of predictors utilized in the model.

$$\bar{R}^2 = 1 - \left(1 - R^2\right)\frac{n-1}{n-p-1} \tag{8.4}$$

with p and n denoting the number of features and samples, respectively.

In a linear regression model, the importance of a feature can be gauged through the examination of its associated *t*-statistic. The *t*-statistic assesses the significance of each feature's coefficient by comparing it to a reference value, often zero. Specifically, the magnitude of the *t*-statistic denotes how many standard deviations the coefficient estimate is away from zero. The formula for calculating the *t*-statistic for a coefficient in a linear regression model is derived from the coefficient estimate, its standard error, and the degrees of freedom. Here is the formula:

$$t_{\tilde{\beta}_j} = \frac{\tilde{\beta}_j}{SE\left(\tilde{\beta}_j\right)} \tag{8.5}$$

where $\tilde{\beta}_j$ represents the coefficient estimate for the jth feature. βnullβnull is the hypothesized value of the coefficient under the null hypothesis (often taken as zero). $SE\left(\tilde{\beta}_j\right)$ denotes the standard error associated with the coefficient estimate $\tilde{\beta}_j$. t is the t-statistic.

The absolute value of the *t*-statistic for a feature's coefficient serves as an indicator of the significance of that feature in relation to predicting the outcome variable. A larger absolute *t*-statistic implies a stronger deviation of the coefficient from zero, indicating that the feature holds more importance in explaining variations in the predicted outcome. Essentially, a higher absolute *t*-statistic suggests that the associated feature is more statistically significant and contributes more substantially to the model's predictive ability [7]. In practical terms, when assessing the importance of features in a linear regression model, features with larger absolute *t*-statistic values (typically above a certain threshold, often 1.96 or 2 in absolute value for a 95% confidence interval) are considered more important, as they indicate a stronger statistical significance and influence on predicting the outcome variable. These features are likely to provide more meaningful information for understanding and predicting the target variable compared to features with smaller absolute *t*-statistic values.

```
import numpy as np
import matplotlib.pyplot as plt
from sklearn.datasets import load_diabetes
from sklearn.linear_model import LinearRegression
from sklearn.model_selection import train_test_split

# Load the diabetes dataset
diabetes = load_diabetes()
X, y = diabetes.data, diabetes.target

# Split the data into training and testing sets
X_train, X_test, y_train, y_test = train_test_split(X, y, test_size=0.2,
random_state=42)

# Initialize and fit the linear regression model
model = LinearRegression()
model.fit(X_train, y_train)

# Visualize the weight (coefficient) of all features
plt.figure(figsize=(10, 6))
plt.barh(diabetes.feature_names, model.coef_, color='skyblue')
plt.xlabel('Weight (Coefficient)')
plt.title('Weight Plot')
plt.show()

# Select one feature for the effect plot (let's say the third feature)
feature_index = 2  # Change this index to visualize different features
X_feature = X_train[:, feature_index].reshape(-1, 1)
X_feature_test = X_test[:, feature_index].reshape(-1, 1)

# Fit the model with the selected feature
model_feature = LinearRegression()
model_feature.fit(X_feature, y_train)

# Create an effect plot (partial dependence plot) for the selected feature
feature_values = np.linspace(np.min(X_feature), np.max(X_feature),
num=100).reshape(-1, 1)
predictions = model_feature.predict(feature_values)

plt.figure(figsize=(8, 6))
plt.scatter(X_feature_test, y_test, color='lightcoral', label='Actual')
plt.plot(feature_values, predictions, color='darkblue', linewidth=2,
label='Predicted')
plt.xlabel(diabetes.feature_names[feature_index])
plt.ylabel('Target')
plt.title('Effect Plot')
plt.legend()
plt.show()
```

8.2.2 Logistic regression

Logistic regression, a fundamental statistical method, holds significance as a versatile tool for binary classification tasks. It differs from linear regression in its application to predict the probability of occurrence of a categorical outcome rather than continuous values. In this section, we delve into the fundamental concepts and principles of logistic regression, elucidating its core components and underlying principles.

Logistic regression operates within a binary classification framework, aiming to model the relationship between one or more independent variables and a categorical dependent variable with two possible outcomes (commonly denoted as 0 and 1). This method is adept at predicting the probability that an observation belongs to a particular class based on input features. At the core of logistic regression lies the logistic function (also known as the sigmoid function), which transforms the linear combination of independent variables into a probability value between 0 and 1. The logistic function is expressed as:

$$P(Y=1|X)=\frac{1}{1+\exp\left(-\left(\beta_0+\beta_1 x_1^{(i)}+\ldots+\beta_p x_p^{(i)}\right)\right)} \tag{8.6}$$

where $P(Y=1|X)$ denotes the probability that the outcome variable Y equals 1 given the input features X. $\beta_0, \beta_1, \beta_2, \ldots, \beta_p$ represent the coefficients associated with the independent variables. $X_1, X_2, \ldots, X_p$ signify the independent variables.

In logistic regression, the interpretation of coefficients involves understanding log-odds and odds ratios, which are fundamental concepts aiding in the analysis of the relationship between independent variables and the log-odds of the outcome occurring. Log-odds, also known as logit, are the logarithm of the odds ratio. Logistic regression predicts log-odds rather than probabilities directly [8]. The log-odds of an event happening (denoted as logit(p)logit(p)) are expressed mathematically as:

$$\ln\left(\frac{P(y=1)}{1-P(y=1)}\right)=\log\left(\frac{P(y=1)}{P(y=0)}\right)=\beta_0+\beta_1 x_1+\ldots+\beta_p x_p \tag{8.7}$$

Here, P represents the probability of the binary outcome (success), and $(1-P)$ represents the probability of the complementary outcome (failure). The log-odds indicate the logarithm of the ratio between the probability of success and the probability of failure. The odds ratio is a measure of association between an independent variable and the outcome variable in logistic regression. It quantifies the change in odds for a one-unit change in the independent variable, assuming all other variables remain constant. The odds ratio for a predictor X_i with coefficient β_i is given by:

$$\text{Odds ratio}=\exp(\beta_j) \tag{8.8}$$

Interpreting the odds ratio involves understanding how a one-unit increase in the predictor variable affects the odds of the outcome variable. For instance, an odds ratio of 1 suggests no change in the odds, an odds ratio greater than 1 implies an increase in the odds of the event occurring with an increase in the predictor, and an odds ratio less than 1 indicates a decrease in the odds of the event with an increase in the predictor.

In interpreting coefficients in logistic regression, the exponentiated coefficients provide insights into the impact of each independent variable on the odds of the outcome variable. The interpretations vary for different types of features within the model, elucidating how alterations in feature values correspond to changes in the estimated odds of the outcome variable. For a numerical feature x_j, a unitary increase in its value results in a change in the estimated odds by a factor of $\exp(\beta_j)$. This implies that the odds of the event occurring alter multiplicatively by the exponentiated coefficient β_j when the numerical feature x_j is incremented by one unit. When dealing with a binary categorical feature, one of the two values serves as the reference category. Altering the feature x_j from the reference category to the other category changes the estimated odds by a factor of $\exp(\beta_j)$. This adjustment signifies a multiplicative change in the odds of the event based on the exponentiated coefficient β_j. For categorical features encompassing more than two categories, employing techniques like one-hot-encoding becomes essential. This encoding method allocates a distinct column for each category, utilizing $L-1$ columns for a categorical feature with L categories to avoid over-parameterization. Here, the Lth category operates as the reference category. The interpretation of each category mirrors that of binary features, akin to the interpretation of a binary categorical feature. The intercept coefficient β_0 denotes the estimated odds when all numerical features hold a value of zero, and categorical features are at the reference category. The calculated odds are represented by $\exp(\beta_j)$. Typically, the interpretation of the intercept weight is not considered notably relevant within the context of the model.

```
import numpy as np
import matplotlib.pyplot as plt
from sklearn.datasets import load_breast_cancer
from sklearn.linear_model import LogisticRegression
from sklearn.model_selection import train_test_split

# Load the breast cancer dataset
breast_cancer = load_breast_cancer()
X, y = breast_cancer.data, breast_cancer.target

# Split the data into training and testing sets
X_train, X_test, y_train, y_test = train_test_split(X, y, test_size=0.2,
random_state=42)

# Initialize and fit the logistic regression model
model = LogisticRegression(max_iter=10000)  # Increase max_iter for
convergence if needed
model.fit(X_train, y_train)

# Visualize the weight (coefficient) of all features
plt.figure(figsize=(10, 6))
plt.barh(breast_cancer.feature_names, model.coef_[0], color='skyblue')
plt.xlabel('Weight (Coefficient)')
plt.title('Weight Plot')
plt.show()
```

8.2.3 Generalized linear model

Generalized linear models (GLMs) stand as a comprehensive extension of traditional linear models, offering a versatile framework to handle a wider range of data distributions and response types. GLMs serve as a powerful statistical modeling approach designed to accommodate diverse data types by extending the conventional linear regression paradigm. Unlike linear regression, which assumes a normally distributed continuous response variable, GLMs encompass various response distributions (e.g., binary, Poisson, and gamma) and allow for non-linear relationships between predictors and the response [9]:

$$g(E_Y(y|,x)) = \beta_0 + \beta_1 x_1 + \ldots \beta_p x_p \quad (8.9)$$

GLMs consist of several key components that differentiate them from traditional linear regression models. Understanding these components is crucial in comprehending how GLMs handle different types of data and distributions. The main components of GLMs include:

- **Systematic component ($X^T\beta$)**: The systematic component in GLMs mirrors the linear predictor in traditional linear models. It comprises a linear combination of predictors, each multiplied by its respective coefficient. However, GLMs encompass a link function that relates the mean of the response variable to the linear predictor, enabling the model to capture non-linear relationships between predictors and the response.
- **Probability distribution (E_Y)**: Unlike linear regression, GLMs do not assume normality of the response variable. Instead, GLMs accommodate a broad spectrum of response distributions, such as binomial, Poisson, gamma, and others. This flexibility allows GLMs to model a diverse array of data types, including binary outcomes, counts, and skewed continuous data.
- **Link function (g)**: A distinguishing feature of GLMs is the inclusion of a link function that connects the systematic component to the expected value of the response variable. The link function bridges the linear predictor and the mean of the response, ensuring a proper transformation to maintain the relationship between predictors and the response variable.

Moreover, link functions play a pivotal role in connecting the linear predictor to the expected value (mean) of the response variable. The choice of a specific link function significantly influences the interpretability of the model and the relationship between predictors and the response variable. Link functions serve as a bridge between the linear predictor, which is a linear combination of predictor variables, and the meaning of the response variable. They provide a mechanism to model non-linear relationships between predictors and the response, allowing for greater flexibility in capturing various patterns in the data. This can have different impact on interpretability:

- **Linearity:** The choice of a link function dictates the form of the relationship between predictors and the mean of the response variable. For instance, the logit link function in logistic regression ensures that changes in predictor variables result in a multiplicative effect on the odds of the event occurring. In contrast, the log link function in Poisson regression models the logarithm of the mean response, implying that predictors have a multiplicative impact on the response variable.

- **Ease of interpretation:** Different link functions offer varying degrees of interpretability. Some link functions, such as the logit and probit in binary response models, allow for straightforward interpretation in terms of odds ratios and probabilities. However, other link functions, like the log link in Poisson regression, may require additional transformation to interpret the coefficients in the context of the response variable.
- **Model flexibility:** Certain link functions provide more flexibility in capturing complex relationships between predictors and the response. A suitable choice of link function enables the model to better represent non-linear relationships and accommodate diverse data distributions, contributing to a more accurate representation of the underlying data structure.

```
import numpy as np
import matplotlib.pyplot as plt
import statsmodels.api as sm
from sklearn.datasets import make_regression

# Generate synthetic data for linear regression
X, y = make_regression(n_samples=100, n_features=1, noise=10,
random_state=42)

# Add intercept term to input features
X = sm.add_constant(X)

# Fit a Generalized Linear Model (OLS - Ordinary Least Squares)
glm = sm.GLM(y, X, family=sm.families.Gaussian())
result = glm.fit()

# Print model summary
print(result.summary())

# Visualize the linear regression line
plt.figure(figsize=(8, 6))
plt.scatter(X[:, 1], y, color='blue', label='Actual')
plt.plot(X[:, 1], result.predict(X), color='red', linewidth=2,
label='Predicted')
plt.xlabel('X')
plt.ylabel('Y')
plt.title('Linear Regression')
plt.legend()
plt.show()
```

Generalized additive models (GAMs) constitute a flexible and powerful extension of GLMs, allowing for nonlinear relationships between predictors and the response variable while accommodating various types of distributions. GAMs offer a versatile framework suitable for modeling complex relationships in multidimensional datasets, enhancing predictive accuracy and interpretability. GAMs adopt an additive approach, where the relationship between the predictors and the response variable is expressed as a sum of smooth functions of the predictors. Unlike traditional linear models, GAMs can incorporate nonlinear relationships and interactions between predictors without requiring explicit functional forms. Mathematically, the formulation of GAMs is given as follows:

$$g(E_Y(y|,x)) = \beta_0 + \beta_1 f_1(x_1) + \ldots \beta_p f_p(x_p) \tag{8.10}$$

where $f_1(X_1), f_2(X_2), ..f_p(X_p)$ denote the smooth functions of predictors capturing their nonlinear relationships with the response The main components of GAMs include:

- **Smooth functions:** GAMs employ smooth functions (often represented using splines) to model the relationship between each predictor and the response. These smooth functions capture nonlinear patterns in the data, allowing for flexibility in modeling complex relationships.
- **Additivity:** The additive nature of GAMs enables the combination of multiple smooth functions for different predictors to create a composite model that accounts for the collective influence of predictors on the response.

8.3 Graphical models

Graphical models encompass various models that utilize graphical representations, such as Bayesian networks or Markov networks, to express dependencies between variables. These models are adept at capturing complex relationships among various factors, making them valuable for explaining probabilistic reasoning and causal relationships. Their graphical structure enables intuitive visualization, aiding in understanding the interconnections within the data and the model's decision rationale [10].

8.3.1 Naive Bayes

The Naive Bayes classifier is a simple yet effective probabilistic model used for classification tasks, particularly in text classification, spam filtering, and recommendation systems. Its foundation lies in Bayes' theorem, combined with the assumption of independence among features, hence the term "naive." At the core of Naive Bayes lies Bayes' theorem, which provides a framework for calculating posterior probabilities based on prior knowledge and observed evidence. It is expressed as:

$$P(Y|X) = \frac{P(X|Y) \cdot P(Y)}{P(X)} \tag{8.11}$$

where $P(Y|X)$ represents the probability of event Y given event X. $P(X|Y)$ is the probability of event X given event Y. $P(Y)$ and $P(X)$ denote the probabilities of events Y and X, respectively.

The Naive Bayes classifier assumes independence among features given the class label. This assumption implies that the presence of one feature is independent of the presence of other features, which simplifies the computation of the conditional probability of observing a set of features given a class label. In other words, the Naive Bayes classifier assumes that the features (predictors) are conditionally independent of each other given the class label. Mathematically, this can be represented as:

$$P(y|, x_1 \ldots x_n) = \frac{P(x_1|y)P(x_2|y)\ldots P(x_n|, y)P(y)}{P(x_1)P(x_2)\ldots P(x_n)} \tag{8.12}$$

where $P(y|, x_1 \ldots x_n)$ is the posterior probability of class y given features. $P(y)$ is the prior probability of class y. $P(x_i|, y)$ represents the conditional probability of feature x_i given class y. This can be expressed as:

$$P(y|x_1 \ldots x_n) = \frac{P(y)\prod_{i=1}^{n} P(x_i|y)}{P(x_1)P(x_2)\ldots P(x_n)} \tag{8.13}$$

Since the denominator remains constant across a specific input, it can be omitted:

$$P(y|x_1 \ldots x_n) \propto P(y)\prod_{i=1}^{n} P(x_i|y) \tag{8.14}$$

In the context of the classifier model, it is crucial to determine the probability of the data across all potential values of y and identify the output with the highest probability, which can be mathematically articulated as:

$$y = \text{argmax}_y P(y)\prod_{i=1}^{n} P(x_i|y) \tag{8.15}$$

Conditional independence assumption in Naive Bayes significantly contributes to the model's interpretability by simplifying the understanding of relationships between predictors. By assuming that the predictors are conditionally independent given the class label, Naive Bayes assumes that the presence of one particular feature is unrelated to the presence of another feature, given the class label. This assumption vastly reduces the complexity of the model, allowing for a more straightforward interpretation of how individual predictors contribute to the classification decision. This simplicity enables straightforward estimations of the likelihood of a class based on the presence of specific features, aiding in understanding the importance and impact of each feature on the classification outcome. For instance, in text classification, the model can distinctly identify and weigh each word's significance independently of others, simplifying the understanding of which words or features contribute more toward determining the class.

Moreover, conditional independence enhances interpretability by enabling a clearer distinction between different predictors' contributions to the classification process. The assumption facilitates the separation of influence among predictors, allowing practitioners to discern the specific effect of each predictor on the class assignment. This clarity aids in understanding the relative importance and relevance of various features in determining the final classification outcome. By assuming independence, Naive Bayes allows analysts to interpret the effect of each feature on the predicted class, providing insights into which features strongly influence the classification decision and which ones might have minimal impact. Consequently, this improved understanding of individual predictors' contributions enhances the model's interpretability and facilitates a more transparent analysis of how each predictor influences the classification results.

```
from sklearn.datasets import load_iris
from sklearn.model_selection import train_test_split
from sklearn.naive_bayes import GaussianNB
from sklearn.metrics import accuracy_score, classification_report

# Load the Iris dataset (example dataset)
iris = load_iris()
X, y = iris.data, iris.target

# Split the data into training and testing sets
X_train, X_test, y_train, y_test = train_test_split(X, y, test_size=0.2,
random_state=42)

# Initialize the Gaussian Naive Bayes classifier
nb_classifier = GaussianNB()

# Train the classifier
nb_classifier.fit(X_train, y_train)

# Make predictions
predictions = nb_classifier.predict(X_test)

# Calculate accuracy
accuracy = accuracy_score(y_test, predictions)
print(f"Accuracy: {accuracy}")

# Classification report
print(classification_report(y_test, predictions))
```

8.3.2 Bayesian network

Bayesian networks, also known as belief networks or causal probabilistic networks, are graphical models that represent probabilistic relationships among variables using directed acyclic graphs (DAGs). They offer a powerful framework for modeling uncertain domains by encoding dependencies and conditional relationships between variables. These networks leverage Bayesian probability theory to infer complex relationships and make predictions based on observed evidence [11]. The graphical representation of Bayesian networks involves depicting variables as nodes and their probabilistic relationships as directed edges between nodes. Each node represents a random variable, and the directed edges signify the probabilistic dependencies or causal relationships between variables. The structure of the graph captures the conditional independence assumptions among variables given their parents in the graph. The nodes in a Bayesian network correspond to the variables being modeled, and each node holds a conditional probability distribution that quantifies the probability of that variable given its parent variables. These conditional probability distributions encapsulate the probabilistic relationships and dependencies among variables.

The directionality of edges in Bayesian networks signifies the causal flow of influence from parent nodes (causes) to child nodes (effects). This structure allows for inferring causal relationships and understanding the impact of one variable on another within the network. For instance, in a medical Bayesian network, a node representing a disease may influence nodes representing symptoms, medications, or test results, reflecting the causal relationship between the disease and its manifestations [12]. Each node in a Bayesian network holds a conditional probability distribution (CPD) that quantifies the probability of that variable given its parent variables. These CPDs encode the strength and nature of the causal relationship between variables. By examining these distributions, one can understand how changes in parent variables affect the probabilities of outcomes in child variables, elucidating the causal impact.

Bayesian networks facilitate counterfactual reasoning, allowing analysts to evaluate hypothetical scenarios by altering the state of variables. By manipulating the states of certain nodes and observing the propagated changes, one can infer the potential outcomes or changes in other variables, shedding light on the causal effects of interventions or changes in variables. Bayesian networks offer interpretability by providing a visual and intuitive representation of causal relationships. Analysts can interpret the graphical structure to identify direct causal paths, indirect effects, and dependencies among variables. This aids in understanding the underlying mechanisms driving the system being modeled. Leveraging causal relationships, Bayesian networks support decision-making by simulating the effects of interventions or decisions on the system. They enable probabilistic predictions by incorporating causal knowledge, enhancing the accuracy and reliability of predictions.

8.4 Tree-based models

Tree-based models, such as decision trees and random forests, operate by partitioning the feature space into a set of hierarchical decision rules. These models offer inherent interpretability by presenting decision paths in a tree-like structure. Decision trees, in particular, are intuitive and easily understandable, making them suitable for explaining classification or regression decisions. Ensemble methods like random forests further enhance interpretability while improving predictive performance [13].

8.4.1 Decision tree

Decision trees are hierarchical, tree-like structures used for both classification and regression tasks in machine learning. They operate by recursively partitioning the input space into smaller regions or segments, aiming to make predictions or assign labels to instances based on the features' values. A decision tree consists of nodes representing points where decisions are made, or information is gathered. These nodes include:

- Root node: The initial node, representing the entire dataset or a subset.
- Internal nodes: Intermediate nodes that split the data based on feature conditions.
- Leaf nodes: Terminal nodes that represent the final predicted outcome or class label.

At each internal node, the decision tree selects the best feature and corresponding threshold (for numerical features) to split the dataset. This process aims to maximize the homogeneity or purity of the resulting subsets based on the target variable. The tree-building process continues recursively, branching into subsets based on feature conditions until certain stopping criteria are met. These criteria could include reaching a maximum depth, achieving minimum samples per leaf, or other measures to prevent overfitting.

Decision trees employ various splitting criteria to determine the best feature for node splitting. The selection of an optimal attribute to split the data at each node significantly impacts the tree's structure and predictive accuracy. Three commonly used splitting criteria are Gini impurity, entropy, and information gain, each offering a distinct perspective on feature selection within decision trees.

Gini impurity measures the degree of impurity or disorder in a dataset by calculating the probability of misclassifying an instance's label. For a given node t containing samples from K classes, the Gini impurity $G(t)$ is computed as:

$$G(t) = 1 - \sum_{i=1}^{K} p(i|,t)^2 \tag{8.16}$$

where $p(i|,t)$ represents the probability of an instance in node t belonging to class i. Lower Gini impurity values indicate higher purity or homogeneity within a node, signifying a better split.

Entropy is a measure of the average amount of information or disorder in a dataset. It quantifies the impurity of a node by computing the entropy $H(t)$ as:

$$H(t) = -\sum_{i=1}^{K} p(i|,t)\log p(i|,t) \tag{8.17}$$

Here, $p(i|,t)$ denotes the probability of an instance in node tt belonging to class ii. Lower entropy values imply higher purity or homogeneity within a node.

Information gain is a metric used to evaluate the usefulness of a feature in reducing uncertainty or disorder in a dataset. It measures the difference in entropy or Gini impurity before and after the split based on a particular feature. The information gain IGIG for a split using a feature AA is calculated as:

$$IG(A) = H(parent) - \sum_{i=1}^{m} \frac{N_i}{m} H(child_i) \tag{8.18}$$

where N is the total number of samples in the parent node, N_i represents the number of samples in the ith child node, mm denotes the number of child nodes, and H ($parent$) and $G(parent)$ refer to the entropy and Gini impurity of the parent node, respectively.

Decision tree algorithms, such as ID3, C4.5, and CART, employ these criteria to select the best feature for splitting at each node. They calculate the information gain, Gini impurity, or entropy for each feature and choose the attribute that maximizes information gain or minimizes impurity as the splitting criterion. This iterative process continues recursively to construct the tree until a stopping criterion is met. Gini impurity and entropy are alternative measures of node impurity, while information gain is the method used to evaluate the effectiveness of a split. Gini impurity tends to be faster to compute and can be used for multi-class classification. Entropy and information gain, rooted in information theory, provide another perspective and are more widely used due to their flexibility and theoretical underpinnings [14].

Tree pruning is a critical step in decision tree construction aimed at reducing overfitting and improving the generalization ability of the model. It involves the deliberate removal or simplification of certain parts of the tree to prevent excessive complexity and enhance its predictive performance. Several pruning techniques are employed to achieve this goal:

- **Cost-complexity pruning:** Also known as minimal cost-complexity pruning or weakest link pruning, is a widely used method to prevent overfitting in decision trees. This technique involves iteratively pruning the tree by assessing the impact of removing individual nodes or branches on the tree's overall complexity and performance. During cost-complexity pruning, a cost parameter, often denoted as $a\alpha$, is introduced to penalize the tree's complexity. The process involves evaluating the cost-complexity measure for different subtrees obtained by progressively pruning the least informative nodes, where the measure is given by:

$$R_{\alpha}(T) = R(T) + \alpha|T| \tag{8.19}$$

Here, *R*(*T*) represents the total misclassification rate or impurity of tree *T*, $|T|$ denotes the number of terminal nodes (leaves), and $\alpha\alpha$ controls the trade-off between model complexity and accuracy. The subtree with the smallest R_{α} value or the most favorable cost-complexity measure, considering various values of $\alpha\alpha$, is selected as the pruned tree.

- **Reduced error pruning:** It is an alternative technique used in decision trees, focusing on minimizing classification error or impurity in the tree nodes. This method involves evaluating the error rate or impurity measure on a validation dataset while iteratively removing nodes to find an optimal tree size. Reduced error pruning assesses the impact of removing individual nodes and their corresponding subtrees on the classification performance using a validation dataset. If the removal of a subtree does not deteriorate performance, the subtree is pruned. This process continues until further pruning does not enhance the model's predictive capability.
- **Pre-pruning and post-pruning: Pre-pruning** involves setting criteria to stop the growth of the tree during construction, such as limiting the maximum depth, minimum samples per leaf, or minimum samples required for a split. In contrast, **post-pruning** involves building a full tree and then applying pruning techniques to trim or simplify the tree structure.

Decision trees are renowned for their inherent interpretability owing to their intuitive structure, which mirrors human decision-making processes. The hierarchical nature of decision trees, resembling a flowchart or a series of if–else statements, offers a transparent and easily understandable representation of decision-making logic. Each node in the tree corresponds to a feature and presents a decision criterion based on which the data is partitioned, leading to subsequent nodes or leaves. This step-by-step branching structure allows practitioners to comprehend how the model arrives at its predictions or classifications, offering explicit insights into the decision-making process. Furthermore, decision trees enable the visualization of the entire decision-making path, allowing analysts to trace and interpret each node's contribution to the final prediction, fostering transparency and aiding in model explanation [15]. Visualizing decision trees provides an additional avenue to enhance interpretability. Techniques such as graphical representations, flowcharts, or tree diagrams offer visual insights into the tree's structure and decision rules. Visualizations facilitate a comprehensive understanding of the tree's hierarchy, splitting criteria, and the importance of various features in determining the outcomes. Feature importance can be visually depicted by showcasing the hierarchy of nodes and the number of instances that pass through each decision point. This visualization aids in identifying influential features and understanding their contributions to the model's predictions, thus assisting in model interpretation for both experts and non-experts. Moreover, visualization tools enable interactive exploration, allowing users to traverse the tree, inspect nodes, and comprehend the decision rules, enhancing the model's interpretability in various applications and domains.

```
from sklearn.datasets import load_iris
from sklearn.model_selection import train_test_split
from sklearn.tree import DecisionTreeClassifier, plot_tree
import matplotlib.pyplot as plt

# Load the Iris dataset (example dataset)
iris = load_iris()
X, y = iris.data, iris.target

# Split the data into training and testing sets
X_train, X_test, y_train, y_test = train_test_split(X, y, test_size=0.2,
random_state=42)

# Initialize the Decision Tree classifier
tree_classifier = DecisionTreeClassifier(random_state=42)

# Train the classifier
tree_classifier.fit(X_train, y_train)

# Visualize the Decision Tree
plt.figure(figsize=(10, 8))
plot_tree(tree_classifier, feature_names=iris.feature_names,
class_names=iris.target_names, filled=True)
plt.title('Decision Tree Visualization')
plt.show()
```

8.4.2 Decision rules

Decision rules are fundamental components derived from decision tree structures or learned directly from the data, representing a series of if-else conditions that guide the classification or prediction process. These rules encapsulate the decision logic employed by decision trees in a human-readable format, offering transparent and interpretable representations of the decision-making criteria. Each rule corresponds to a path within the decision tree, starting from the root node and traversing through intermediate nodes until reaching a leaf node. At each node, a specific feature and a threshold (for continuous variables) or a category (for categorical variables) are considered, establishing the conditions for data partitioning. As a result, decision rules are formulated based on the splitting criteria determined by these nodes, reflecting the criteria used to classify or predict instances in the tree structure. The formulation of decision rules involves a sequence of conditions resembling if-else statements, where each condition represents a decision node in the tree. These rules delineate the criteria for assigning instances to particular classes or categories based on the values of specific features. By following these if–else conditions sequentially, data instances can be assigned to respective classes or outcomes, offering a clear pathway to understand and interpret how the decision-making process unfolds within the model. Decision rules provide a simplified representation of the decision tree's logic, facilitating interpretability and offering insights into the model's decision boundaries. Additionally, these rules can be extracted from decision trees and from other machine-learning models, providing valuable explanations of the models' predictions or classifications in various domains and applications [16].

Rule-based systems leverage decision rules as fundamental components to perform classification and aid in decision-making processes across diverse domains. These systems rely on a set of if-else conditions or logical statements derived from decision tree structures, expert knowledge, or machine-learning algorithms. Decision rules extracted from decision trees or learned directly from the data are utilized within rule-based systems to categorize instances or guide actions based on specific conditions. The rules encompass the logic underlying the decision-making process, where each rule represents a distinct pathway or criterion that determines the assignment of instances to particular classes or outcomes. Rule-based systems play a crucial role in various fields such as medicine, finance, robotics, and natural language processing, utilizing decision rules to automate decision-making, diagnose diseases, predict trends, or control systems in a transparent and interpretable manner.

The architecture of rule-based systems typically involves a knowledge base containing a collection of decision rules, a rule interpreter or reasoning engine that processes these rules, and an inference mechanism for decision-making. When presented with new instances or data, the rule interpreter evaluates the conditions specified by the decision rules and determines the appropriate action or classification. The modularity and transparency of rule-based systems facilitate human understanding and enable domain experts to incorporate their expertise into the system by defining or refining decision rules. Moreover, these systems often offer flexibility for rule modification, allowing adaptations to changing environments or evolving datasets. By employing decision rules as a foundation, rule-based systems serve as powerful tools for automated reasoning, decision support, and classification tasks, contributing to advancements in various industries and scientific domains [17]. Techniques for extracting decision rules from decision trees play a pivotal role in facilitating human-readable explanations of a model's behavior, enhancing interpretability, and providing transparent insights into the decision-making process. One common approach to extract decision rules involves traversing the decision tree structure and capturing the conditions at each node, creating a set of if–else statements that represent the logic embedded within the tree. Starting from the root node and following each branch to the leaf nodes, these if–else conditions delineate the criteria for classifying or predicting instances based on specific feature thresholds or categories. By systematically capturing these conditions, decision rules are formed, presenting a sequence of logical statements that guide the model's predictions. This extraction process offers a clear and structured representation of the decision-making pathways employed by the model, enabling human-readable explanations of how input features lead to particular outcomes or classifications.

Another technique for extracting decision rules involves pruning the decision tree to its optimal size or simplifying its structure while preserving predictive accuracy. Post-pruning methodologies, such as reduced error pruning or cost-complexity pruning, aim to trim the tree by removing less informative branches or nodes. As a result of this simplification, decision rules are generated, capturing the essential decision criteria without compromising the model's predictive performance. These extracted rules form a compact and interpretable representation of the original decision tree, providing concise explanations of the model's behavior and aiding in understanding the relationship between input features and predicted outcomes. By employing these extraction techniques, decision trees can offer human-understandable explanations, ensuring transparency and interpretability, particularly in domains where comprehensible model explanations are crucial for informed decision-making.

Utilizing decision rules as a standalone model presents distinct differences, advantages, and limitations compared to employing decision trees. Decision rules, derived from decision tree structures or machine-learning algorithms, offer a simplified representation of the decision-making process, providing human-readable explanations in a straightforward if–else format. These rules present a clear and concise set of conditions that guide predictions or classifications, offering interpretability and transparency in understanding the model's behavior. One notable advantage of decision rules lies in their simplicity and ease of comprehension, making them highly interpretable for both domain experts and non-experts. Additionally, decision rules allow for greater modifiability and interpretability, enabling experts to adjust or refine rules based on evolving knowledge or changing requirements. However, the use of decision rules alone may result in oversimplification, limiting their ability to capture complex relationships present in the data, especially in cases where interactions between features are essential for accurate predictions. In contrast, decision trees encompass decision rules within a hierarchical and branching structure, allowing for more nuanced and multifaceted modeling of interactions among features. Decision trees excel in capturing complex decision boundaries and interactions between predictors, offering a comprehensive representation of the decision-making process. One significant advantage of decision trees lies in their ability to handle both categorical and continuous variables while automatically determining feature importance during tree construction. However, decision trees may tend toward overfitting, particularly when growing excessively complex trees that perfectly fit the training data but fail to generalize well to unseen data. Moreover, decision trees might lack the level of interpretability provided by decision rules, as the hierarchical structure might be more challenging to decipher, especially in larger and deeper trees. Balancing between interpretability and predictive power is crucial when choosing between decision rules and decision trees, considering the trade-offs between simplicity and capturing intricate relationships in the data. Detailed comparison between decision trees and decision rules is given in Table 8.1.

Table 8.1 Comparison between decision rules and decision trees

Aspect	Decision rules	Decision trees
Structure	Simplified, comprised of if–else conditions	Hierarchical, tree-like structure
Representation	Condensed, concise representation	Detailed, comprehensive structure
Interpretability	Highly interpretable, human-readable	Moderately interpretable, requires visualization
Complexity	Lower complexity, easier to comprehend	Higher complexity, may require pruning
Explainability	Offers clear explanations of individual rules	Offers insights via visualization of tree
Flexibility	May lacks complexity, less adaptable	More adaptable to capture complex relationships
Predictive accuracy	Potential loss of predictive power	Generally robust predictive performance
Overfitting	Less prone to overfitting	Prone to overfitting, requires pruning
Handling noise	May struggle with noise or outlier sensitivity	Robustness to noise, can handle outliers

8.5 K-nearest neighbors

K-nearest neighbors (KNN) and related algorithms, known for their simplicity and intuitive nature, also contribute to the taxonomy of Intrinsically XAI. KNN determines the classification of a data point by analyzing the majority class of its nearest neighbors. While inherently less structured than other models, KNN's simplicity allows for easy interpretation and explanation of predictions, particularly in scenarios where identifying similar instances or patterns is vital for IoT security analysis [18].

KNN is a non-parametric and intuitive machine-learning algorithm used for classification and regression tasks. It operates on the principle of similarity, where an instance is classified by a majority vote of its neighbors. KNN assumes that similar instances are located close to each other in the feature space, making predictions based on the proximity of the query instance to its nearest neighbors in the training dataset. The algorithm assigns the class label or predicts the value for a new data point based on the classes or values of its KNN, determined by a predefined distance metric [19]. The key components of the KNN algorithm involve two primary parameters: "*k*," representing the number of neighbors used to make predictions, and the distance metric, typically Euclidean distance, to measure the proximity between instances in the feature space. The selection of "*k*" influences the algorithm's bias-variance trade-off, where smaller "*k*" values result in more flexible models prone to noise, while larger "*k*" values lead to smoother decision boundaries but may miss local patterns. The distance metric determines how the algorithm calculates distances between data points, influencing the neighborhood definition. Additionally, the training dataset, comprising feature vectors and corresponding labels or target values, forms the foundation for the algorithm to identify the nearest neighbors during the prediction phase based on similarity measures in the feature space.

The KNN algorithm involves several steps for classification. Here is a detailed explanation of each step:

Step 1. Select the *k* value: The process begins by selecting the value of "*k*," the number of neighbors to consider when making predictions. The choice of "*k*" influences the model's performance and generalization. A smaller "*k*" may result in more flexible but noisy predictions, while a larger "*k*" could lead to smoother decision boundaries but may miss local patterns.
Step 2. Calculate the distance between the point and known category data: For a new data point seeking classification, the algorithm calculates the distance (typically Euclidean distance) between the new point and each point in the training dataset, representing known categories. This step establishes the proximity or similarity between the new instance and the existing data points.
Step 3. Sort distances in increasing order: The calculated distances are sorted in ascending order, placing the nearest neighbors at the beginning of the sorted list. This sorting facilitates the selection of the "*k*" nearest neighbors for the prediction process.
Step 4. Select *k* points with the smallest distance: From the sorted list, the algorithm selects the "*k*" points with the smallest distances to the new data point. These "*k*" nearest neighbors are crucial in determining the predicted class or label for the new instance.
Step 5. Determine the frequency of occurrence of categories: Among the "*k*" selected nearest neighbors, the algorithm counts the frequency of occurrence for each category or class they belong to. This step assesses which category is more prevalent among the "*k*" neighbors.
Step 6. Return the predicted classification: Finally, based on the majority vote or the highest frequency of occurrence among the categories of the "*k*" nearest neighbors, the algorithm assigns the predicted classification or label to the new data point. The category with the highest occurrence among the "*k*" neighbors becomes the predicted class for the new instance.

The KNN algorithm possesses inherent interpretability, making it a valuable tool for comprehensible and transparent decision-making. One primary reason for its interpretability lies in its simplicity and intuitive nature. KNN operates on the principle of similarity, where predictions are determined based on the classes of neighboring data points. This straightforward concept enables users to grasp the rationale behind predictions: an instance is classified by considering the majority vote of its nearest neighbors in the feature space. This simplicity makes KNN easily understandable, even for individuals without extensive machine learning expertise [20]. Additionally, KNN's interpretability stems from its explicit representation of decision boundaries. Unlike complex models with intricate decision boundaries, KNN's decision boundaries are delineated by the clustering of data points. In visualizations, decision boundaries are often smooth and correspond to natural clusters in the data, which align with human perception. This characteristic allows analysts to visually inspect and comprehend the logic behind predictions. Additionally, since KNN does not assume any underlying functional form, its predictions are directly tied to the data points themselves, providing direct insight into the features that influence a prediction. By observing which features contribute to nearby points in the feature space, users can infer the importance of these features in the decision-making process. Moreover, KNN's transparency lies in its ability to provide explanations for individual predictions. For a given instance, the "*k*" nearest neighbors that influence the prediction can be identified and examined [20]. This allows analysts to inspect specific instances and understand why a certain prediction was made, providing a clear rationale for the model's decision. This property of KNN is highly beneficial, especially in critical domains like healthcare or finance, where interpretability and justification of decisions are paramount.

```
from sklearn.datasets import load_iris
from sklearn.model_selection import train_test_split
from sklearn.neighbors import KNeighborsClassifier
import matplotlib.pyplot as plt
from matplotlib.colors import ListedColormap

# Load the Iris dataset (example dataset)
iris = load_iris()
X, y = iris.data[:, :2], iris.target  # Considering only the first two
features for visualization

# Split the data into training and testing sets
X_train, X_test, y_train, y_test = train_test_split(X, y, test_size=0.2,
random_state=42)

# Initialize the K-Nearest Neighbors classifier
knn_classifier = KNeighborsClassifier(n_neighbors=5)

# Train the classifier
knn_classifier.fit(X_train, y_train)

# Visualize the decision boundaries
cmap_light = ListedColormap(['#FFAAAA', '#AAFFAA', '#AAAAFF'])
cmap_bold = ListedColormap(['#FF0000', '#00FF00', '#0000FF'])

# Plot the decision surface
x_min, x_max = X[:, 0].min() - 1, X[:, 0].max() + 1
y_min, y_max = X[:, 1].min() - 1, X[:, 1].max() + 1
xx, yy = np.meshgrid(np.arange(x_min, x_max, 0.02), np.arange(y_min,
y_max, 0.02))
Z = knn_classifier.predict(np.c_[xx.ravel(), yy.ravel()])
Z = Z.reshape(xx.shape)

plt.figure(figsize=(8, 6))
plt.pcolormesh(xx, yy, Z, cmap=cmap_light)
```

```
# Plot the training points
plt.scatter(X_train[:, 0], X_train[:, 1], c=y_train, cmap=cmap_bold,
edgecolor='k', s=20, label='Train')
# Plot the testing points
plt.scatter(X_test[:, 0], X_test[:, 1], c=y_test, cmap=cmap_bold,
edgecolor='k', s=50, marker='x', label='Test')
plt.xlim(xx.min(), xx.max())
plt.ylim(yy.min(), yy.max())
plt.title("K-Nearest Neighbors (KNN) Visualization")
plt.xlabel('Feature 1')
plt.ylabel('Feature 2')
plt.legend()
plt.show()
```

Overall, KNN's simplicity, explicit decision boundaries, and ability to provide individual-level explanations contribute significantly to its inherent interpretability, making it a valuable and accessible algorithm for understanding and explaining predictions in various applications. Table 8.2 summarizes the key properties that contribute to the interpretability of the KNN algorithm.

Table 8.2 Summary of interpretability-enhancing properties in KNN algorithm with associated benefits

Properties	Description	Benefits
Intuitive concept	Operates on the principle of similarity, where predictions are based on the majority vote of neighbors	Easily understood by non-experts
Explicit decision boundaries	Decision boundaries align with clusters in data, visually clear and understandable	Offers visual interpretability
Feature importance clarity	Directly linked to the features of nearby instances, indicating their influence on predictions	Identifies influential features
Individual-level explanations	Identifies and provides insights into the specific neighbors influencing each prediction	Enables explanation for each instance·
Visual interpretation	Offers visual clarity in decision boundaries, aiding in understanding the model's behavior	Facilitates insights from data plots
Simplicity	Simple methodology without complex mathematical functions, enhancing ease of comprehension	Reduces complexity for users
Human-readable explanations	Provides explanations in a way that is understandable and transparent to non-experts	Enhances model transparency
Sensitivity to local structure	Adapts well to local data patterns and captures intricate local structures	Suitable for nonlinear relation-ships
Distance metric flexibility	Accommodates various distance metrics, allowing customization based on data characteristics	Adaptable to different data types

References

[1] Kök, I., Okay, F. Y., Muyanlı, Ö., and Özdemir, S. (2023). Explainable artificial intelligence (XAI) for internet of things: a survey. *IEEE Internet of Things Journal*, *10*(16), 14764–14779.

[2] Masud, M. T., Keshk, M., Moustafa, N., Linkov, I., and Emge, D. K. (2024). Explainable Artificial Intelligence for resilient security applications in the Internet of Things. *IEEE Open Journal of the Communications Society*, 99, 1. DOI:10.1109/OJCOMS.2024.3413790.

[3] Mi, J. X., Jiang, X., Luo, L., and Gao, Y. (2024). Toward explainable artificial intelligence: A survey and overview on their intrinsic properties. *Neurocomputing*, *563*, 126919.

[4] Ai, Q., and Narayanan. R, L. (2021, October). Model-agnostic vs. model-intrinsic interpretability for explainable product search. In *Proceedings of the 30th ACM International Conference on Information & Knowledge Management* (pp. 5–15).

[5] Ilic, I., Görgülü, B., Cevik, M., and Baydoğan, M. G. (2021). Explainable boosted linear regression for time series forecasting. *Pattern Recognition*, *120*, 108144.

[6] Aras, S., and Lisboa, P. J. (2022). Explainable inflation forecasts by machine learning models. *Expert Systems with Applications*, *207*, 117982.

[7] Witten, D., and Tibshirani, R. A comparison of fold-change and the t-statistic for microarray data analysis. *Analysis 1776* (2007): 58–85.

[8] Norton, E. C., and Dowd, B. E. (2018). Log odds and the interpretation of logit models. *Health Services Research*, *53*(2), 859–878.

[9] Hastie, T. J., and Pregibon, D. (2017). Generalized linear models. *In Statistical Models in S* (pp. 195–247). Routledge.

[10] Lauritzen, S. L. (2003). Some modern applications of graphical models. *Oxford Statistical Science Series*, 13–32.

[11] Drury, B., Valverde-Rebaza, J., Moura, M. F., and de Andrade Lopes, A. (2017). A survey of the applications of Bayesian networks in agriculture. *Engineering Applications of Artificial Intelligence*, *65*, 29–42.

[12] Scanagatta, M., Salmerón, A., and Stella, F. (2019). A survey on Bayesian network structure learning from data. *Progress in Artificial Intelligence*, *8*(4), 425–439.

[13] Fletcher, S., and Islam, M. Z. (2019). Decision tree classification with differential privacy: A survey. *ACM Computing Surveys (CSUR)*, *52*(4), 1–33.

[14] Kotsiantis, S. B. (2013). Decision trees: A recent overview. *Artificial Intelligence Review*, *39*, 261–283.

[15] Fawcett, T. W., Fallenstein, B., Higginson, A. D., *et al.* (2014). The evolution of decision rules in complex environments. *Trends in Cognitive Sciences*, *18*(3), 153–161.

[16] Apté, C., and Weiss, S. (1997). Data mining with decision trees and decision rules. *Future Generation Computer Systems*, *13*(2-3), 197–210.

[17] Samet, H. (2007). K-nearest neighbor finding using MaxNearestDist. *IEEE Transactions on Pattern Analysis and Machine Intelligence*, *30*(2), 243–252.

[18] Huai, M., Miao, C., Li, Y., Suo, Q., Su, L., and Zhang, A. (2020). Learning distance metrics from probabilistic information. *ACM Transactions on Knowledge Discovery from Data (TKDD)*, *14*(5), 1–33.

[19] Laaksonen, J., and Oja, E. (1996, June). Classification with learning k-nearest neighbors. In *Proceedings of International Conference on Neural Networks (ICNN'96)* (Vol. 3, pp. 1480-1483). IEEE.

[20] Mucherino, A., Papajorgji, P. J., Pardalos, P. M., Mucherino, A., Papajorgji, P. J., and Pardalos, P. M. (2009). K-nearest neighbor classification. *Data Mining in Agriculture*, 83–106.

Chapter 9
Model-agnostic methods for globally interpretable machine learning

9.1 Introduction

In Internet-of-Things (IoT) systems, deploying sophisticated artificial intelligence (AI) algorithms has significantly enhanced functionality but concurrently introduced complex black-box models, raising concerns about their interpretability and trustworthiness. The importance of globally explainable AI methodologies within IoT security cannot be overstated. With the increasing interconnectivity of devices and the sensitive nature of data exchanged, ensuring transparency and interpretability in AI models becomes a pivotal aspect of fortifying IoT systems against vulnerabilities and adversarial attacks. Embracing model-agnostic techniques offers a profound solution, allowing stakeholders and users to comprehend how these AI systems arrive at decisions, fostering trust and enabling efficient mitigation of potential risks inherent in interconnected IoT environments. Moreover, the implications of globally explainable AI stretch far beyond mere interpretability. These methods serve as crucial tools to decipher and scrutinize the intricate interactions between variables and model outputs, elucidating potential biases, vulnerabilities, or unexpected behaviors that could compromise the security and reliability of IoT systems. By harnessing model-agnostic approaches like partial dependence plots (PDPs), individual conditional expectation (ICE), and feature interaction analyses, among others, stakeholders gain insights into the inner workings of AI models, enabling proactive identification and rectification of vulnerabilities. Consequently, this fosters robustness, resilience, and accountability within IoT ecosystems, bolstering confidence among users, regulators, and industry players as they navigate the evolving landscape of interconnected devices and data privacy concerns.

Ensuring the security and transparency of IoT applications stands as a paramount concern in an increasingly interconnected world. Model-agnostic methods serve as a cornerstone in achieving these critical objectives by providing a pathway toward comprehensibility and trust in the functioning of AI models within IoT ecosystems. Unlike model-specific techniques that cater to individual algorithms or systems, model-agnostic approaches offer a versatile toolkit that transcends specific model architectures. This versatility enables stakeholders to decipher the decision-making processes of various AI models, regardless of their complexity or diversity. In the context of IoT applications, where diverse devices, protocols, and data formats coalesce, the adoption of model-agnostic methods emerges as a unifying strategy to ensure transparency and interpretability across this heterogeneous landscape. Furthermore, the applicability of model-agnostic techniques extends beyond mere transparency; it plays a pivotal role in fortifying the security posture of IoT applications. The ability to comprehend and scrutinize the inner workings of AI models becomes instrumental in identifying potential security vulnerabilities, biases, or anomalous behaviors that might jeopardize the integrity and confidentiality of IoT systems. By embracing model-agnostic methods such as PDPs, permutation feature importance, or global surrogate models, stakeholders can delve into the underlying complexities of AI models, unveiling critical insights into potential security loopholes. This proactive approach empowers developers, administrators, and regulatory bodies to address vulnerabilities before they are exploited, thereby safeguarding sensitive data, enhancing resilience, and fostering trust in IoT applications [1].

9.2 PDP

This section delves into an in-depth exploration of PDPs, elucidating their theoretical underpinnings, practical applications, and relevance within the context of fortifying the security posture of interconnected IoT environments. PDPs serve as indispensable visual tools in elucidating the intricate associations between individual features and the target variable within machine-learning models. Fundamentally, a PDP illustrates the marginal effect of a specific feature on the model's predictions while keeping other features fixed at certain values. In the context of securing IoT systems, PDPs offer a methodical depiction of how alterations in a particular feature influence the model's output, thereby enabling stakeholders to comprehend the significance and impact of each feature on the overall predictive behavior. This visualization technique operates by systematically varying the values of the chosen feature across its entire range while maintaining other features' values constant at their average or selected levels. Consequently, the resulting plot showcases the average prediction outcome across these varied feature values, providing a clear and interpretable representation of the feature's influence on the target variable within the AI model deployed in the IoT security domain. Moreover, PDPs facilitate a comprehensive understanding of non-linear relationships between features and the target variable, transcending the limitations of simple correlation measures. By presenting a graphical depiction of how changes in a single feature affect the model's predictions, PDPs unveil insights into the complex interactions and dependencies within AI algorithms used in IoT security applications. This comprehensive visualization aids in uncovering potential complexities that might not be immediately discernible through conventional statistical analyses, allowing stakeholders to discern nuances in feature importance, identify potential biases, and unveil unexpected behaviors within the AI model. As a result, PDPs stand as a pivotal asset in the arsenal of model-agnostic methods, enabling a deeper comprehension of the relationship between features and the target variable, crucial for fortifying the transparency and security of IoT systems [2].

In the realm of securing IoT systems, the application of PDPs manifests as a pivotal strategy in deciphering and fortifying the security infrastructure of interconnected devices. One notable instance where PDPs significantly contribute to enhancing IoT security lies in anomaly detection within sensor networks. By employing PDPs to analyze the behavior of individual sensors or sensor clusters, security analysts gain valuable insights into how alterations in sensor readings impact the predictive outcomes of anomaly detection models. For instance, in a smart home environment, PDPs enable the visualization of how deviations in temperature, humidity, or motion sensor data influence the model's identification of anomalous activities. Such insights aid in fine-tuning anomaly detection algorithms, ensuring robust and accurate identification of potential security breaches or abnormal behaviors within the IoT ecosystem. Furthermore, PDPs find application in bolstering the security of IoT devices through intrusion detection systems. By employing these plots to scrutinize the relationship between network traffic patterns and the likelihood of intrusion, security practitioners gain a comprehensive understanding of how variations in network features affect the predictive capabilities of intrusion detection models. For instance, in a smart city infrastructure, PDPs can illustrate the impact of different network traffic characteristics on the model's identification of suspicious activities or potential cyber-attacks [3]. Through these visualizations, security analysts can identify critical network features that significantly influence intrusion detection, allowing for targeted optimizations and robust defenses against security threats within the intricate web of interconnected IoT devices and systems (Figure 9.1).

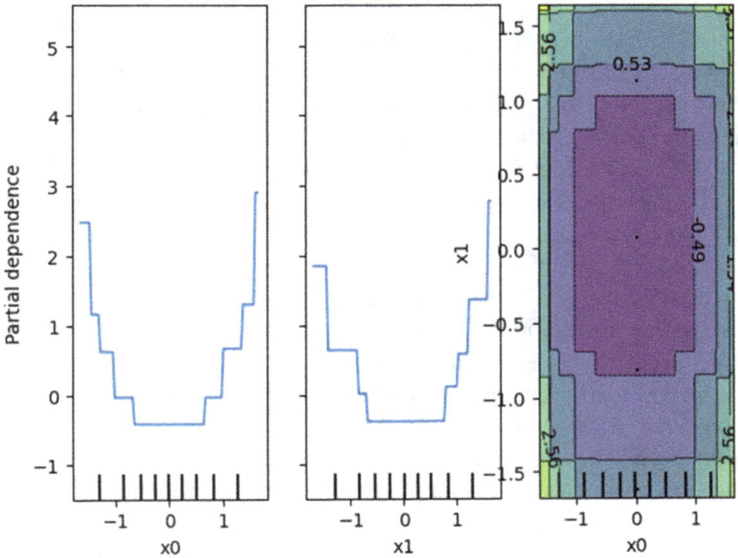

Figure 9.1 Global ALE plot for a global explanation of predictions from GradientBoostingClassifier

```
from sklearn.datasets import make_hastie_10_2
from sklearn.ensemble import GradientBoostingClassifier
from sklearn.inspection import PartialDependenceDisplay

X, y = make_hastie_10_2(random_state=0)
clf = GradientBoostingClassifier(n_estimators=100, learning_rate=1.0,
    max_depth=1, random_state=0).fit(X, y)
features = [0, 1, (0, 1)]
PartialDependenceDisplay.from_estimator(clf, X, features)
```

9.3 ICE plots

This section embarks on a comprehensive exploration of ICE plots, delineating their theoretical foundations, practical applications, and relevance within the context of fortifying the security posture of interconnected IoT environments. ICE plots serve as a powerful elucidative tool within the domain of interpretable artificial intelligence, offering a nuanced perspective on how individual instances or data points respond to changes in a specific feature within machine-learning models. Unlike aggregate visualization methods such as PDPs, ICE plots unravel the intricate relationships between a single feature and predictions by presenting a series of curves, each representing the model's predictions for different instances as the values of the chosen feature are varied. In the context of securing IoT systems, ICE plots facilitate a deeper understanding of the impact of diverse IoT-related features, such as sensor readings or network behaviors, on the model's predictions, enabling stakeholders to discern the varied responses of individual instances to changes in these features and uncover potential vulnerabilities or unexpected behaviors within the AI models deployed in IoT security applications [4]. Moreover, ICE plots offer a granularity that transcends the limitations of global feature importance measures, providing a fine-grained analysis of feature impacts on individual predictions. This granularity becomes particularly crucial in comprehending the intricate dynamics present within IoT systems, where diverse and heterogeneous data sources contribute to complex decision-making processes. For instance, within a smart healthcare IoT environment, ICE plots enable a detailed exploration of how variations in patient vital signs or health parameters affect the model's predictions for individual patients, allowing healthcare practitioners to discern specific patterns or anomalies that might elude traditional aggregate analyses. By zooming into individual instance-level predictions, ICE plots empower stakeholders to detect subtle variations, biases, or potential security threats within IoT systems, thereby bolstering the transparency, interpretability, and ultimately, the security resilience of interconnected devices and applications.

In securing IoT devices and networks, the application of ICE plots emerges as a pivotal strategy in comprehending and fortifying the security fabric of interconnected systems. One prominent instance where ICE plots significantly contribute to enhancing IoT security lies in anomaly detection across heterogeneous sensor networks. By employing ICE plots to scrutinize individual sensor readings and their impact on anomaly detection models, security analysts gain a nuanced understanding of how deviations in specific sensor data influence the model's predictions for anomaly occurrences [3–5]. For example, in an industrial IoT setting, ICE plots enable the visualization of how variations in temperature, pressure, or vibration sensor readings affect the identification of anomalous behaviors, thereby aiding in fine-tuning anomaly detection algorithms. Such insights facilitate the precise detection of potential security breaches or irregularities within IoT networks, enhancing the overall security posture. Furthermore, ICE plots play a crucial role in strengthening IoT device security through intrusion detection systems. By leveraging ICE plots to investigate the relationship between diverse network traffic features and the probability of intrusion, security practitioners attain a granular understanding of how alterations in specific network characteristics impact the predictive capabilities of intrusion detection models. For instance, in smart city deployments, ICE plots allow for the visualization of how variations in network traffic patterns affect the model's identification of potential cyber-attacks or malicious activities. This level of scrutiny aids in identifying critical network features that significantly influence intrusion detection, enabling targeted enhancements and robust defenses against security threats within the intricate web of interconnected IoT devices and networks [5].

For a global view, we can create ICE plots for multiple instances and overlay them to see general trends across the dataset. By plotting ICE curves for several instances, you can visualize the average behavior of the model across the feature space. In the following, we provide a Python example on visualization of local ICE explanation.

```
import pandas as pd
import numpy as np
import matplotlib.pyplot as plt
from sklearn.ensemble import GradientBoostingRegressor
import scipy as sp

# Generating the random data
N = 1000
df = pd.DataFrame(sp.stats.uniform.rvs(-1, 2, size=(N, 3)), columns=['x1',
'x2', 'x3'])
noise = sp.stats.norm.rvs(size=N)
y = 0.2 * df.x1 - 5 * df.x2 + 10 * df.x2 * (df.x3 >= 0) + noise
gbm = GradientBoostingRegressor()
gbm.fit(df.values, y)

# Global ICE plot function
def plot_global_ICE(feature, data, model, num_instances=10):
    plt.figure(figsize=(8, 6))
    feature_values = np.linspace(data[feature].min(), data[feature].max(),
100)

    for i in range(num_instances):
        instance_idx = np.random.randint(0, len(data))
        instance = data.iloc[instance_idx]
```

```
        predictions = []
        for value in feature_values:
            modified_instance = instance.copy()
            modified_instance[feature] = value
            prediction = model.predict(modified_instance.values.reshape(1,
-1))
            predictions.append(prediction[0])

        plt.plot(feature_values, predictions, label=f'Instance
{instance_idx}')

    plt.xlabel(feature)
    plt.ylabel('Prediction')
    plt.title(f'Global Individual Conditional Expectation (ICE) Plot for
{num_instances} instances')
    plt.legend()
    plt.show()


# Testing global ICE plot
plot_global_ICE('x2', df, gbm, num_instances=5)
```

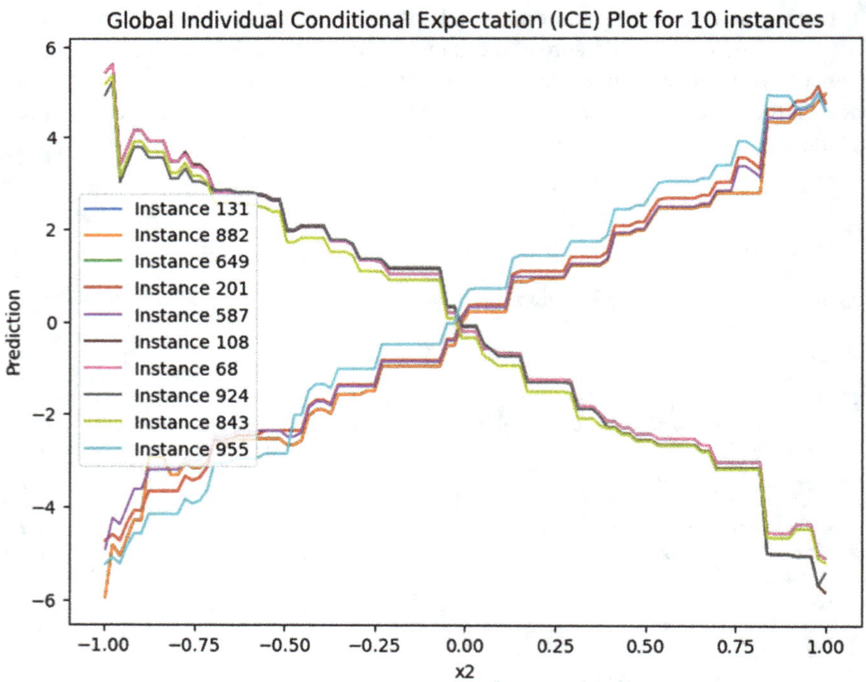

Figure 9.2 Global ICE Plot for "x2" feature across 10 instances

In Figure 9.2, the model's predictive responses are depicted across the "x2" feature space, showcasing this relationship across multiple instances (five instances selected randomly). This plot offers an overview of how the model behaves concerning "x2" for different instances, revealing common trends or variations in the model's predictions as "x2" values fluctuate across the dataset. By overlaying ICE curves for various instances, it provides insights into the broader behavior of the model concerning "x2" and highlights any consistent patterns or differences in predictions across these instances.

9.4 Automated dependence plots

Automated dependence plots (ADPs) represent a significant advancement in the realm of interpretable artificial intelligence, offering an automated approach to unraveling complex dependencies between features and predictions within machine-learning models. This section explores the ADPs for global explanation, elucidating their underlying methodology, practical purposes, and significance within the context of fortifying the security posture of interconnected IoT environments [6].

ADPs represent a sophisticated methodology in interpretable artificial intelligence, offering a systematic and automated approach to visualizing complex feature interactions within machine-learning models. Unlike traditional manual methods that struggle to handle multifaceted feature dependencies, ADPs provide a structured and systematic way to explore the interplay between multiple features and their collective influence on model predictions. By leveraging ADPs, stakeholders gain a comprehensive understanding of how various features interact with each other, allowing for the visualization of intricate relationships and dependencies present within the model. These plots systematically showcase how changes in one feature, in conjunction with alterations in other related features, collectively affect the model's predictions. In securing IoT systems, where diverse sensor data, network behaviors, and environmental variables converge, ADPs offer a robust framework to systematically unveil the interdependencies among these features, enabling stakeholders to decipher complex patterns and potential vulnerabilities that might otherwise remain concealed. Moreover, ADPs facilitate the identification of non-linear and higher-order feature interactions, enabling a holistic view of the complex relationships embedded within machine-learning models [7]. By systematically visualizing feature interactions across various dimensions, ADPs empower stakeholders to discern nuanced behaviors and unexpected correlations that might significantly impact the model's predictions in IoT security applications. For instance, within smart home environments, ADPs enable the visualization of how interactions between multiple devices' data, such as temperature sensors, motion detectors, and energy consumption metrics, collectively influence the model's predictions of security breaches or anomalous activities. This systematic exploration aids in uncovering critical feature interactions, fostering a deeper comprehension of IoT system dynamics and contributing to fortifying the security infrastructure against potential threats and vulnerabilities [8].

In securing IoT environments, ADPs serve as instrumental tools in fortifying the security posture against potential vulnerabilities and threats. One significant instance where ADPs significantly contribute to enhancing IoT security is in anomaly detection across heterogeneous sensor networks. By leveraging ADPs to analyze complex feature interactions across multiple sensors, security analysts gain insights into how combinations of sensor readings collectively impact anomaly detection models. For instance, within an industrial IoT setting, ADPs enable the visualization of intricate relationships between temperature, pressure, and vibration sensor data, offering a systematic exploration of how changes in these interdependent features collectively influence the detection of abnormal behaviors or potential security breaches. This systematic exploration aids in fine-tuning anomaly detection algorithms, enabling more accurate and robust identification of irregularities within the IoT network, ultimately strengthening its overall security infrastructure [9]. Furthermore, ADPs play a pivotal role in bolstering IoT device security through intrusion detection systems. By utilizing ADPs to dissect complex feature interactions within network traffic patterns, security practitioners attain a comprehensive understanding of how variations in multiple network characteristics collectively impact the predictive capabilities of intrusion detection models. For instance, within smart city deployments, ADPs facilitate the visualization of intricate relationships among diverse network parameters, aiding in the identification of critical feature interactions influencing the model's predictions of potential cyber-attacks or malicious activities. This deeper insight enables stakeholders to develop targeted strategies and deploy robust defenses against security threats within interconnected IoT devices and networks, significantly enhancing the security resilience of the IoT ecosystem [10].

9.5 Accumulated local effect plots

Accumulated local effect (ALE) plots represent a pivotal advancement in the domain of interpretable artificial intelligence, offering a nuanced approach to comprehend the cumulative effects of features on machine-learning model predictions. Within the context of securing IoT systems, where transparency and interpretability are indispensable for ensuring robustness, the adoption of ALE plots emerges as a crucial strategy.

ALE plots stand as a sophisticated methodology within interpretable artificial intelligence, providing a comprehensive depiction of the cumulative impact of individual features on machine-learning model predictions. Unlike traditional methods that solely focus on isolating individual feature effects, ALE plots offer a nuanced approach by capturing the cumulative effects across the entire range of a specific feature. ALE plots achieve this by computing the average change in model predictions as the values of the chosen feature vary. By systematically dissecting the feature's values into smaller intervals, ALE plots depict how alterations in the feature within each interval contribute to the changes in model predictions. The cumulative effects are then visualized by summing up the average changes across intervals, presenting a holistic view of how the feature collectively influences the model's predictions [11]. In the domain of securing IoT systems, where understanding the collective impact of diverse features is critical, ALE plots provide stakeholders with a nuanced understanding of how changes in individual features, such as sensor readings or network behaviors, cumulatively affect the model's predictions, thereby unraveling complex relationships and potential security vulnerabilities within the AI models deployed in IoT security applications. Moreover, ALE plots facilitate the identification of non-linear and heterogeneous feature effects, enabling a more comprehensive assessment of feature impact on predictions. By capturing the cumulative effects of features on model predictions across their respective ranges, ALE plots overcome the limitations of linear assumptions, allowing stakeholders to discern nuanced and non-monotonic relationships between features and predictions. This level of granularity becomes particularly crucial in understanding the intricacies present within IoT systems, where diverse data sources and complex interactions influence decision-making processes [12]. For example, within a smart city IoT infrastructure, ALE plots enable the visualization of how cumulative effects of various environmental and traffic-related features collectively influence predictions regarding traffic congestion or environmental hazards. This in-depth understanding aids in uncovering critical feature dependencies, contributing to fortifying the security infrastructure against potential threats and vulnerabilities within interconnected IoT environments.

Within the realm of securing IoT systems, the application of ALE plots presents a robust methodology to fortify the security infrastructure against potential vulnerabilities and threats. One significant instance where ALE plots significantly contribute to enhancing IoT security lies in anomaly detection across diverse sensor networks. By leveraging ALE plots to analyze cumulative effects of multiple sensor readings on anomaly detection models, security analysts gain insights into how collective changes in sensor data influence the model's predictions for detecting anomalies or security breaches. For instance, within industrial IoT setups, ALE plots enable a systematic exploration of how variations in temperature, pressure, or humidity sensor readings cumulatively affect the identification of abnormal behaviors, allowing for fine-tuning anomaly detection algorithms. This detailed scrutiny aids in improving the accuracy and robustness of anomaly detection within the IoT network, bolstering its security resilience. Additionally, ALE plots play a pivotal role in enhancing IoT device security through intrusion detection systems. By employing ALE plots to visualize cumulative effects of diverse network traffic features on intrusion detection model predictions, security practitioners attain a comprehensive understanding of how combined variations in multiple network characteristics impact the model's ability to detect potential cyber-attacks or intrusions [13]. For instance, within smart city deployments, ALE plots facilitate a nuanced examination of how cumulative effects of various network parameters influence predictions regarding cyber-threats or malicious activities. This deep understanding empowers stakeholders to identify critical feature interactions, aiding in the development of targeted strategies to strengthen defenses against security threats within interconnected IoT devices and networks, ultimately contributing to a more secure IoT ecosystem (Figure 9.3).

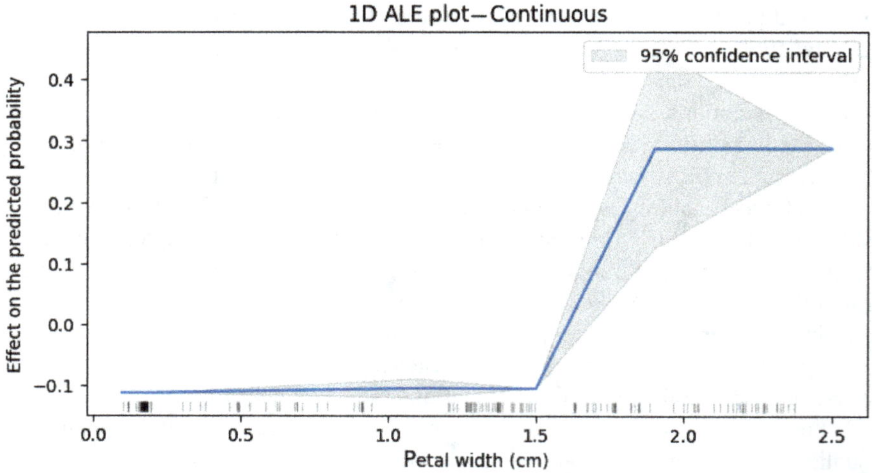

Figure 9.3 Global ALE plot for global explanation of predictions from RandomForestClassifier

```
import pandas as pd
import numpy as np
from sklearn.datasets import load_iris
from sklearn.ensemble import RandomForestClassifier
import matplotlib.pyplot as plt
from PyALE import ale

# Load the Iris dataset and prepare the data
iris = load_iris()
X = pd.DataFrame(iris.data, columns=iris.feature_names)
y = pd.Categorical.from_codes(iris.target, iris.target_names)

# Create a RandomForestClassifier
clf = RandomForestClassifier(n_estimators=5, random_state=873)
clf.fit(X, y)

# Predict probabilities for a subset of samples
subset_probabilities = clf.predict_proba(X)[110:120, :]

# Define a custom class to mimic model predictions for a single class
class clf_dummy():
    def predict(df):
        return(clf.predict_proba(df)[:, 2])

# Calculate the accumulated local effect (ALE) for the 'petal width (cm)'
feature
feat_eff = ale(X=X, model=clf_dummy, feature=['petal width (cm)'],
grid_size=5)

# Plot the ALE plot
plt.ylabel('Effect on the predicted probability')
plt.show()
```

9.6 Feature interaction

Feature interaction (H-statistic) serves as a fundamental measure within the domain of interpretable artificial intelligence, providing a quantitative assessment of the interactions between features in machine-learning models. In the context of securing IoT systems, where the complex interplay among diverse features influences decision-making processes, the adoption of H-statistic emerges as a pivotal approach. Unlike traditional methods that often struggle to capture higher-order and non-linear feature interactions, the H-statistic offers a robust and systematic means to quantify the degree of interaction between features, shedding light on the complexities inherent in AI models deployed within interconnected IoT environments [14].

The H-statistic represents a critical metric within interpretable artificial intelligence, designed to quantify the degree of interaction between features in machine-learning models. Unlike traditional methods that solely focus on assessing individual feature importance, the H-statistic delves into the intricate relationships among features by measuring their joint contributions to model predictions. This metric encapsulates the essence of higher-order and non-linear interactions between features, providing a comprehensive understanding of how multiple variables collectively impact the model's output. In the context of securing IoT systems, where diverse and interconnected features shape decision-making processes, the H-statistic offers a nuanced lens to scrutinize the complexities inherent in AI models. By quantifying the interactions among features, the H-statistic aids in identifying critical feature pairs or combinations that significantly influence model predictions, thereby facilitating a more comprehensive comprehension of the interplay among diverse IoT-related features, such as sensor data, network behaviors, or environmental parameters. The H-statistic measures feature interactions by evaluating deviations from the model's expected behavior when considering individual features against their joint effects [15]. It quantifies the discrepancy between the observed model output and the predicted output when accounting for interactions between specific features. This metric calculates the mutual information between features and the model predictions, effectively capturing the non-additive effects arising from interactions among multiple variables. In IoT security applications, the H-statistic acts as a powerful diagnostic tool, revealing the presence and magnitude of feature interactions that might not be discernible through conventional linear or additive models. By uncovering intricate relationships between features, this metric aids in identifying hidden dependencies and vulnerabilities within AI models deployed in IoT systems, ultimately contributing to fortifying the security infrastructure against potential threats and risks [16].

The H-statistic holds paramount significance in unraveling the complex relationships ingrained within IoT security frameworks. In the intricate landscape of securing interconnected IoT systems, understanding the interplay among diverse features, such as sensor readings, network behaviors, or environmental data, stands as a critical factor in fortifying the security posture. The H-statistic, with its ability to capture higher-order and non-linear interactions between features, serves as a beacon of insight into the intricate dynamics of AI models deployed in IoT security applications. By quantifying feature interactions, this metric offers stakeholders a deeper comprehension of hidden dependencies, anomalies, or vulnerabilities that might elude conventional analysis. Within IoT security frameworks, the H-statistic facilitates the identification of critical feature combinations or relationships that significantly influence model predictions, aiding in the proactive detection and mitigation of potential security threats or anomalous behaviors within the interconnected IoT ecosystem [1]. Moreover, the H-statistic's role extends beyond conventional linear assessments, providing a nuanced understanding of complex relationships that dictate decision-making processes within IoT security frameworks. By unveiling higher-order interactions and non-linear dependencies among features, this metric empowers security analysts to decipher intricate patterns and behaviors present within AI models. For instance, in smart home or industrial IoT settings, the H-statistic facilitates the identification of subtle yet impactful feature interactions that might signal potential security breaches or anomalous activities. This in-depth understanding enables stakeholders to proactively address vulnerabilities, strengthen defense mechanisms, and foster a more resilient security infrastructure within interconnected IoT environments. Thus, the H-statistic emerges as an indispensable tool, offering a sophisticated means to navigate the complexities of feature interactions within IoT security frameworks, ultimately contributing to a more robust and secure IoT ecosystem [4].

9.7 Functional decomposition

Functional decomposition stands as a pivotal methodology in the realm of interpretable artificial intelligence, offering a systematic approach to dissecting the complex behaviors of machine-learning models. Within the context of securing IoT systems, where transparent and interpretable AI models are imperative, the adoption of functional decomposition emerges as a critical strategy. Unlike traditional model inspection techniques that often struggle to comprehend complex model behaviors, functional decomposition disentangles the intricate relationships between input features and model predictions, presenting a structured breakdown of how individual features contribute to the overall model output. This section embarks on an extensive exploration of functional decomposition, elucidating its theoretical underpinnings, practical applications, and significance within the context of fortifying the security posture of interconnected IoT environments [5].

Functional decomposition stands as an indispensable technique within interpretable artificial intelligence, enabling the deconstruction of intricate model predictions into simpler and interpretable components. This methodological approach achieves a systematic breakdown of complex predictions by unraveling the contribution of individual input features to the model's output. Unlike conventional black-box models that obscure the decision-making process, functional decomposition dissects the underlying mechanisms, providing a structured breakdown of how each feature influences the model's predictions. By employing techniques such as Shapley values or additive feature attribution methods, functional decomposition assigns importance scores to individual features, elucidating their contributions in shaping the final model output. In the context of securing IoT systems, where comprehending the influences of various sensors, network behaviors, or environmental variables on predictions is crucial, functional decomposition offers a precise lens to disentangle the intricate relationships among these features, thereby fostering transparency and interpretability within AI models deployed in IoT security applications. Moreover, functional decomposition not only dissects the contributions of individual features but also reveals the interactions and dependencies among these features, providing a holistic understanding of the model's decision-making process [9]. By breaking down complex predictions into simpler components, functional decomposition reveals how the interactions between features collectively influence the model's output. This comprehensive breakdown offers insights into the combined effects and interactions among multiple features, allowing stakeholders to discern subtle relationships or unexpected behaviors that might impact the security framework within IoT systems. For instance, in smart city deployments or industrial IoT environments, functional decomposition enables the visualization of how diverse sensor readings or network parameters collectively contribute to predictions regarding security threats or anomalous activities. Thus, functional decomposition is a powerful methodology, offering a detailed understanding of feature interactions and their collective influence on complex model predictions within IoT security frameworks [20].

Functional decomposition is pivotal in elucidating and fortifying IoT security measures by providing a transparent and granular understanding of the model's decision-making process. In the context of securing IoT systems, where the dynamics among diverse features significantly impact security measures, functional decomposition serves as a powerful tool to explain the influences of various sensor data, network behaviors, or environmental parameters on security predictions. By decomposing complex predictions into interpretable components, this methodology enables stakeholders to comprehend the contributions and interactions of individual features in the context of security threats or anomalous behaviors. For instance, in industrial IoT settings, functional decomposition facilitates the elucidation of how variations in sensor readings, such as temperature or pressure, contribute to predictions of potential security breaches. This level of transparency aids in identifying critical features or feature combinations that might indicate vulnerabilities, thereby empowering stakeholders to fortify security measures and proactively mitigate potential risks within interconnected IoT environments. Furthermore, functional decomposition supports the fortification of IoT security measures by providing actionable insights into improving security infrastructure. By systematically breaking down complex predictions, this approach reveals feature importance and interactions, guiding the optimization of security strategies and defenses. For example, in smart city deployments, functional decomposition enables stakeholders to discern how combinations of network parameters or environmental data collectively impact security predictions, thereby identifying areas for improvement or reinforcement in security protocols. This detailed comprehension aids in fine-tuning security measures, enhancing anomaly detection algorithms, and deploying targeted strategies to strengthen defenses against potential cyber-attacks or intrusions. Thus, functional decomposition serves as a cornerstone in fortifying IoT security measures by fostering transparency, enabling actionable insights, and supporting the development of robust security frameworks within interconnected IoT environments [17].

9.8 Permutation feature importance

This section embarks on a comprehensive exploration of permutation feature importance, delineating its theoretical foundations, practical applications, and significance within the context of fortifying the security posture of interconnected IoT environments.

Permutation feature importance serves as a crucial metric within interpretable artificial intelligence, quantifying the impact of individual features on the predictive performance of machine-learning models. This methodology evaluates feature importance by assessing the change in model performance when the values of a particular feature are randomly shuffled or permuted, disrupting the relationship between the feature and the target variable. By systematically permuting each feature and observing the resulting change in model performance metrics such as accuracy, F1 score, or mean squared error, permutation feature importance gauges the extent to which each feature contributes to the model's predictive capacity. Features that, when permuted, cause a substantial decrease in model performance indicate their significance in model predictions. In the context of securing IoT systems, where numerous sensors, network behaviors, or environmental variables contribute to decision-making processes, permutation feature importance offers a means to quantify the impact of each feature, aiding in the identification of critical features influencing security predictions within the complex web of interconnected IoT devices and systems. Moreover, permutation feature importance provides a model-agnostic approach, making it applicable to various machine-learning algorithms deployed in IoT security applications. This methodology's versatility allows for the assessment of feature importance across different models without being tied to specific model architectures or characteristics. By evaluating feature importance independent of model intricacies, permutation feature importance becomes instrumental in comprehending the contribution of features regardless of the underlying algorithm employed. In IoT security frameworks, where a variety of machine-learning models may be used due to the diverse nature of data sources and applications, the model-agnostic nature of permutation feature importance enables a standardized assessment of feature significance, offering insights into critical feature contributions and aiding in the development of robust security strategies within interconnected IoT environments [18].

Permutation feature importance holds substantial promise in fortifying security measures for IoT devices and networks by providing actionable insights into feature importance. In securing interconnected IoT systems, where identifying critical features influencing security predictions is paramount, permutation feature importance offers a systematic means to quantify the impact of individual features. By discerning the relative importance of various sensors, network behaviors, or environmental parameters in influencing security predictions, this methodology aids in prioritizing features crucial for effective anomaly detection or threat identification. For instance, within smart home environments, permutation feature importance enables stakeholders to gauge the significance of different sensor data, such as motion sensors, temperature readings, or camera feeds, in predicting potential security breaches. Understanding the impact of these features assists in allocating resources for securing and monitoring critical aspects of IoT networks, enhancing overall security resilience against potential threats or vulnerabilities. Furthermore, the permutation feature importance contributes to the iterative enhancement of security measures within IoT environments by facilitating continuous model improvement. By systematically evaluating feature importance across different machine-learning models or iterations, this methodology guides the refinement of anomaly detection algorithms, intrusion detection systems, or predictive models used in IoT security applications. As IoT ecosystems evolve dynamically, permutation feature importance aids in adapting security strategies to changing environments by identifying essential features that influence evolving security threats. For example, in industrial IoT deployments, this approach assists in identifying critical network behaviors or sensor readings that evolve over time and significantly impact security predictions. Consequently, permutation features are important in supporting the development of adaptive and resilient security frameworks within interconnected IoT environments, fortifying defenses against emerging threats and ensuring the sustained security of IoT devices and networks.

9.9 Global surrogate models

This section embarks on an extensive exploration of global surrogate models, elucidating their theoretical foundations, practical applications, and significance within the context of fortifying the security posture of interconnected IoT environments.

Global surrogate models serve as indispensable tools in approximating and understanding the decision-making processes of complex machine-learning models. Often, black-box models such as deep neural networks or ensemble methods might exhibit exceptional predictive performance but lack interpretability due to their inherent complexity. Surrogate models offer a solution by acting as interpretable approximations of these complex models, providing insights into their underlying decision rules without requiring a comprehensive understanding of the original model's intricacies. These surrogate models are constructed to mimic the behavior of the black-box model by using a more interpretable algorithm, allowing stakeholders to gain a transparent view of the relationships between input features and predictions. In the context of securing IoT systems, where model transparency is crucial for understanding security predictions derived from intricate machine-learning models, surrogate models serve as invaluable tools to unravel the decision logic behind these complex models. By offering a simplified representation, surrogate models facilitate a clearer comprehension of critical decision rules and feature contributions within AI models deployed in IoT security applications. Moreover, surrogate models play a pivotal role in addressing the trade-off between model complexity and interpretability, which is often encountered in securing IoT systems. While complex models might achieve high accuracy, they often lack transparency, hindering their interpretability and comprehensibility. Global surrogate models bridge this gap by providing simplified yet interpretable representations that approximate the decision boundaries of these complex models. These surrogate models, built using more transparent algorithms such as decision trees or linear models, capture the essential patterns and interactions between features, enabling stakeholders to decipher the underlying decision-making process. In IoT security frameworks, where diverse and heterogeneous data sources contribute to decision-making, surrogate models offer a means to distill complex model behaviors into understandable rules, aiding in the identification of critical feature contributions and potential vulnerabilities within interconnected IoT environments. As a result, surrogate models stand as crucial assets, enabling a balance between model accuracy and interpretability, essential for fortifying security measures within complex IoT systems [19].

Global surrogate models play a pivotal role in bolstering the transparency and security of IoT-based AI systems by providing comprehensible insights into the decision-making processes of complex machine-learning models. In the realm of securing interconnected IoT systems, where understanding the inner workings of AI models is critical for fortifying security measures, surrogate models serve as indispensable tools. These surrogate models offer a simplified representation of complex black-box models used in IoT security applications, enabling stakeholders to grasp crucial decision rules and feature importance without delving into the intricacies of the original complex models. By providing interpretable approximations of these black-box models, global surrogate models aid in elucidating how various sensor data, network behaviors, or environmental parameters contribute to security predictions. This understanding empowers stakeholders to identify critical features, discern potential vulnerabilities, and fortify security measures within the multifaceted IoT ecosystem. Furthermore, global surrogate models support the advancement of secure IoT-based AI systems by enabling a deeper comprehension of model behavior and facilitating proactive security measures. These surrogate models act as interpretable proxies for complex models, offering a transparent view of the model's decision logic and predictions. In IoT security frameworks, this transparency allows stakeholders to detect anomalies, identify unexpected behaviors, and address potential security risks before they escalate. For instance, within smart city IoT deployments, global surrogate models aid in comprehending how interactions between various environmental sensors or network behaviors influence security predictions, allowing stakeholders to preemptively reinforce security protocols. The interpretable nature of these surrogate models assists in establishing robust security strategies, enhancing anomaly detection, and ensuring a more resilient IoT-based AI system that proactively mitigates security threats within interconnected IoT environments. Hence, global surrogate models stand as instrumental assets in fortifying the security posture and transparency of AI-driven IoT systems.

9.10 Prototypes and criticisms

This section embarks on an in-depth exploration of prototypes and criticisms, elucidating their theoretical underpinnings, practical applications, and significance within the context of fortifying the security posture of interconnected IoT environments.

Prototypes serve as pivotal elements in elucidating model decisions within interpretable artificial intelligence. These representative instances are characterized by their proximity to the decision boundaries or areas of significant model uncertainty, encapsulating the essence of complex decision-making processes. In the context of securing IoT systems, where understanding the rationale behind model predictions is crucial for fortifying security measures, prototypes offer invaluable insights into the inner workings of machine-learning models [3]. By identifying instances that exhibit model sensitivity or are pivotal in determining specific predictions, prototypes shed light on the features and patterns that significantly influence security-related decisions. For instance, within IoT security frameworks, prototypes may represent data points or instances that pose potential security threats or anomalies, aiding stakeholders in comprehending the features contributing to such predictions. By focusing on these representative instances, stakeholders gain a deeper understanding of the decision rules and feature importance, facilitating the development of targeted strategies to address security vulnerabilities and enhance the robustness of AI-driven security measures within interconnected IoT environments. Furthermore, prototypes contribute significantly to enhancing the interpretability and transparency of machine-learning models deployed in IoT security applications. These instances act as tangible examples that highlight the intricacies of model behavior, allowing stakeholders to scrutinize the decision boundaries and areas of uncertainty. By analyzing prototypes, stakeholders can discern the underlying patterns or features that lead to specific model predictions, fostering a clearer comprehension of how various sensor data, network behaviors, or environmental parameters influence security-related decisions. This transparent view into the decision-making process aids in fortifying security measures by enabling stakeholders to identify potential weaknesses or areas requiring reinforcement within IoT-based AI systems. For example, in smart home IoT deployments, prototypes may represent instances where anomalous behaviors occur, providing insights into the critical features driving such predictions and guiding the implementation of targeted security measures [7]. Thus, prototypes stand as indispensable tools in explicating model decisions, contributing to the transparency and interpretability of AI-driven security frameworks in the complex landscape of interconnected IoT environments.

Despite their significant contributions to enhancing model interpretability, model-agnostic methods encounter several criticisms and limitations within the domain of interpretable artificial intelligence. One common criticism concerns the potential computational overhead associated with model-agnostic techniques, especially when dealing with large-scale, high-dimensional datasets commonly found in IoT applications. These methods might demand substantial computational resources and time, hindering their feasibility for real-time or resource-constrained IoT environments. Moreover, model-agnostic techniques may sometimes lack the granularity to capture intricate interactions or complex patterns present in certain machine-learning models, leading to oversimplification of model explanations. Additionally, the interpretability provided by these methods might not always meet domain-specific needs, as they offer generalized explanations that might not align perfectly with the intricate requirements of diverse IoT applications. To address these limitations, several potential solutions have been proposed. One approach involves the development of optimized algorithms and computational techniques explicitly tailored for handling large-scale IoT datasets efficiently. By optimizing computational workflows and utilizing parallel processing techniques, the computational overhead associated with model-agnostic methods can be alleviated, making them more applicable in resource-constrained IoT environments. Additionally, advancements in model-agnostic methodologies aim to enhance their capacity to capture more intricate interactions and nuanced patterns within machine-learning models. Techniques incorporating advanced feature engineering or ensemble approaches can enhance the granularity of explanations provided by model-agnostic methods, offering more detailed and accurate insights into complex model behaviors. Furthermore, tailoring model-agnostic methods to specific domain requirements through customization and integration with domain-specific knowledge and constraints can help address the issue of interpretability alignment with the intricacies of IoT applications. By fine-tuning explanations to better suit the requirements of diverse IoT domains, model-agnostic methods can offer more contextually relevant and actionable insights, thereby mitigating some of the criticisms and limitations encountered within interpretable AI frameworks [18–20].

References

[1] Gianfagna, L., and Di Cecco, A. (2021). Model-agnostic methods for XAI. In *Explainable AI with Python* (pp. 81–113). Cham: Springer International Publishing.

[2] Moosbauer, J., Herbinger, J., Casalicchio, G., Lindauer, M., and Bischl, B. (2021). Explaining hyperparameter optimization via partial dependence plots. *Advances in Neural Information Processing Systems*, *34*, 2280–2291.

[3] Shi, H., Yang, N., Yang, X., and Tang, H. (2023). Clarifying relationship between pm2. 5 concentrations and spatiotemporal predictors using multi-way partial dependence plots. *Remote Sensing, 15*(2), 358.

[4] Goldstein, A., Kapelner, A., Bleich, J., and Pitkin, E. (2015). Peeking inside the black box: Visualizing statistical learning with plots of individual conditional expectation. *Journal of Computational and Graphical Statistics*, *24* (1), 44–65.

[5] Yeh, A., and Ngo, A. (2021, September). Bringing a ruler into the black box: Uncovering feature impact from individual conditional expectation plots. In *Joint European Conference on Machine Learning and Knowledge Discovery in Databases* (pp. 34–48). Cham: Springer International Publishing.

[6] Inouye, D., Leqi, L., Kim, J. S., Aragam, B., and Ravikumar, P. (2020, August). Automated dependence plots. *In Conference on Uncertainty in Artificial Intelligence* (pp. 1238–1247). PMLR.

[7] Molnar, C., Freiesleben, T., König, G., *et al.* (2023, July). Relating the partial dependence plot and permutation feature importance to the data generating process. *In World Conference on Explainable Artificial Intelligence* (pp. 456–479). Cham: Springer Nature Switzerland.

[8] Loftus, J. R., Bynum, L. E., and Hansen, S. (2023). Causal Dependence Plots. *arXiv preprint arXiv:2303.04209*.

[9] Muschalik, M., Fumagalli, F., Jagtani, R., Hammer, B., and Hüllermeier, E. (2023, July). iPDP: On partial dependence plots in dynamic modeling scenarios. In *World Conference on Explainable Artificial Intelligence* (pp. 177–194). Cham: Springer Nature Switzerland.

[10] Freiesleben, T., Molnar, C., König, G., *et al.* (2023). Relating the partial dependence plot and permutation feature importance to the data generating process. *What Does Explainable AI Explain?*, pp. 456–479.

[11] Gkolemis, V., Dalamagas, T., and Diou, C. (2023, April). DALE: Differential accumulated local effects for efficient and accurate global explanations. In *Asian Conference on Machine Learning* (pp. 375–390). PMLR.

[12] Danesh, T., Ouaret, R., Floquet, P., and Negny, S. (2022). Interpretability of neural networks predictions using Accumulated Local Effects as a model-agnostic method. In *Computer Aided Chemical Engineering* (Vol. 51, pp. 1501–1506). Elsevier.

[13] Gkolemis, V., Dalamagas, T., Ntoutsi, E., and Diou, C. (2023). Rhale: Robust and heterogeneity-aware accumulated local effects. *In ECAI 2023* (pp. 859–866). IOS Press.

[14] Oh, S. (2022). Predictive case-based feature importance and interaction. *Information Sciences*, *593*, 155–176.

[15] Havrylenko, Y., and Heger, J. (2024). Detection of interacting variables for generalized linear models via neural networks. *European Actuarial Journal*, *14*(2), 551–580.

[16] Rajbahadur, G. K., Wang, S., Oliva, G. A., Kamei, Y., and Hassan, A. E. (2021). The impact of feature importance methods on the interpretation of

defect classifiers. *IEEE Transactions on Software Engineering*, *48*(7), 2245–2261.

[17] Zupan, B., Bratko, I., Bohanec, M., and Demšar, J. (1999). Function decomposition in machine learning. In *Advanced Course on Artificial Intelligence* (pp. 71–101). Berlin, Heidelberg: Springer Berlin Heidelberg.

[18] Kaneko, H. (2022). Cross-validated permutation feature importance considering correlation between features. *Analytical Science Advances*, *3*(9-10), 278–287.

[19] Fumagalli, F., Muschalik, M., Hüllermeier, E., and Hammer, B. (2023). Incremental permutation feature importance (iPFI): Towards online explanations on data streams. *Machine Learning*, *112*(12), 4863–4903.

[20] Das, A., and Rad, P. (2020). Opportunities and challenges in explainable artificial intelligence (xai): A survey. *arXiv preprint arXiv:2006.11371*.

Chapter 10

Model-agnostic methods for locally explainable AI to secure IoT system

10.1 Introduction

The successful integration of IoT devices in various domains has accelerated the potential for sophisticated functionalities and improved efficiency. However, this proliferation of interconnected devices has introduced complex models and algorithms that operate in diverse environments, raising concerns about their transparency and reliability. In the context of securing IoT systems, the significance of local explainability emerges as a critical factor in ensuring trust, transparency, and accountability. Local explainability techniques facilitate the comprehension of individual predictions made by ML models within IoT systems, providing insights into the reasoning behind specific outcomes. With local explainability, stakeholders gain a clearer understanding of how and why decisions are made, empowering them to detect and mitigate potential biases, errors, or security vulnerabilities. Moreover, in the context of IoT systems, where data privacy, security, and safety are paramount, local explainability a fundamental tool in ensuring compliance with regulatory standards and ethical guidelines. The ability to generate understandable and context-specific explanations for individual predictions allows stakeholders, including users, developers, and regulators, to scrutinize and validate the behavior of AI models deployed in IoT devices. Local explanations enable the identification of potential vulnerabilities or instances where sensitive information might be at risk, thus aiding in the identification and rectification of potential security threats or data breaches. By enabling the scrutiny of specific instances rather than merely offering global model insights, local explainability techniques empower stakeholders to validate model behavior against established security standards, fostering a more robust and secure IoT ecosystem that adheres to ethical norms and regulatory frameworks, ultimately ensuring the safety and privacy of end-users interacting with IoT systems.

In the IoT security landscape, ML models have become ubiquitous, offering sophisticated solutions for data analysis, predictive maintenance, and anomaly detection. However, the diverse nature of IoT environments with varying data distributions, device types, and operational conditions presents challenges in deploying and interpreting ML models effectively. Model-agnostic methods stand as a pivotal paradigm in IoT security, transcending the constraints imposed by specific model architectures or algorithms. These methods focus on generating explanations for model predictions without relying on internal model structures, thereby offering a universal approach applicable across different types of ML models. This universality and versatility in providing interpretability make model-agnostic methods indispensable in IoT security. They enable stakeholders to comprehend the decisions made by diverse ML models employed in IoT devices, allowing for the identification of potential vulnerabilities, biases, or errors irrespective of the underlying model complexities. As such, the adaptability of model-agnostic methods ensures comprehensive and uniform interpretability, fostering transparency and accountability in IoT systems, which is critical for mitigating security risks and ensuring robustness across various IoT applications and industries.

The significance of model-agnostic methods in IoT security extends beyond interpretability to address broader challenges related to trust, fairness, and ethical considerations. In the IoT ecosystem, where devices interact in real-time, decisions made by ML models directly impact user experiences, safety, and privacy. Model-agnostic methods offer a pathway to enhance transparency and fairness by enabling stakeholders to scrutinize individual predictions made by diverse models operating in IoT environments. This scrutiny aids in uncovering potential biases or discriminatory patterns, facilitating the identification and rectification of ethical concerns or unintended consequences embedded within these systems. By allowing for the generation of explanations that are independent of specific model architectures or functionalities, model-agnostic methods enable a systematic examination of decision-making processes in IoT devices, fostering a culture of accountability and responsible AI deployment. Consequently, adopting model-agnostic methods in IoT security ensures robust interpretability and aligns with ethical imperatives, contributing to the development of trustworthy and ethically aligned IoT systems that prioritize user well-being and societal values.

In the pursuit of enhancing transparency and interpretability within the intricate landscape of IoT systems, this chapter seeks to explore several model-agnostic explanation techniques as pivotal tools in unraveling the decision-making processes of ML models deployed across diverse IoT environments. One such technique, individual conditional expectation (ICE) plots, focuses on visualizing the relationship between individual feature values and model predictions, providing localized insights into how changes in specific input variables affect the model's output for a particular instance. Another prominent technique, local interpretable model-agnostic explanations (LIMEs), operates by approximating complex model behaviors using interpretable surrogate models at a local level. Moreover, the landscape of model-agnostic explanation techniques in securing IoT systems encompasses the SHapley Additive exPlanations (SHAP) method, renowned for its ability to assign attributions to each feature contributing to a model's prediction. PyExplainer, a local rule-based model-agnostic technique, operates by generating simplified, rule-based models that closely approximate the behavior of complex ML models at a local level. Contrastive explanations, another significant technique, highlight the differences in model predictions between two instances, shedding light on the factors leading to disparate outcomes within IoT contexts. Counterfactual explanations offer hypothetical instances that, when fed into the model, would have led to a different prediction, aiding in understanding model behaviors within IoT devices. Lastly, the Anchor method generates interpretable and sufficient conditions that ensure model predictions remain unchanged within predefined bounds, contributing to robust explanations in the context of securing IoT systems.

10.2 ICE plots

Within IoT security, the interpretability and transparency of ML models play a fundamental role in ensuring robustness and trustworthiness. As this section unfolds, the focus is directed toward unraveling the principles, functionalities, and applications of ICE plots in enhancing local interpretability within IoT security contexts. The discussion encapsulates the methodology behind ICE plots, elucidating their significance in scrutinizing model predictions, detecting biases, and identifying potential vulnerabilities specific to individual instances in IoT scenarios.

ICE plots serve as a pivotal model-agnostic technique in elucidating the localized behavior of ML models deployed in the intricate landscape of IoT systems. The fundamental principle underlying ICE plots revolves around unveiling the relationship between individual feature values and the corresponding model predictions for specific instances within an IoT environment. This technique generates a series of conditional predictions while varying a single feature's value for a given instance, effectively illustrating how changes in that particular feature influence the model's output. Visualized as a collection of individual curves or lines, each representing the model's predictions with respect to a particular feature's values, ICE plots offer a comprehensive and interpretable depiction of feature influence on model predictions within IoT contexts. The functionality of ICE plots lies in their ability to present a clear and localized view of a model's behavior, enabling stakeholders to grasp how alterations in input variables affect the model's output for specific instances encountered in IoT systems. Interpretation of ICE plots involves observing the shape, slope, or variation in the individual curves, allowing for insights into the feature's impact on predictions. Flat or erratic curves may indicate minimal or inconsistent influence. In contrast, steep or fluctuating curves may signal strong feature dependence, empowering stakeholders to discern the relevance and contribution of specific features toward individual predictions within IoT security field.

Furthermore, the interpretability and practical application of ICE plots in securing IoT systems lie in their ability to provide nuanced insights into the decision-making processes of ML models operating in diverse IoT environments. By facilitating a granular understanding of feature dependencies within individual instances, ICE plots aid in uncovering complex model behaviors that might go unnoticed in global model analyses. This technique allows stakeholders, including developers, security analysts, and end-users, to grasp how variations in specific input features influence the model's predictions in real-world IoT scenarios. Such localized interpretability proves invaluable in identifying potential biases, anomalies, or vulnerabilities specific to certain instances or contexts within IoT systems. Additionally, ICE plots enable the identification of non-linear relationships or interactions between features and model predictions, fostering a deeper comprehension of the intricate decision boundaries of ML models deployed in IoT devices. As a result, ICE plots serve as a powerful tool for enhancing transparency and aiding in the detection of security threats or irregularities, ultimately contributing to fortifying the robustness and reliability of IoT systems in an ever-evolving technological landscape [1].

The utilization of ICE plots presents a myriad of applications and profound relevance in explicating the behavior of ML models deployed within the intricate framework of IoT systems. Within the realm of IoT security, ICE plots find applicability in various domains, including anomaly detection, predictive maintenance, and personalized IoT services. In anomaly detection, ICE plots offer stakeholders a granular understanding of how individual features contribute to specific predictions, enabling the identification of unusual behaviors or deviations in IoT device functionalities. For local explanations, ICE plots display the model's predictions for each instance in the dataset as the feature of interest changes. Let us say we are focusing on a single instance, and we want to understand how its prediction changes with variations in one specific feature. In the following, we provide a Python example on visualization of local ICE explanation.

```
import pandas as pd
import numpy as np
import matplotlib.pyplot as plt
from sklearn.ensemble import GradientBoostingRegressor
import scipy as sp

# Generating the random data
N = 1000
df = pd.DataFrame(sp.stats.uniform.rvs(-1, 2, size=(N, 3)), columns=['x1',
'x2', 'x3'])
noise = sp.stats.norm.rvs(size=N)
y = 0.2 * df.x1 - 5 * df.x2 + 10 * df.x2 * (df.x3 >= 0) + noise
gbm = GradientBoostingRegressor()
gbm.fit(df.values, y)

# Local ICE plot function
def plot_ICE(feature, instance_idx, data, model):
    instance = data.iloc[instance_idx]
    feature_values = np.linspace(data[feature].min(), data[feature].max(),
100)

    predictions = []
    for value in feature_values:
        modified_instance = instance.copy()
        modified_instance[feature] = value
        prediction = model.predict(modified_instance.values.reshape(1, -
1))
        predictions.append(prediction[0])
```

```
    plt.figure(figsize=(8, 6))
    plt.plot(feature_values, predictions, label='ICE')
    plt.xlabel(feature)
    plt.ylabel('Prediction')
    plt.title(f'Individual Conditional Expectation (ICE) Plot for instance
{instance_idx}')
    plt.legend()
    plt.show()


# Testing local ICE plot
plot_ICE('x1', 5, df, gbm)
```

The generated local ICE plot in Figure 10.1 showcases the predictive behavior of the model for the "x1" feature concerning a specific instance (Instance 5). This plot illustrates how the model's predictions change as the "x1" feature varies while holding other features constant for this particular data point. By observing the curve, we can discern the model's sensitivity to alterations in "x1" for this singular instance, shedding light on the local impact of "x1" on the model's predictions.

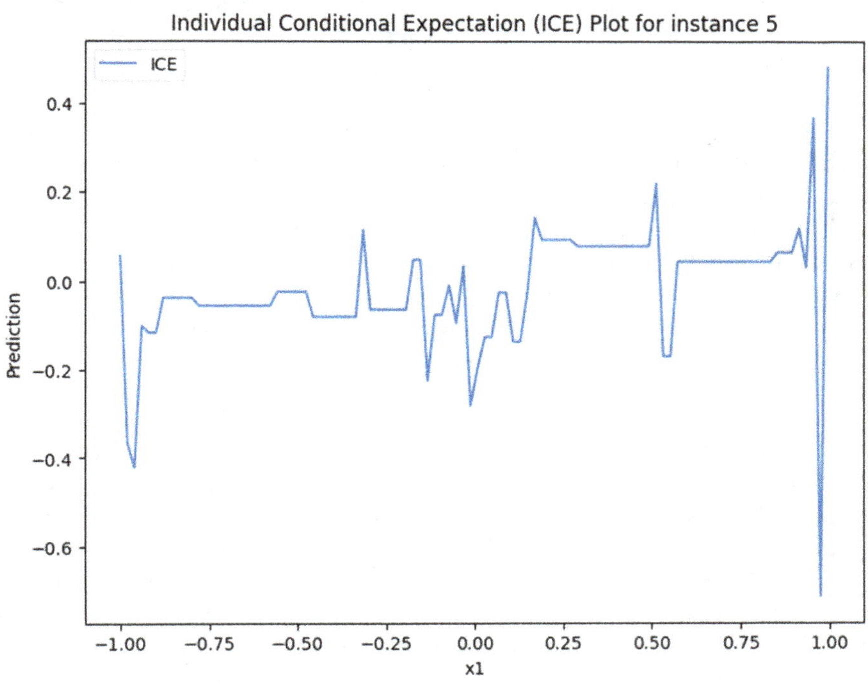

Figure 10.1 ICE plot for "x1" feature of instance 5"

10.3 Local interpretable model-agnostic explanations

This section delves into the exploration of LIME, a prominent and versatile model-agnostic explanation technique designed to provide transparent insights into the decision-making processes of ML models within the intricate landscape of IoT systems. LIME is a foundational methodology aimed at generating easily understandable and locally faithful explanations for individual predictions made by ML models operating in diverse IoT environments.

LIME stands as a pivotal technique in model-agnostic explanations, offering a comprehensive framework to unravel the decision-making processes of ML models within the complex landscape of IoT systems. At its core, LIME operates on the principle of generating interpretable surrogate models that approximate the behavior of complex ML models at a local level. This methodology enables stakeholders to scrutinize and comprehend the reasoning behind individual predictions made by diverse models employed in IoT devices, transcending the black-box nature of these models. The underlying principle of LIME revolves around creating local fidelity models by perturbing input instances and observing the changes in model predictions, thereby approximating the local decision boundaries of the original model. By simplifying the intricate relationships between input features and predictions within a specified vicinity of an instance, LIME strives to provide transparent and understandable explanations for individual model outputs encountered in IoT contexts. This approach ensures that stakeholders can grasp the reasoning behind specific predictions, enabling them to identify potential biases, vulnerabilities, or inconsistencies that may impact the security and reliability of IoT systems.

The working mechanism of LIME involves a multi-step process designed to elucidate the decision-making process of ML models for individual predictions in IoT scenarios. Initially, LIME focuses on selecting a particular instance or sample for which an explanation is sought within the IoT dataset. Subsequently, LIME perturbs this instance by generating a diverse set of similar instances by introducing slight variations in the feature values around the original sample. These perturbed instances are then presented to the underlying ML model, and the resulting predictions are recorded. Following this, LIME constructs a locally interpretable surrogate model, often using linear models or decision trees, by incorporating the perturbed instances along with their corresponding predictions. The surrogate model's simplicity allows for a transparent understanding of how individual features influence predictions within the vicinity of the selected instance. By emphasizing feature importance and their contributions to the model's output in the local context, LIME generates understandable explanations for stakeholders, shedding light on the reasoning behind specific predictions made by ML models in IoT environments. This methodology ensures stakeholders gain insights into the model's decision-making rationale for individual instances, contributing to enhanced transparency and interpretability crucial for fortifying the security and trustworthiness of IoT systems [2].

The implementation of LIME within the realm of explaining IoT models represents a significant stride toward fortifying the transparency and interpretability of ML models deployed in diverse IoT environments. One practical implementation involves the application of LIME in IoT devices used for predictive maintenance in industrial settings. In this scenario, various sensors collect data related to machinery health and performance. LIME facilitates the comprehension of how specific sensor readings influence the predictions made by ML models regarding potential equipment failures or maintenance needs. By utilizing LIME to generate locally interpretable explanations, stakeholders discern the correlations between sensor readings and predictions, pinpointing crucial indicators of machinery malfunctions. This localized understanding empowers maintenance personnel to take proactive measures, conduct timely inspections, or apply necessary interventions, thereby preventing potential failures and ensuring the continuous operation of industrial IoT systems. Thus, the implementation of LIME in explaining IoT models aids in uncovering critical insights into the reasoning behind individual predictions, supporting informed decision-making and proactive interventions in IoT systems, ultimately contributing to bolstering their reliability and security.

Furthermore, the benefits derived from implementing LIME in explaining IoT models extend beyond enhanced interpretability to significantly impact the security posture of IoT ecosystems. In IoT security contexts, where the reliability and integrity of ML models are paramount, LIME plays a pivotal role in elucidating potential security vulnerabilities or risks embedded within these models. By providing transparent insights into the decision-making rationale behind individual predictions, LIME enables stakeholders to detect and rectify biases, inconsistencies, or unexpected behaviors exhibited by ML models in specific IoT scenarios. For instance, in IoT applications involving user authentication or intrusion detection systems, LIME aids in scrutinizing how input features, such as behavioral patterns or network data, contribute to the model's predictions. This scrutiny facilitates the identification of potential adversarial attacks or anomalies that might evade traditional security measures. Additionally, by highlighting the influential factors behind predictions through LIME explanations, security analysts can refine and adapt security protocols, enhancing the resilience of IoT systems against potential cyber threats or adversarial manipulations. Consequently, the implementation of LIME not only amplifies interpretability but also fortifies the security posture of IoT ecosystems by enabling the proactive identification and mitigation of potential security risks or vulnerabilities embedded within ML models deployed in IoT devices [3].

One compelling use case showcasing the practical application of LIME in bolstering IoT security involves anomaly detection in a network of IoT devices employed for environmental monitoring within a smart city infrastructure. In this scenario, various sensors embedded across the city collect data pertaining to air quality, temperature, humidity, and noise levels. These sensors feed into ML models tasked with identifying abnormal environmental conditions or potential hazards. LIME is employed to provide interpretable explanations for individual predictions made by these models. By employing LIME, stakeholders gain localized insights into the relationships between sensor readings and the predictions of anomalies. For instance, LIME reveals how variations in air quality readings strongly influence predictions related to environmental anomalies. By scrutinizing these LIME-generated explanations, anomalies such as pollution spikes or unusual noise patterns in specific areas are elucidated, allowing authorities to address potential environmental hazards or security threats promptly. This utilization of LIME not only enhances the interpretability of ML models within IoT systems but also facilitates proactive measures to ensure public safety and security within the smart city infrastructure, showcasing the efficacy of LIME in providing actionable insights for IoT security applications.

10.4 SHapley Additive exPlanations

This section examines SHAP, a robust and adaptable model-agnostic explanation technique designed to uncover the sophisticated decision-making processes of ML models used in IoT systems. We explore the fundamental principles, methodology, and applications of SHAP, clarifying its crucial role in offering clear insights into the complex connections between input data and predictions in the context of IoT security.

Shapley values represent a concept derived from cooperative game theory, offering a systematic and principled approach to attribute contributions to individual players within a coalition or group. In the context of ML interpretability, Shapley values serve as a fundamental methodology for assigning fair and consistent feature importance to input variables within predictive models. The theoretical basis of Shapley values originates from the work of Lloyd Shapley in the 1950s, primarily introduced to allocate payouts fairly among contributing players in a cooperative game based on their marginal contributions to the coalition's overall payoff. In the domain of ML, particularly in the interpretation of complex models deployed within IoT systems, Shapley values have been adapted to estimate the impact of each input feature on model predictions. By considering all possible combinations and permutations of input features, Shapley values quantify the average marginal contribution of each feature across all possible feature subsets, enabling a comprehensive understanding of each feature's significance in influencing model predictions.

The computation of Shapley values involves a combinatorial approach that considers all possible permutations of features and evaluates their individual contributions to the model's output. Specifically, Shapley values quantify the marginal importance of a particular feature by examining its inclusion or exclusion in every possible subset of features alongside other features. This methodology necessitates evaluating the model's predictions for each subset containing different combinations of features and comparing these predictions to those made without the specific feature. The difference in predictions serves as the basis for calculating the marginal contribution of the feature in question. Shapley values ensure fairness and consistency by considering each feature's contribution to the model predictions across various feature permutations, thereby attributing accurate importance scores to each input variable. In the context of interpreting ML models within IoT systems, Shapley values offer a coherent and principled framework to discern the impact of individual features on model predictions, fostering transparency and interpretability crucial for ensuring the security and reliability of IoT ecosystems [4].

SHAP stands as a pivotal advancement in the domain of model interpretability, offering a principled and versatile framework for understanding the contributions of individual input features to the predictions of ML models deployed within the complex landscape of IoT systems. At its core, SHAP is rooted in Shapley values from cooperative game theory, adapted to the realm of ML to provide comprehensive and consistent feature importance scores. The fundamental principle underlying SHAP revolves around computing Shapley values for each input feature, quantifying the average marginal contribution of a feature across all possible feature combinations. This unique approach ensures fairness and coherence in attributing importance to features by considering their varying interactions and dependencies within the model. SHAP achieves this by assessing the impact of each feature's absence or presence in different combinations, thereby elucidating the relative influence of individual features on model predictions within IoT systems. By providing a unified and principled framework for computing feature importance, SHAP transcends the limitations of black-box models, offering stakeholders a clear and interpretable understanding of feature contributions, thus bolstering transparency and trustworthiness within IoT ecosystems [5].

Functionally, SHAP operates by leveraging Shapley values to attribute importance scores to individual input features based on their contributions to the model's predictions. This involves computing the marginal contributions of each feature across all possible subsets of features and averaging these contributions, ensuring consistent and unbiased estimations of feature importance. SHAP's functionality extends to offering a local interpretation of model predictions, allowing stakeholders to examine the relative importance of features for specific instances encountered within IoT environments. This local interpretability empowers users to scrutinize the reasoning behind individual predictions, thereby enabling the identification of potential biases, anomalies, or security vulnerabilities specific to certain IoT scenarios. Significantly, SHAP's capability to provide locally faithful explanations aids stakeholders in comprehending how changes in input variables influence predictions for individual instances, contributing to informed decision-making, robustness assessment, and proactive measures to fortify the security and reliability of IoT systems. As such, SHAP stands as a powerful tool for enhancing local explanations within IoT contexts, elucidating the intricate relationships between input features and model predictions crucial for ensuring the integrity and trustworthiness of ML models deployed in IoT devices.

The adoption of SHAP serves as a pivotal strategy in elucidating and fortifying the security aspects of IoT systems by providing transparent insights into the decision-making processes of ML models. SHAP facilitates a comprehensive understanding of the relative importance and interactions among input features within complex models deployed in IoT environments. In explaining IoT systems, SHAP enables stakeholders to scrutinize the contributions of individual features to model predictions, thereby enhancing interpretability and transparency. For instance, in an IoT application where sensors collect data for predictive maintenance in industrial machinery, SHAP aids in identifying critical features influencing predictions related to potential equipment failures. By quantifying the impact of each feature on predictions, SHAP explanations uncover significant indicators or anomalies in sensor readings, empowering maintenance teams to take preemptive actions, conduct timely inspections, or initiate necessary interventions. This proactive approach ensures the robustness and reliability of industrial IoT systems by preventing potential downtimes or failures, showcasing SHAP's utility in explicating model decisions and fortifying the security of IoT applications across various domains. Moreover, SHAP plays a pivotal role in securing IoT systems by offering insights into the underlying mechanisms of ML models, thereby aiding in the identification and mitigation of potential security vulnerabilities or risks. In IoT security contexts, SHAP's ability to provide locally faithful explanations proves invaluable in detecting anomalies or irregularities specific to certain instances encountered within IoT ecosystems. For instance, in a smart home IoT environment employing predictive models for user behavior analysis, SHAP elucidates the contributions of various factors such as occupancy patterns or device usage in predicting user actions. This enables stakeholders to discern potential security threats or abnormal behaviors that might evade traditional detection mechanisms. By revealing the underlying factors driving model predictions, SHAP empowers security analysts to refine intrusion detection algorithms, fortify authentication systems, or implement enhanced security measures, thereby bolstering the resilience of IoT systems against potential cyber threats or adversarial attacks [6]. Consequently, the use of SHAP in explaining and securing IoT systems not only enhances interpretability but also aids in proactive risk mitigation, contributing to the integrity, resilience, and security posture of IoT ecosystems across diverse applications and industries.

10.5 PyExplainer: a local rule-based model-agnostic technique

This section emphasizes the exploration of PyExplainer, a notable local rule-based model-agnostic technique designed to offer transparent and interpretable explanations for ML models within the intricate landscape of IoT systems. PyExplainer stands as a versatile and intuitive methodology aimed at generating simplified and comprehensible rule-based models that approximate the behavior of complex ML models at a local level. This section delves into the foundational principles, functionalities, and applications of PyExplainer, elucidating its pivotal role in unraveling the decision-making processes of ML models within IoT security contexts.

PyExplainer represents a significant advancement in the domain of local rule-based model-agnostic techniques, offering a streamlined framework designed to unravel the decision-making processes of ML models deployed within IoT systems. The design of PyExplainer is rooted in its ability to generate simplified and transparent rule-based models that approximate the behavior of complex ML models, specifically tailored for local explanations. Fundamentally, PyExplainer functions by analyzing individual predictions and creating intuitive, human-readable rules that mirror the decision boundaries of the underlying model for specific instances encountered within IoT environments. This technique synthesizes the intrinsic relationships between input features and predictions, providing stakeholders with interpretable explanations. PyExplainer operates by leveraging a combination of optimization algorithms and local rule induction to generate simple rule sets. These rules aim to accurately capture the decision-making rationale of ML models in IoT contexts, thereby fostering transparency and interpretability crucial for enhancing the security and trustworthiness of IoT systems. The functionality of PyExplainer extends to offering clear, concise, and interpretable rules that represent the decision logic of the ML models at a local level, aiding stakeholders in comprehending the factors influencing specific predictions within diverse IoT applications [7].

The applications of PyExplainer span across various domains within IoT security, encompassing anomaly detection, predictive maintenance, and personalized services. For instance, in the realm of anomaly detection, PyExplainer aids in elucidating the decision boundaries of ML models analyzing sensor data within industrial IoT setups. By generating intuitive rules, PyExplainer unveils critical conditions or thresholds within sensor readings that trigger predictions of potential anomalies or equipment failures. This transparent elucidation of decision rules enables proactive measures, such as preemptive maintenance or timely interventions, fortifying the reliability and security of industrial IoT systems. Moreover, in personalized IoT services like smart homes, PyExplainer offers comprehensible rules that represent the decision-making rationale of models predicting user preferences or behavior. This facilitates tailored services, allowing stakeholders to understand the factors influencing recommendations or actions generated by IoT devices. PyExplainer's applications underscore its significance in providing interpretable and contextually relevant explanations, empowering stakeholders to make informed decisions and fortify the security and reliability of diverse IoT applications [8].

PyExplainer plays a pivotal role in augmenting local interpretability within IoT systems by offering transparent and human-understandable rule-based explanations for ML models deployed in diverse IoT applications. At the core of its functionality, PyExplainer contributes to fortifying the security of IoT devices by providing stakeholders with accessible and interpretable insights into the decision-making processes of ML models operating within these systems. By generating simplified rule-based models that approximate the behavior of complex ML models at a local level, PyExplainer aids in deciphering the rationale behind individual predictions for specific instances encountered in IoT environments. This localized interpretability enables stakeholders, including developers, security analysts, and end-users, to comprehend the critical factors influencing predictions made by ML models within IoT systems. Consequently, PyExplainer facilitates the identification of potential biases, anomalies, or security vulnerabilities that may impact the reliability or integrity of IoT devices, thereby empowering stakeholders to take proactive measures to fortify the security posture of IoT ecosystems [9].

Moreover, PyExplainer's role in local interpretability within IoT security contexts extends to its ability to provide comprehensible and contextually relevant explanations for individual predictions. This empowers stakeholders to scrutinize and verify the reasoning behind specific predictions made by ML models, aiding in the detection and mitigation of potential threats or irregularities within IoT systems. For instance, in an IoT application involving predictive maintenance in industrial settings, PyExplainer elucidates intuitive rules that mirror the decision logic of ML models analyzing sensor data. These rules unveil critical thresholds or conditions within sensor readings that trigger predictions of potential equipment failures or malfunctions. This transparent elucidation of decision rules empowers maintenance personnel to proactively identify and address potential vulnerabilities or risks, thus reinforcing the security and reliability of industrial IoT systems. In essence, PyExplainer's provision of clear and interpretable rules aids in local interpretability, enabling stakeholders to comprehend the decision-making processes of ML models in IoT applications, ultimately contributing to the enhancement of security measures and the fortification of IoT systems against potential threats or vulnerabilities [10].

One illustrative scenario showcasing PyExplainer's significance in bolstering IoT security involves anomaly detection within a network of IoT devices used for environmental monitoring in a smart city infrastructure. In this context, various sensors collect data on environmental parameters such as air quality, temperature, humidity, and pollution levels. ML models analyze this data to predict environmental anomalies or hazards. PyExplainer plays a crucial role by providing transparent and interpretable rule-based explanations for individual predictions made by these models. For instance, PyExplainer generates human-readable rules that mirror the decision boundaries of ML models, offering insights into critical thresholds or conditions within sensor readings that trigger predictions of potential environmental anomalies. These explanations enable city administrators to swiftly identify areas experiencing abnormal pollution spikes or environmental hazards, empowering them to take proactive measures, such as targeted interventions or corrective actions. By offering interpretable explanations, PyExplainer aids in fortifying the security and reliability of environmental monitoring systems within the smart city infrastructure, showcasing its effectiveness in elucidating the decision-making processes of ML models and facilitating proactive responses to potential threats. Furthermore, PyExplainer proves instrumental in enhancing IoT security by providing transparent insights into the decision logic of ML models within healthcare IoT applications. Consider a scenario where wearable medical devices collect physiological data to predict health-related events or anomalies. PyExplainer aids healthcare professionals by generating human-interpretable rules that approximate the decision-making rationale of ML models for predicting critical health events. These rules offer insights into the factors influencing predictions, enabling healthcare providers to identify and understand anomalies or irregular patterns in patients' physiological data. For example, PyExplainer elucidates the influential factors driving predictions of potential health emergencies, empowering healthcare professionals to promptly detect and respond to abnormal health patterns or critical events. By providing interpretable explanations, PyExplainer enhances the interpretability of ML models within healthcare IoT systems, facilitating timely interventions, and ultimately bolstering the security and reliability of healthcare applications in IoT ecosystems. Thus, PyExplainer's role in offering transparent rule-based explanations contributes significantly to fortifying the security posture of IoT applications across various domains, aiding stakeholders in proactive risk mitigation and informed decision-making [11].

10.6 Contrastive explanations

This section delves into the exploration of contrastive explanations, a significant methodology within the realm of model interpretability, specifically tailored to unravel the decision-making processes of ML models deployed in the complex landscape of IoT systems. This section delves into the foundational principles, methodologies, and applications of contrastive explanations, elucidating its pivotal role in enhancing interpretability and transparency within IoT security.

Contrastive explanations serve as a fundamental methodology within the domain of model interpretability, specifically designed to elucidate the decision-making processes of ML models in complex environments such as IoT systems. At its core, contrastive explanations focus on highlighting and explicating the differences in model predictions between instances or observations. The fundamental principle underlying contrastive explanations revolves around providing insights into why an ML model reaches different decisions for similar instances within a dataset. This methodology aims to shed light on the factors or features that significantly contribute to divergent predictions, emphasizing the crucial differences that influence the model's decision-making. Contrastive explanations operate by comparing and contrasting the features or attributes of instances that lead to differing predictions, thereby unveiling the discriminative elements pivotal in steering the model's output. By pinpointing the distinctive attributes between instances that generate divergent predictions, contrastive explanations offer stakeholders a clear understanding of the specific factors driving the model's decisions, enhancing transparency and interpretability crucial for securing and fortifying IoT systems against potential risks or vulnerabilities [12].

Functionally, contrastive explanations work by systematically comparing the features or attributes of instances that result in varying predictions from an ML model. This process involves selecting instances where the model's predictions significantly differ and subsequently identifying the feature variations that contribute to the contrasting outcomes. Contrastive explanations operate by altering or perturbing features between instances to understand the impact of these variations on the model's predictions. By systematically examining and manipulating the features that lead to divergent predictions, contrastive explanations facilitate the identification of critical attributes or conditions influencing the model's decisions within specific contexts encountered in IoT environments. This methodology ensures a focused analysis of feature discrepancies, enabling stakeholders to comprehend the nuanced differences in feature contributions that drive disparate model predictions. Ultimately, contrastive explanations serve as a powerful tool for providing context-specific insights into the decision-making processes of ML models in IoT applications, contributing to enhanced interpretability and facilitating proactive measures to fortify the security and reliability of IoT systems.

Contrastive explanations play a pivotal role in augmenting local interpretability within IoT systems by offering nuanced insights into the decision-making processes of ML models, thereby fortifying the security measures deployed in these complex environments. Their contribution lies in elucidating why ML models produce divergent predictions for similar instances encountered within IoT ecosystems. By pinpointing the specific features or attributes that lead to disparate model outputs, contrastive explanations empower stakeholders to understand the underlying factors influencing predictions at a local level. In IoT security contexts, this localized interpretability aids in the identification and mitigation of potential risks or vulnerabilities within ML models. For instance, in predictive maintenance applications where sensors collect data on industrial machinery health, contrastive explanations highlight specific sensor readings or conditions that trigger differing predictions related to potential equipment failures. This transparency allows maintenance teams to proactively address critical sensor readings or anomalies, fortifying the reliability and security of industrial IoT systems. This methodology operates by identifying the key feature variations between instances that result in contrasting model outputs. This clarity enables stakeholders to detect potential anomalies, irregularities, or biases in model behavior specific to certain instances encountered within IoT deployments. For example, in user authentication systems where models predict user behaviors, contrastive explanations unveil distinctive behavioral patterns or input characteristics leading to differing authentication outcomes. This facilitates the identification of potential security threats or vulnerabilities, enabling the refinement of authentication mechanisms or the implementation of enhanced security protocols within IoT systems. Thus, contrastive explanations significantly contribute to local interpretability, empowering stakeholders to comprehend and address specific challenges or discrepancies within ML models deployed in IoT environments, thereby fortifying their security and reliability [13].

One compelling use case that demonstrates the effectiveness of contrastive explanations in bolstering IoT security involves anomaly detection within a network of IoT devices employed for infrastructure monitoring in smart cities. In this scenario, sensors gather data on various parameters such as traffic flow, environmental conditions, and infrastructure health. ML models analyze this data to predict potential anomalies or hazards. Contrastive explanations play a pivotal role by unveiling the specific attributes or conditions within sensor readings that lead to contrasting predictions for similar instances. For instance, contrastive explanations elucidate the distinctive patterns or sensor anomalies triggering differing predictions related to infrastructure faults or potential hazards. This transparent revelation empowers city administrators to swiftly detect and address critical issues, such as unusual traffic patterns or infrastructure irregularities, thus facilitating timely interventions or corrective actions. By offering interpretable insights into the specific attributes driving divergent predictions, contrastive explanations aid in fortifying the reliability and security of smart city infrastructure, showcasing their efficacy in elucidating model decisions and enabling proactive responses to potential threats. Moreover, contrastive explanations prove instrumental in enhancing IoT security within healthcare applications, particularly in patient monitoring systems utilizing wearable devices. In this use case, wearable medical devices collect physiological data to predict health-related events or anomalies. Contrastive explanations offer transparent insights into the specific attributes or physiological patterns that result in differing predictions for similar instances. For example, contrastive explanations reveal distinct physiological readings or patterns triggering varying predictions related to potential health emergencies or irregularities. This elucidation empowers healthcare practitioners to swiftly identify and respond to critical health events or anomalies in patient data, facilitating prompt interventions and medical attention. By providing interpretable insights into the localized differences influencing predictions, contrastive explanations contribute significantly to fortifying the security and reliability of healthcare IoT systems, enabling proactive measures to ensure patient safety and well-being. Thus, the practical use cases underscore the effectiveness of contrastive explanations in offering transparent and interpretable insights, aiding stakeholders in proactive risk mitigation and informed decision-making within diverse IoT applications, ultimately contributing to enhancing IoT security [14,15].

10.7 Counterfactual explanations

Herin, we embark on an exploration of counterfactual explanations, an influential methodology within the domain of model interpretability, specifically tailored to elucidate the decision-making processes of ML models deployed in the intricate landscape of IoT systems. Counterfactual explanations stand as a notable technique aimed at providing insightful and contextually relevant explanations by exploring the "what–if" scenarios that lead to different model predictions. This section delves into the foundational principles, methodologies, and applications of counterfactual explanations, elucidating its pivotal role in enhancing interpretability and transparency within IoT security contexts. By generating explanations that outline the minimal changes required in input features to alter a model's prediction, counterfactual explanations offer stakeholders a deeper understanding of the factors influencing ML model outputs within IoT ecosystems. The discussion within this section encapsulates the functionality, interpretability, and significance of counterfactual explanations in unraveling the decision-making processes of ML models within IoT environments, ultimately contributing to fortifying the security, trustworthiness, and reliability of IoT systems in an increasingly interconnected and data-driven technological landscape.

Counterfactual explanations represent a vital technique in the realm of model interpretability, specifically aimed at elucidating the decision-making mechanisms of ML models deployed in the complex realm of IoT systems. The fundamental concept underlying counterfactual explanations revolves around exploring "what–if" scenarios, seeking to comprehend the minimal changes in input features necessary to alter a model's prediction for a specific instance. This methodology aims to provide contextually relevant explanations by generating hypothetical instances that deviate minimally from the original input while causing a shift in the model's output. The purpose of counterfactual explanations lies in offering insights into the decision boundaries of ML models by constructing alternate scenarios that would lead to different predictions, thereby enhancing interpretability and transparency within IoT security contexts. Interpreting counterfactual explanations involves understanding the generated hypothetical instances, which outline the perturbations in input features required to alter a model's prediction. By elucidating the specific feature modifications that influence the model's output, counterfactual explanations empower stakeholders to comprehend the factors driving model decisions within IoT environments. This methodology operates by generating counterfactual instances that are as close as possible to the original input while causing a change in the model's prediction, thereby shedding light on the salient features that significantly impact model outcomes [16,17].

Functionally, counterfactual explanations offer a framework to interpret the rationale behind ML model predictions within IoT systems by outlining the minimum changes necessary in input features to trigger a different prediction outcome. The interpretation of counterfactual explanations involves assessing the perturbations in feature values, which, when applied to the original instance, lead to a change in the model's output. This technique enables stakeholders to discern the critical attributes or conditions influencing the model's decision-making process for a specific instance encountered within IoT deployments. For instance, in a predictive maintenance scenario within industrial IoT setups, counterfactual explanations provide insights into the minimal changes required in sensor readings to provoke varying predictions related to potential equipment failures. By generating counterfactual instances that demonstrate the subtle alterations in sensor data influencing predictions, stakeholders gain a deeper understanding of the decision boundaries of ML models. This enhances interpretability by revealing the specific feature modifications that trigger diverse predictions, thus empowering stakeholders to proactively identify and address potential risks or vulnerabilities within IoT systems. Ultimately, the interpretation of counterfactual explanations aids in comprehending the nuanced factors steering model predictions in IoT environments, contributing to enhanced transparency and informed decision-making in securing IoT systems [18].

Counterfactual explanations serve as a significant tool for enhancing local interpretability within IoT systems, specifically tailored to fortify the security measures deployed in these intricate environments. Their application revolves around providing contextually relevant insights into the decision-making processes of ML models at a local level within IoT contexts. By generating hypothetical instances that depict minimal changes in input features leading to different model predictions, counterfactual explanations empower stakeholders to comprehend the specific factors influencing predictions for individual instances. In the realm of IoT security, the application of counterfactual explanations aids in the identification and mitigation of potential risks or vulnerabilities within ML models. For instance, in anomaly detection scenarios where ML models predict irregularities in sensor data, counterfactual explanations elucidate the slight modifications in sensor readings required to trigger varying predictions. This localized interpretability enables stakeholders to understand the critical sensor readings or conditions leading to different predictions, thereby facilitating proactive measures to address potential anomalies or security threats within IoT systems.

Functionally, counterfactual explanations contribute to local interpretability by showcasing the minimal feature alterations needed to influence the decision boundaries of ML models for specific instances encountered within IoT deployments. This methodology aids stakeholders, including developers and security analysts, in comprehending the discriminative attributes or conditions steering different predictions for similar instances. For instance, in predictive maintenance scenarios within industrial IoT setups, counterfactual explanations offer insights into the minute changes in sensor data required to provoke divergent predictions related to equipment failures. By generating counterfactual instances that demonstrate subtle alterations in sensor readings impacting predictions, stakeholders gain a deeper understanding of the localized decision-making rationale of ML models [19]. This knowledge enhances interpretability by revealing the specific feature modifications triggering diverse predictions, empowering stakeholders to proactively detect and mitigate potential risks or vulnerabilities within IoT systems. Consequently, the application of counterfactual explanations significantly contributes to local interpretability, aiding in the identification and mitigation of specific challenges or discrepancies within ML models deployed in IoT environments, ultimately fortifying their security and reliability.

An illustrative example showcasing the role of counterfactual explanations in fortifying IoT systems involves anomaly detection within a network of sensors used for predictive maintenance in manufacturing. In this scenario, ML models predict potential equipment failures based on sensor data collected from industrial machinery. Counterfactual explanations play a pivotal role by generating hypothetical instances that elucidate minimal changes in sensor readings leading to divergent predictions. For instance, in the context of predicting machinery failures, counterfactual explanations provide insights into the subtle modifications in sensor data required to trigger varying predictions. These explanations unveil the critical sensor readings or conditions that influence predictions of potential equipment breakdowns. By comprehending the localized changes in sensor data driving divergent predictions, maintenance teams gain actionable insights to proactively address and prevent potential machine failures, thereby enhancing the reliability and security of industrial IoT systems.

Moreover, a study showcasing the efficacy of counterfactual explanations within healthcare IoT applications involves patient monitoring systems utilizing wearable devices. In this study, ML models predict health-related events or anomalies based on physiological data collected from patients. Counterfactual explanations offer transparent insights into the minimal changes in physiological readings necessary to provoke differing predictions for similar instances. For example, in the context of predicting health emergencies, counterfactual explanations elucidate the subtle variations in physiological data triggering diverse predictions related to critical health events. These explanations empower healthcare practitioners to swiftly detect and respond to potential health anomalies or emergencies. By understanding the localized changes in patient data influencing diverse predictions, healthcare professionals can take timely interventions, ensuring patient safety and well-being within IoT-enabled healthcare systems. These examples and studies underscore the pivotal role of counterfactual explanations in providing interpretable insights, aiding stakeholders in proactive risk mitigation and informed decision-making within diverse IoT applications, ultimately contributing to enhancing IoT security and reliability.

10.8 Anchor

This section delves into an exploration of Anchor, a significant methodology within the domain of model interpretability, specifically designed to elucidate the decision-making processes of ML models deployed in the intricate landscape of IoT systems. The discussion within this section encapsulates the functionality, interpretability, and significance of Anchor in unraveling the decision-making processes of ML models within IoT environments, ultimately contributing to fortifying the security, trustworthiness, and reliability of IoT systems in an increasingly interconnected and data-driven technological landscape.

The Anchor method represents a vital technique in the domain of model interpretability, focused on providing clear and locally faithful explanations for ML models operating within the intricate framework of IoT systems. At its core, the Anchor method functions by defining succinct rules or conditions that "anchor" the model's predictions to specific feature ranges or values, ensuring stability, and enhancing interpretability. This methodology aims to identify the most influential features or attributes pivotal in a model's decision-making for individual instances encountered within IoT environments. The functioning of the Anchor method involves generating rule-based explanations that capture the range of feature values critical to a model's predictions, thereby facilitating a deeper understanding of the decision boundaries. By defining these rules, the Anchor method offers stakeholders transparent insights into the specific conditions under which a model's predictions hold true, fostering interpretability within complex IoT systems. The benefits of the Anchor method lie in its ability to provide locally accurate and understandable explanations, aiding stakeholders in comprehending the crucial features driving model predictions within IoT contexts. However, the method's limitations include potential challenges in scalability or complexity when dealing with high-dimensional or intricate models, which might hinder its application in certain IoT deployments, requiring careful consideration and adaptation in such scenarios.

The Anchor method's value resides in its capacity to offer transparent and contextually relevant explanations, elucidating the rationale behind ML model predictions in IoT systems. By defining rules that anchor a model's predictions to specific feature values or ranges, the Anchor method aids stakeholders in interpreting the decision-making processes of ML models within IoT environments. Additionally, this method facilitates better trust and reliance on model predictions by establishing conditions under which predictions are reliable, contributing to enhanced transparency and reliability within IoT systems. However, the Anchor method's limitations encompass potential challenges in its applicability to highly complex or high-dimensional models commonly encountered in IoT scenarios, which might restrict its effectiveness in certain contexts. Despite these limitations, the Anchor method remains a valuable approach in providing interpretable explanations for ML models within IoT applications, contributing to fortifying the security and reliability of IoT systems by offering transparent insights into model decisions while requiring careful consideration of its adaptability and scalability within diverse IoT environments.

The Anchor method significantly enhances local explanations in securing IoT devices or systems by offering precise and transparent insights into the decision-making processes of ML models operating in these intricate environments. Its application revolves around providing clear and locally faithful explanations that capture the critical features driving a model's predictions for specific instances within IoT ecosystems. Anchor achieves this by defining concise rules or conditions that anchor the model's predictions to specific feature ranges or values, enabling stakeholders to comprehend the factors influencing predictions at a local level. In the context of IoT security, Anchor aids in the identification and mitigation of potential risks or vulnerabilities within ML models by offering interpretable explanations. For instance, in anomaly detection scenarios where models predict irregularities in sensor data, Anchor elucidates the feature ranges or conditions critical to triggering specific predictions. This localized interpretability enables stakeholders to understand the specific features influencing predictions, facilitating proactive measures to address potential anomalies or security threats within IoT systems [20–22].

Functionally, Anchor enhances local explanations within IoT security by providing concise and interpretable rules that delineate the feature values or conditions essential for a model's predictions in specific instances. This method aids stakeholders, including developers and security analysts, in comprehending the discriminative attributes or conditions steering predictions for individual instances within IoT deployments. For example, in predictive maintenance scenarios within industrial IoT setups, Anchor offers insights into the critical sensor readings or conditions that trigger diverse predictions related to equipment failures. By defining these rule-based explanations, stakeholders gain a deeper understanding of the decision boundaries of ML models, enabling them to proactively detect and mitigate potential risks or vulnerabilities within IoT systems. Consequently, the application of Anchor significantly contributes to enhancing local explanations, aiding stakeholders in comprehending specific challenges or discrepancies within ML models deployed in IoT environments, ultimately fortifying their security and reliability.

References

[1] Lundberg, S. M., Erion, G., Chen, H., *et al.* (2020). From local explanations to global understanding with explainable AI for trees. *Nature Machine Intelligence*, *2*(1), 56–67.

[2] Zafar, M. R., and Khan, N. (2021). Deterministic local interpretable model-agnostic explanations for stable explainability. *Machine Learning and Knowledge Extraction*, *3*(3), 525–541.

[3] Zafar, M. R., and Khan, N. M. (2019). *DLIME:* A deterministic local interpretable model-agnostic explanations approach for computer-aided diagnosis systems. arXiv preprint *arXiv:1906.10263.*

[4] Antwarg, L., Miller, R. M., Shapira, B., and Rokach, L. (2021). Explaining anomalies detected by autoencoders using Shapley Additive Explanations. *Expert Systems with Applications*, *186*, 115736.

[5] Nohara, Y., Matsumoto, K., Soejima, H., and Nakashima, N. (2022). Explanation of machine learning models using shapley additive explanation and application for real data in hospital. *Computer Methods and Programs in Biomedicine*, *214*, 106584.

[6] Albini, E., Long, J., Dervovic, D., and Magazzeni, D. (2022, June). Counterfactual shapley additive explanations. In *Proceedings of the 2022 ACM Conference on Fairness, Accountability, and Transparency* (pp. 1054-1070).

[7] Pornprasit, C., Tantithamthavorn, C., Jiarpakdee, J., Fu, M., and Thongtanunam, P. (2021, November). Pyexplainer: Explaining the predictions of just-in-time defect models. In *2021 36th IEEE/ACM International Conference on Automated Software Engineering (ASE)* (pp. 407-418). IEEE.

[8] Yu, J., Fu, M., Ignatiev, A., Tantithamthavorn, C., and Stuckey, P. (2024). A formal explainer for just-in-time defect predictions. *ACM Transactions on Software Engineering and Methodology*, 33(7), 1–31.

[9] Yang, F., Zeng, G., Zhong, F., Xiao, P., Zheng, W., and Qiu, F. (2024). CfExplainer: Explainable just-in-time defect prediction based on counterfactuals. *Journal of Systems and Software*, *218*, 112182.

[10] Awal, M. A., and Roy, C. K. (2024). EvaluateXAI: A framework to evaluate the reliability and consistency of rule-based XAI techniques for software analytics tasks. *Journal of Systems and Software*, *217*, 112159.

[11] Awal, M. D., Rochan, M., and Roy, C. K. (2024). Investigating Adversarial Attacks in Software Analytics via Machine Learning Explainability. *arXiv preprint arXiv:2408.04124.*

[12] Dhurandhar, A., Chen, P. Y., Luss, R., Tu, C. C., Ting, P., Shanmugam, K., and Das, P. (2018). Explanations based on the missing: Towards contrastive explanations with pertinent negatives. *Advances in Neural Information Processing Systems, 31.*

[13] Jacovi, A., Swayamdipta, S., Ravfogel, S., Elazar, Y., Choi, Y., and Goldberg, Y. (2021). Contrastive explanations for model interpretability. *arXiv preprint arXiv:2103.01378.*

[14] Karimi, A. H., Barthe, G., Schölkopf, B., and Valera, I. (2022). A survey of algorithmic recourse: Contrastive explanations and consequential recommendations. *ACM Computing Surveys*, *55*(5), 1–29.

[15] Prabhushankar, M., Kwon, G., Temel, D., and AlRegib, G. (2020, October). Contrastive explanations in neural networks. *In 2020 IEEE International Conference on Image Processing (ICIP)* (pp. 3289–3293). IEEE.

[16] Guidotti, R. (2024). Counterfactual explanations and how to find them: Literature review and benchmarking. *Data Mining and Knowledge Discovery*, *38*(5), 2770–2824.

[17] Verma, S., Dickerson, J., and Hines, K. (2020). Counterfactual explanations for machine learning: A review. *arXiv preprint arXiv:2010.10596*, 2, 1.

[18] Goyal, Y., Wu, Z., Ernst, J., Batra, D., Parikh, D., and Lee, S. (2019, May). Counterfactual visual explanations. In *International Conference on Machine Learning* (pp. 2376–2384). PMLR.

[19] Verma, S., Boonsanong, V., Hoang, M., Hines, K., Dickerson, J., and Shah, C. (2024). Counterfactual explanations and algorithmic recourses for machine learning: A review. *ACM Computing Surveys*, *56*(12), 1–42.

[20] Ribeiro, M. T., Singh, S., and Guestrin, C. (2018, April). Anchors: High-precision model-agnostic explanations. *In Proceedings of the AAAI Conference on Artificial Intelligence* (Vol. 32, No.1).

[21] Delaunay, J., Galárraga, L., and Largouët, C. (2020, October). Improving anchor-based explanations. *In Proceedings of the 29th ACM International Conference on Information & Knowledge Management* (pp. 3269–3272).

[22] Yan, L., Hwang, A., Wu, Z., and Head, A. (2024, May). Ivie: Lightweight anchored explanations of just-generated code. *In Proceedings of the CHI Conference on Human Factors in Computing Systems* (pp. 1–15).

Chapter 11

Explainability evaluation metrics for explainable security intelligence in the Internet of Things

11.1 Introduction

In the rapidly evolving landscape of AI applied within the IoT, the deployment of complex AI models demands a thorough and systematic evaluation to ensure their reliability, interpretability, and ethical adherence. The significance of evaluation metrics in XAI within IoT ecosystems cannot be overstated. As IoT devices continue to proliferate across various domains, encompassing healthcare, smart homes, transportation, and industrial automation, the need for transparent and accountable AI algorithms becomes vital. Evaluation metrics serve as critical tools that allow practitioners and stakeholders to gauge the performance, fairness, robustness, and interpretability of these AI systems in real-world scenarios. Particularly within IoT environments, where resource constraints, dynamic data streams, and edge computing prevail, tailored evaluation metrics play a pivotal role in not only ensuring the trustworthiness of AI decision-making but also fostering user confidence, addressing ethical considerations, and enhancing the overall adoption of AI technologies in IoT applications [1]. Moreover, in the intricate network of IoT devices generating copious amounts of data, the interpretability of AI models is central to deriving actionable insights and facilitating human understanding. Evaluation metrics tailored for XAI within IoT settings provide a systematic framework to assess the fidelity of explanations offered by AI models. These metrics help validate the consistency and reliability of explanations, aiding in comprehending the decisions made by AI systems operating within diverse IoT environments. By leveraging specialized evaluation metrics, practitioners can ascertain the degree to which AI models in IoT elucidate their decision-making processes, enabling stakeholders to make informed judgments, diagnose system errors, and instill user trust. Furthermore, these evaluation metrics contribute to mitigating biases, ensuring fairness, and upholding ethical standards, aligning with the imperative need to cultivate responsible AI practices within the intricate fabric of IoT, where AI-enabled devices and systems increasingly influence and shape our daily lives [2].

Within the realm of XAI, the reliance on sophisticated algorithms and complex models underscores the criticality of evaluation metrics to guarantee the integrity and efficacy of AI systems across multifaceted domains. Evaluation metrics constitute a cornerstone in pursuing fairness, faithfulness, robustness, localization, complexity, randomization, and axiomatic properties within XAI frameworks. The importance of these metrics lies in their ability to assess and validate the ethical, interpretative, and functional facets of AI systems, particularly in scenarios where decision-making impacts human lives, societal structures, and critical infrastructures. By quantifying and qualifying the performance of XAI models, these metrics not only ensure equitable treatment and prevent biases but also uphold the fidelity of model explanations, bolster system resilience against adversarial attacks, facilitate local interpretations of decisions, gauge the complexity of models for better understanding, inject randomness for security enhancements, and verify compliance with foundational axiomatic principles. In scientific parlance, these metrics serve as indispensable instruments enabling a comprehensive and rigorous examination of XAI systems, fostering transparency, accountability, and ethical governance in their deployment across diverse applications and industries [3].

Integrating evaluation metrics to navigate the nuanced terrain of fairness, faithfulness, robustness, and axiomatic considerations within XAI systems is pivotal to bridging. These metrics serve as guiding beacons, offering systematic methodologies and quantifiable indices to gauge the extent to which AI models adhere to ethical norms, provide accurate explanations, endure adversarial pressures, pinpoint localized decision rationales, exhibit understandable complexity levels, resist malicious intrusions through randomization techniques, and comply with fundamental axiomatic principles. By elucidating the performance benchmarks and evaluation thresholds, these metrics lay the foundation for principled AI development and deployment, ensuring that XAI systems transcend the realms of mere functionality to embrace ethical responsibility and societal inclusivity. In scientific discourse, these metrics form the bedrock for scrutinizing and fine-tuning AI algorithms, fostering a paradigm shift toward accountable and transparent AI systems that earn trust, ensure reliability, and navigate the intricate interplay between technological innovation and ethical stewardship in our increasingly AI-infused world.

11.2 Foundations of evaluation metrics

The assessment of XAI models is a pivotal juncture in the trajectory of AI advancement, necessitating the integration of comprehensive evaluation metrics to ascertain their quality, reliability, and suitability for deployment across multifaceted domains. Evaluation metrics are foundational in discerning the nuanced attributes that encapsulate the quality of XAI models. These metrics serve as indispensable tools that quantitatively and qualitatively gauge the performance, fairness, robustness, interpretability, and ethical adherence of AI systems, transcending mere algorithmic functionality to encompass a spectrum of attributes critical to their acceptance and integration within real-world scenarios. In the scientific domain, these metrics provide a systematic framework for rigorous scrutiny, enabling researchers, practitioners, and stakeholders to assess the strengths and limitations of XAI models, identify potential biases, verify the faithfulness of explanations, ascertain their robustness against adversarial attacks, discern the complexity for comprehensibility, introduce randomness for security, and validate adherence to foundational axiomatic principles. Thus, evaluation metrics stand as the linchpin in the continual refinement and validation of XAI models, facilitating a deeper understanding and trust in their decision-making processes, thereby propelling the evolution of AI systems toward greater transparency, accountability, and ethical efficacy [4].

The essence of evaluation metrics in assessing the quality of XAI models extends beyond mere performance metrics, embracing a holistic evaluation paradigm that harmonizes technological innovation with ethical imperatives. These metrics function as a conduit, enabling a rigorous examination of XAI models not solely based on their predictive accuracy or efficiency but also on their ethical soundness, interpretative clarity, and societal implications. The quality assessment of XAI models becomes a multidimensional endeavor by employing a diverse array of evaluation metrics, ranging from fairness and robustness assessments to considerations of model complexity and adherence to axiomatic principles. In scientific discourse, these metrics elevate the discourse surrounding AI development, fostering an environment where responsible AI innovation converges with ethical and societal considerations. Moreover, the judicious application of evaluation metrics in assessing XAI model quality serves as a cornerstone in the pursuit of accountable and transparent AI systems, engendering trust, fostering societal acceptance, and paving the way for the responsible integration of AI technologies in our interconnected world [5].

In the endeavor to systematically comprehend the multifaceted landscape of XAI evaluation, the categorization of evaluation metrics into distinct groups stands as an imperative pursuit. These metrics, pivotal in assessing the performance, interpretability, fairness, robustness, and ethical adherence of XAI models, can be systematically classified based on diverse criteria. One such criterion entails bifurcating evaluation metrics by their focus on either model-specific or model-agnostic attributes. Model-specific metrics are tailored to particular AI architectures or algorithms, offering insights into the nuanced intricacies of a specific model's behavior. Conversely, model-agnostic metrics transcend the boundaries of individual models, providing a broader assessment framework that applies uniformly across various AI models. Another critical criterion for classification lies in discerning between quantitative and qualitative evaluation metrics. Quantitative metrics encompass numerical indices that offer measurable criteria, such as accuracy, fairness ratios, or fidelity scores, enabling a quantitative assessment of XAI model performance. Conversely, qualitative metrics involve human-centered assessments, encompassing user studies, surveys, or qualitative judgments regarding the interpretability and trustworthiness of model explanations [6].

Moreover, the taxonomy of evaluation metrics can also be delineated based on their granularity of evaluation. This criterion differentiates between instance-level metrics, which focus on individual predictions or instances, and feature-level metrics that assess the contribution and impact of specific features in generating explanations.

Instance-level metrics delve into the fidelity and accuracy of explanations provided by XAI models for each prediction, aiding in scrutinizing model behavior at a granular level. In contrast, feature-level metrics provide insights into the significance and influence of individual features on the model's decision-making, thus contributing to a deeper understanding of the model's inner workings. Additionally, the classification of evaluation metrics can also extend to encompass metrics focused on interpretability versus transparency, elucidating the degree to which users can comprehend and trust model explanations versus the openness and clarity in the model's functioning. These categorizations form a comprehensive framework, enabling a structured understanding of the diverse spectrum of evaluation metrics and their roles in appraising XAI models across various dimensions.

Within the realm of XAI, a comprehensive understanding of basic terminologies, methodologies, and the fundamental rationale underpinning the utilization of evaluation metrics forms the bedrock of scientific inquiry and advancement. Basic terminologies in the context of XAI evaluation metrics encapsulate a spectrum of attributes crucial for model assessment, including fairness, interpretability, robustness, fidelity, and ethical adherence. These terminologies serve as foundational pillars, delineating the multifaceted dimensions along which XAI models are scrutinized and appraised. Methodologies employed in the evaluation of XAI models encompass a diverse array of quantitative and qualitative approaches, leveraging numerical indices, algorithms, user studies, and domain-specific assessments. These methodologies enable a rigorous examination of AI model behaviors, facilitating an in-depth analysis of their performance, reliability, and ethical compliance. Moreover, the rationale behind employing evaluation metrics in XAI stems from the imperative need to ensure the trustworthiness, transparency, and accountability of AI systems. The utilization of these metrics serves as a systematic and structured approach to quantify and qualify the inherent attributes of XAI models, facilitating their acceptance and integration into real-world applications while mitigating risks associated with biases, interpretability limitations, and ethical concerns [7].

The scientific rationale for employing evaluation metrics in XAI is multifaceted, aiming to address the intricate challenges inherent in AI decision-making processes. Evaluation metrics act as standardized measures that facilitate comparative analysis across various models, aiding in benchmarking and identifying best practices within the domain. Furthermore, these metrics play a pivotal role in fostering interpretability and transparency within AI systems, allowing stakeholders to understand, scrutinize, and validate the decisions made by complex AI algorithms. The overarching goal of employing evaluation metrics in XAI is to navigate the delicate balance between technological advancement and ethical considerations, ensuring that AI systems not only deliver accurate predictions but also adhere to ethical norms, engender trust, and contribute positively to societal well-being. As the field of XAI continues to evolve, the judicious utilization of evaluation metrics remains integral, guiding the trajectory of AI development toward responsible, accountable, and trustworthy systems.

11.3 Fairness metrics

In the landscape of XAI, the concept of fairness stands as a cornerstone imperative, underlining the ethical and moral responsibility of AI systems. Fairness in XAI pertains to the equitable treatment of individuals or groups, ensuring that AI models' decisions are devoid of biases, discrimination, or unfairness based on sensitive attributes like race, gender, or ethnicity. The importance of fairness in XAI transcends mere algorithmic accuracy, extending into the realms of social justice, ethical governance, and equitable access to opportunities. Addressing fairness concerns within AI models is paramount to fostering societal trust, mitigating potential harms, and upholding fundamental human rights. Evaluation metrics play a pivotal role in ensuring fairness within AI models by offering systematic methodologies to quantify and qualify the existence of biases, enabling the identification and mitigation of discriminatory patterns or tendencies. These metrics encompass a spectrum of fairness indices, including group unawareness, demographic parity, equal opportunity, and disparate impact assessments. They provide quantitative measures to scrutinize the distribution of outcomes among different demographic groups, ensuring that AI decisions are not disproportionately impacting any particular group, thereby promoting fairness and inclusivity in AI applications [7].

The use of evaluation metrics to assure fairness in AI models is based on the necessity to reduce biases and disparities that are inherent in data-driven decision-making processes. Evaluation metrics enable a systematic evaluation methodology that examines AI models for fairness across different protected traits, such as race, gender, or socioeconomic position. Metrics such as demographic parity evaluate whether AI models generate evenly distributed outputs across various demographic groupings, guaranteeing equal treatment of individuals regardless of their background. Disparate impact measurements assess the unequal treatment of different groups, determining whether certain groups face negative effects more frequently than others. These measurements serve the purpose of both identifying potential biases and assisting in the development of mitigation techniques and adjustments to models, to guarantee equitable treatment and fairness. Using evaluation metrics, XAI practitioners and stakeholders can measure fairness in AI systems. This allows them to identify and correct biases, ensuring that decision-making remains unbiased, fair, and respectful of individual dignity. By doing so, societal trust is fostered and ethical AI practices are promoted, with the goal of achieving fairness and justice [8].

11.3.1 Group unawareness metric

The group unawareness metric is a fundamental tool in evaluating the fairness of AI models, particularly in the context of XAI. This metric operates on the principle of treating individuals equally regardless of their group membership, ensuring that sensitive attributes such as race, gender, or ethnicity do not influence the outcomes produced by AI algorithms. The working principle of the group unawareness metric revolves around assessing the distribution of model predictions or decisions across different demographic groups without the model having access to or utilizing information related to these groups. Mathematically, the group unawareness metric quantifies the disparity in outcomes between various groups by measuring the difference in the rates of positive or negative predictions or decisions. A formulation of this metric involves calculating the statistical parity difference, which evaluates the absolute difference between the proportion of positive outcomes predicted for different groups. For instance, if the rate of positive predictions for one demographic group differs significantly from another, it indicates potential bias or unfair treatment in the AI model. Application scenarios for the group unawareness metric span various domains where fairness in decision-making processes is paramount, including hiring practices, lending decisions, criminal justice systems, and healthcare. In employment, for instance, this metric can scrutinize whether job recommendations or evaluations from AI systems exhibit disparities across diverse demographic groups, ensuring unbiased and equitable treatment for all individuals regardless of their background [9].

The group unawareness metric serves as a critical tool in assessing and mitigating biases within AI models, aligning with ethical and legal imperatives to ensure fairness and non-discrimination. By evaluating the parity in outcomes among different demographic groups, this metric aims to identify and rectify biases that might be embedded in the decision-making processes of AI systems. Its application is particularly salient when AI-powered systems significantly impact individuals' lives, influencing opportunities, access to resources, or potential risks associated with incorrect predictions. For instance, within healthcare, the group unawareness metric can analyze whether diagnostic or treatment recommendations are distributed equitably across diverse patient populations, ensuring that the AI system's decisions are not influenced by factors irrelevant to the medical condition. Moreover, the group unawareness metric serves as a foundational component in the broader spectrum of fairness evaluation metrics, complementing other metrics to provide a comprehensive assessment framework for ensuring fair and unbiased AI systems across diverse societal contexts. Its utilization underscores the commitment to ethical AI development, promoting transparency and accountability while fostering trust and equity in deploying AI technologies.

11.3.2 Demographic parity metric

The demographic parity metric stands as a pivotal evaluation tool in assessing fairness within AI systems, particularly in the domain of XAI. This metric operates on the foundational principle of ensuring equal representation and treatment across various demographic groups, aiming to mitigate biases and disparities within AI models (refer to Figure 11.1). The working principle of the demographic parity metric revolves around comparing the proportion of positive outcomes predicted by the AI model for different demographic groups, ensuring that these proportions are similar across all groups. Mathematically, the demographic parity metric quantifies

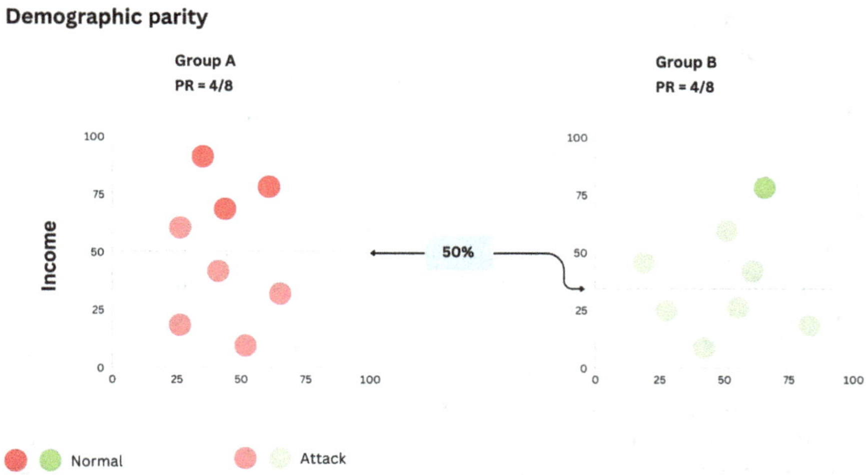

Figure 11.1 *Comparison of predictive performance across demographic groups: demographic parity assessment*

the equality of positive outcome rates by evaluating the difference in these rates across groups. One of the common mathematical formulations of this metric involves computing the absolute difference between the positive outcome rates for each demographic group, aiming for parity or similarity in these rates across all groups. For instance, in predictive tasks such as loan approvals or job recommendations, the demographic parity metric scrutinizes whether the rates of positive outcomes (e.g., loan approvals) are comparable across diverse demographic groups, regardless of their race, gender, or other sensitive attributes. Ensuring similar positive outcome rates indicates a lack of bias or discrimination within the AI model, contributing to fair and equitable decision-making processes [10].

11.3.3 Disparate impact metric

The disparate impact metric serves as a crucial evaluation tool in assessing the fairness of AI models, particularly within the realm of XAI. This metric operates on the fundamental principle of scrutinizing whether the outcomes or decisions made by AI models disproportionately impact certain demographic groups, highlighting potential biases or discrimination within the decision-making process. The working principle of the disparate impact metric involves comparing the rates of favorable outcomes or decisions between different demographic groups and identifying any substantial disparities. Mathematically, the disparate impact metric quantifies the ratio between the favorable outcome rates for different groups, typically focusing on protected attributes like race, gender, or ethnicity. A common mathematical formulation for this metric includes computing the ratio of favorable outcomes for a specific group divided by the favorable outcomes for another group. For instance, in scenarios such as loan approvals or hiring decisions, the disparate impact metric examines whether the acceptance rates for loans or job offers significantly differ among demographic groups, ensuring fairness and non-discrimination. An outcome where there exists a substantial difference in acceptance rates indicates potential biases within the AI model, necessitating corrective actions to ensure equitable decision-making [11].

11.3.4 Equal opportunity metric

The equal opportunity metric is a pivotal tool in assessing fairness within AI models, particularly in the realm of XAI. This metric operates on the fundamental principle of ensuring that the rates of positive outcomes predicted by AI models are consistent across various demographic groups, specifically focusing on the parity in the true positive rates (refer to Figure 11.2). The working principle of the equal opportunity metric revolves around evaluating whether the true positive rates, representing the accuracy of positive predictions for relevant instances (e.g., correctly identifying loan defaulters), are comparable across different demographic groups. Mathematically, the equal opportunity metric quantifies the equality of true positive rates by computing the difference in these rates among groups while focusing on protected attributes like race, gender, or ethnicity. A typical mathematical formulation involves calculating the absolute difference between the true positive rates for each demographic group, aiming for equal opportunities by ensuring minimal disparities in the accuracy of positive predictions. For instance, in scenarios such as loan approvals or criminal justice risk assessments, the equal opportunity metric scrutinizes whether the true positive rates for predicting relevant outcomes are similar across diverse demographic groups, ensuring fairness and equal opportunities in the AI model's predictions [12].

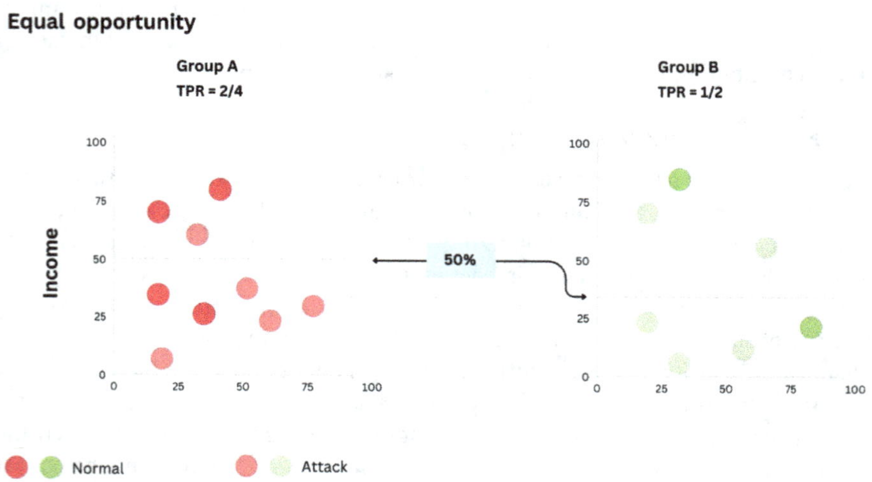

Figure 11.2 Assessing equal opportunity in predictive models: metric evaluation across subgroups

11.3.5 Equal odds metric

The equal odds metric constitutes a pivotal evaluation criterion within the domain of XAI, focusing on ensuring fairness and equitable treatment across diverse demographic groups. This metric operates on the fundamental principle of assessing the parity in both true positive rates and false positive rates (FPR) among different demographic groups (refer to Figure 11.3). The working principle of the equal odds metric revolves around evaluating whether AI models exhibit comparable accuracy in predicting both positive outcomes (true positives) and incorrect predictions of positive outcomes (false positives) across various demographic groups. Mathematically, the equal odds metric quantifies the equality of true positive and FPRs by computing the differences in these rates among groups, typically with respect to protected attributes like race, gender, or ethnicity. One common mathematical formulation involves measuring the absolute differences between the true positive rates and FPRs for each demographic group. For instance, in scenarios such as credit scoring or medical diagnosis, the equal odds metric scrutinizes whether AI models produce similar rates of correct and incorrect positive predictions across diverse demographic groups, ensuring fairness and equal treatment in the model's predictions.

11.3.6 PPV-parity metric

The PPV-parity metric, standing for positive predictive value parity, represents a critical evaluation measure within XAI, specifically focusing on fairness assessment in AI models. This metric operates on the principle of evaluating the parity in PPVs across various demographic groups (refer to Figure 11.4). The working

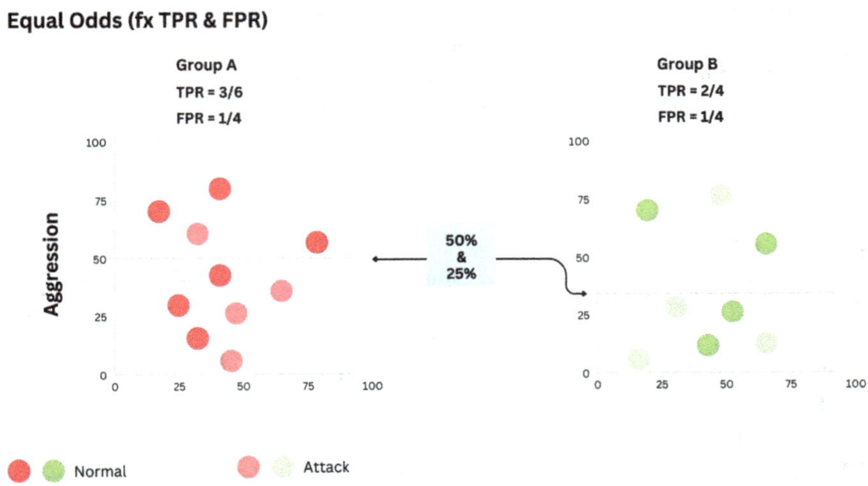

Figure 11.3 Evaluation of equal odds: model fairness analysis based on equal treatment of groups

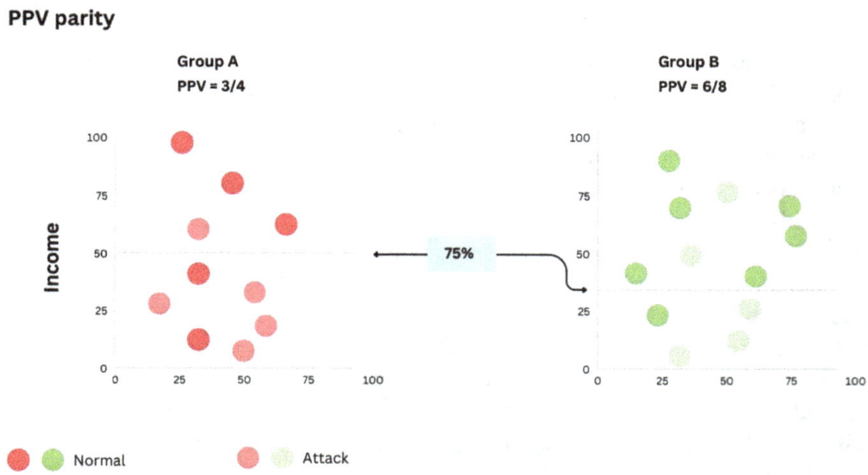

Figure 11.4 Precision parity evaluation: fairness assessment based on positive predictive value (PPV) parity

principle of the PPV-parity metric revolves around assessing whether the accuracy of positive predictions, specifically the likelihood of correct positive predictions among instances predicted as positive by AI models, remains consistent across different demographic groups. Mathematically, the PPV-parity metric quantifies the equality of PPVs by computing the differences in these values among demographic groups, often focusing on protected attributes like race, gender, or ethnicity. A common mathematical formulation involves measuring the absolute differences between the PPVs for each demographic group. For example, in scenarios such as disease diagnosis or credit scoring, the PPV-parity metric scrutinizes whether AI models yield similar rates of accurately predicting positive instances across diverse demographic groups, ensuring fairness and equitable treatment in the model's predictions [13,14].

11.3.7 FPR-parity metric

The FPR-parity metric, which stands for false positive rate parity, plays a crucial role in assessing fairness within AI models, especially in the realm of XAI. This metric operates on the principle of evaluating the parity in FPRs across diverse demographic groups (refer to Figure 11.5). The working principle of the FPR-parity metric revolves around scrutinizing whether the rates of incorrectly predicting positive outcomes (false positives) are consistent among different demographic groups. Mathematically, the FPR-parity metric quantifies the equality of FPRs by computing the differences in these rates among demographic groups, often concerning protected attributes such as race, gender, or ethnicity. A common mathematical formulation involves measuring the absolute differences between the FPRs for each demographic group. For instance, in scenarios like criminal risk assessments or credit scoring, the FPR-parity metric assesses whether AI models produce

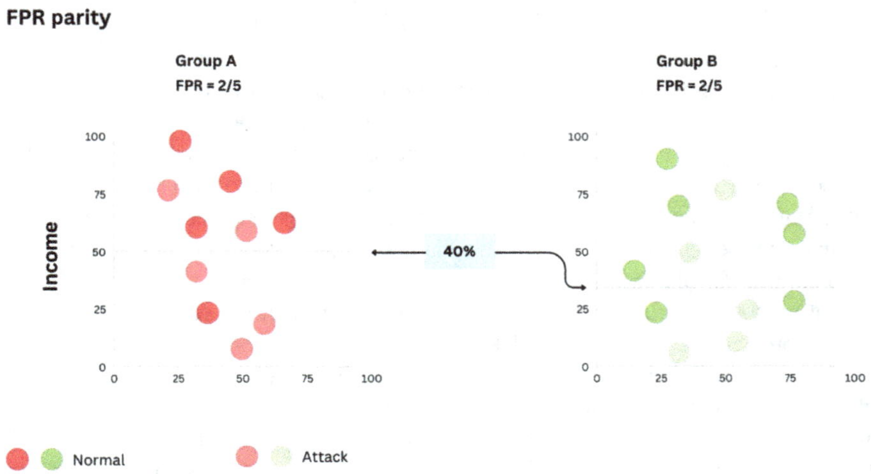

Figure 11.5 FPR parity: fairness metric examination across different groups

similar rates of incorrectly predicting positive outcomes across diverse demographic groups, aiming for fairness and equitable treatment in the model's predictions [15].

11.4 Faithfulness metrics

Faithfulness metrics serve as indispensable tools within XAI, ensuring that explanations offered by AI models accurately align with the model's actual behavior. These metrics play a pivotal role in evaluating the fidelity and reliability of explanations provided by AI systems, ensuring that these explanations are trustworthy, consistent, and aligned with the underlying decision-making processes of the model. The fundamental principle behind faithfulness metrics involves quantifying the correspondence between the explanations offered by the AI model and its actual internal workings. They employ various methodologies such as correlation assessments, pixel-flipping techniques, region segmentation analysis, and sensitivity testing to scrutinize the model's explanations. For instance, metrics like faithfulness correlation examine the coherence between the explanations provided by the model and its predictions, ensuring that the rationale behind each decision aligns with the model's learned features. Moreover, techniques like pixel-flipping or region segmentation evaluate the impact of changing specific inputs or regions on the model's output, validating the faithfulness of the explanations in reflecting how the model processes information. By employing these diverse metrics, XAI practitioners ensure that the explanations generated by AI models are not only interpretable but also faithfully represent the model's actual decision-making mechanisms, instilling trust, understanding, and reliability in AI-driven decisions across various applications and domains.

11.4.1 Faithfulness correlation metric

The faithfulness correlation metric stands as a critical evaluation measure within XAI, aimed at assessing the alignment between the explanations provided by AI models and their actual decision-making processes. This metric operates on the fundamental principle of quantifying the correlation or coherence between the explanations generated by the model and its predictions. The working principle of the faithfulness correlation metric revolves around evaluating the degree of correspondence between the model's explanations and its internal mechanisms. Mathematically, this metric typically employs correlation coefficients, such as Pearson's correlation coefficient or Spearman's rank correlation coefficient, to quantify the relationship between the explanations and the model's predictions. A higher correlation value signifies a stronger alignment between the explanations and the model's decision-making, indicating a higher level of faithfulness. For instance, in applications like image classification or medical diagnostics, the faithfulness correlation metric assesses whether the explanations provided by the model align with the crucial features influencing the model's predictions, ensuring that the explanations accurately capture the rationale behind the model's decisions [15,16]. In the following, we provide an example of evaluating the faithfulness estimate metric for explanation generated from the image classification problem.

```
import numpy as np
import matplotlib.pyplot as plt
import torch
import torchvision
import quantus
from typing import Union

# Load pre-trained MobileNet V3 model.
model = torchvision.models.models.mobilenet_v3_small(pretrained=True)
model = model.to(device)

# Generate Saliency Map of the first batch of the test set.
a_batch = quantus.explain(model, x_batch, y_batch, method='Saliency')
# Save x_batch and y_batch as numpy arrays that will be used to call
metric instances.
x_batch, s_batch, y_batch = x_batch.cpu().numpy(), s_batch.cpu().numpy(),
y_batch.cpu().numpy()
# Make sure s_batch is of correct shape.
s_batch = s_batch.reshape(len(x_batch), 1, 224, 224)
# Quick assert.
assert [isinstance(obj, np.ndarray) for obj in [x_batch, y_batch, s_batch,
a_batch]]

# Return faithfulness correlation scores in an one-liner - by calling the
metric instance.
quantus.FaithfulnessCorrelation(
    nr_runs=100,
    subset_size=224,
    perturb_baseline='black',
    perturb_func=quantus.perturb_func.baseline_replacement_by_indices,
    similarity_func=quantus.similarity_func.correlation_pearson,
    abs=False, return_aggregate=False,) (model=model, x_batch=x_batch,
y_batch=y_batch, a_batch=a_batch, device=device)
```

11.4.2 Faithfulness estimate metric

The faithfulness estimate metric holds significant importance within the domain of XAI, focusing on evaluating the extent to which the explanations provided by AI models accurately reflect the model's behavior. This metric operates on the fundamental principle of estimating the reliability and faithfulness of explanations by assessing how well these explanations align with the model's internal workings. The working principle of the faithfulness estimate metric revolves around quantifying the accuracy and trustworthiness of the explanations generated by the AI model. This metric often employs various statistical and computational techniques to estimate the faithfulness of explanations. Mathematical formulations for the faithfulness estimate metric can vary, but they commonly include methodologies such as error rates, similarity measures, or divergence calculations between the explanations and the model's learned features [17]. For instance, in natural language processing applications, this metric might utilize techniques like cosine similarity to measure the similarity between the explanations provided by the model and the most influential words or features in the model's decision-making process.

In the following, we provide an example of evaluating the faithfulness estimate metric for explanation generated from the image classification problem.

```
import numpy as np
import matplotlib.pyplot as plt
import torch
import torchvision
import quantus
from typing import Union

# Load pre-trained MobileNet V3 model.
model = torchvision.models.models.mobilenet_v3_small(pretrained=True)
model = model.to(device)

# Generate Saliency Map of the first batch of the test set.
a_batch = quantus.explain(model, x_batch, y_batch, method='Saliency')
# Save x_batch and y_batch as numpy arrays that will be used to call
metric instances.
x_batch, s_batch, y_batch = x_batch.cpu().numpy(), s_batch.cpu().numpy(),
y_batch.cpu().numpy()
# Make sure s_batch is of correct shape.
s_batch = s_batch.reshape(len(x_batch), 1, 224, 224)
# Quick assert.
assert [isinstance(obj, np.ndarray) for obj in [x_batch, y_batch, s_batch,
a_batch]]

# Return faithfulness estimate scores in an one-liner - by calling the
metric instance.
quantus.FaithfulnessEstimate(
    perturb_func=quantus.perturb_func.baseline_replacement_by_indices,
    similarity_func=quantus.similarity_func.correlation_pearson,
    features_in_step=224,
    perturb_baseline='black',
)(model=model,
   x_batch=x_batch,
   y_batch=y_batch,
   a_batch=a_batch,
   device=device,)
```

11.4.3 Pixel-flipping metric

The pixel-flipping metric serves as a fundamental tool within the realm of XAI, focusing on assessing the faithfulness of explanations provided by AI models by perturbing individual input features or pixels and observing the resultant impact on the model's predictions. This metric operates on the core principle of evaluating the sensitivity and robustness of the explanations generated by the model concerning perturbations in the input features. The working principle of the pixel-flipping metric involves systematically altering individual pixels or input features within the data and observing the consequent changes in the model's output or explanations. The metric quantifies the faithfulness by measuring how sensitive the model's explanations are to changes in specific features or pixels. Mathematically, this metric often involves algorithms or calculations that perturb individual features or pixels, such as altering pixel values in images or modifying specific attributes in datasets, and then analyzing the impact of these changes on the model's generated explanations. For instance, in image recognition tasks, the pixel-flipping metric might perturb individual pixels within an image to ascertain the significance of each pixel in the model's explanation for its classification decision, thereby assessing the reliability and robustness of the provided explanations.

In the following, we provide an example of evaluating the pixel-flipping metric for explanation generated from the image classification problem.

```
import numpy as np
import matplotlib.pyplot as plt
import torch
import torchvision
import quantus
from typing import Union

# Load pre-trained MobileNet V3 model.
model = torchvision.models.models.mobilenet_v3_small(pretrained=True)
model = model.to(device)

# Generate Saliency Map of the first batch of the test set.
a_batch = quantus.explain(model, x_batch, y_batch, method='Saliency')
# Save x_batch and y_batch as numpy arrays that will be used to call
metric instances.
x_batch, s_batch, y_batch = x_batch.cpu().numpy(), s_batch.cpu().numpy(),
y_batch.cpu().numpy()
# Make sure s_batch is of correct shape.
s_batch = s_batch.reshape(len(x_batch), 1, 224, 224)
# Quick assert.
assert [isinstance(obj, np.ndarray) for obj in [x_batch, y_batch, s_batch,
a_batch]]

# Create the pixel-flipping experiment.
pixel_flipping = quantus.PixelFlipping(
    features_in_step=224,
    perturb_baseline='black',
    perturb_func=quantus.perturb_func.baseline_replacement_by_indices,
)
```

```
# Call the metric instance to produce scores.
scores = pixel_flipping(model=model,
                        x_batch=x_batch,
                        y_batch=y_batch,
                        a_batch=a_batch,
                        device=device,)

# Plot example!
pixel_flipping.plot(y_batch=y_batch, scores=scores)
```

11.4.4 Monotonicity metric

The monotonicity metric represents a crucial evaluation measure within XAI, focused on assessing the monotonic relationship between input features and the model's predictions, ensuring that the model's behavior aligns with expected trends or patterns. This metric operates on the foundational principle of evaluating the monotonicity or consistency in the relationship between individual input features and the model's output predictions. The working principle of the monotonicity-Arya metric involves examining whether the model's predictions exhibit a consistent directional relationship with changes in specific input features. Mathematically, this metric might employ various statistical or computational techniques to assess the monotonic relationship, such as rank-order correlation methods or curve-fitting algorithms, to quantify the degree of monotonicity between input features and model predictions [18].

In the following, we provide an example of evaluating the monotonicity metric for explanation generated from the image classification problem.

```
import numpy as np
import matplotlib.pyplot as plt
import torch
import torchvision
import quantus
from typing import Union

# Load pre-trained MobileNet V3 model.
model = torchvision.models.models.mobilenet_v3_small(pretrained=True)
model = model.to(device)

# Generate Saliency Map of the first batch of the test set.
a_batch = quantus.explain(model, x_batch, y_batch, method='Saliency')
# Save x_batch and y_batch as numpy arrays that will be used to call
metric instances.
x_batch, s_batch, y_batch = x_batch.cpu().numpy(), s_batch.cpu().numpy(),
y_batch.cpu().numpy()
# Make sure s_batch is of correct shape.
s_batch = s_batch.reshape(len(x_batch), 1, 224, 224)
# Quick assert.
assert [isinstance(obj, np.ndarray) for obj in [x_batch, y_batch, s_batch,
a_batch]]
```

```
# Return monotonicity scores in an one-liner - by calling the metric
instance.
quantus.Monotonicity(
    features_in_step=224,
    perturb_baseline='black',
    perturb_func=quantus.perturb_func.baseline_replacement_by_indices,
)(model=model,
   x_batch=x_batch,
   y_batch=y_batch,
   a_batch=a_batch,
   device=device,)
```

11.4.5 Monotonicity correlation metric

The monotonicity correlation metric was proposed by Nguyen and focused on assessing and quantifying the degree of correlation between the monotonic behavior of interpretable layers within neural networks and their relationship to the output layer. The metric aims to ensure and evaluate the consistency of the monotonic behavior between selected interpretable layers and the ultimate output, providing a means to measure the alignment of their directional changes. The formulation emphasized the construction of neural network layers with defined monotonic properties. Specifically, the metric evaluates the relationships between layers within the network, particularly focusing on ensuring monotonicity, either increasing or decreasing, between the interpretable layers and the output layer. This evaluation is based on definitions and constraints formulated within the paper, such as verifying if chosen interpretable layers exhibit monotonic behaviors concerning specific components or features of the output layer.

The proposed methodology involves constructing neural network layers that guarantee monotonic relationships using activation functions and constraints on weight matrices. The metric assesses the correlation of these designated layers' monotonicity with the overall output by ensuring that certain layers exhibit monotonic behaviors concerning specific output components. In essence, the monotonicity correlation metric aims to quantitatively measure the degree of alignment or correlation between the monotonic behavior of interpretable layers and the output layer within neural networks. It evaluates the extent to which these selected layers exhibit desired monotonic properties concerning specific output components, providing a means to ensure consistent directional changes in the interpretable layers that correspond to the model's output.

In the following, we provide an example of evaluating the monotonicity correlation metric for explanation generated from the image classification problem.

```
import numpy as np
import matplotlib.pyplot as plt
import torch
import torchvision
import quantus
from typing import Union
```

```
# Load pre-trained MobileNet V3 model.
model = torchvision.models.models.mobilenet_v3_small(pretrained=True)
model = model.to(device)

# Generate Saliency Map of the first batch of the test set.
a_batch = quantus.explain(model, x_batch, y_batch, method='Saliency')
# Save x_batch and y_batch as numpy arrays that will be used to call
metric instances.
x_batch, s_batch, y_batch = x_batch.cpu().numpy(), s_batch.cpu().numpy(),
y_batch.cpu().numpy()
# Make sure s_batch is of correct shape.
s_batch = s_batch.reshape(len(x_batch), 1, 224, 224)
# Quick assert.
assert [isinstance(obj, np.ndarray) for obj in [x_batch, y_batch, s_batch,
a_batch]]

# Return monotonicity scores in an one-liner - by calling the metric
instance.
quantus.MonotonicityCorrelation(
    nr_samples=10,
    features_in_step=3136,
    perturb_baseline='uniform',
    perturb_func=quantus.perturb_func.baseline_replacement_by_indices,
    similarity_func=quantus.similarity_func.correlation_spearman)(model=mo
del,
   x_batch=x_batch,
   y_batch=y_batch,
   a_batch=a_batch,
   device=device,)
```

11.4.6 Selectivity metric

The selectivity metric operates as a pivotal evaluation criterion within XAI, aiming to measure the extent of feature selectivity or discriminative behavior exhibited by AI models. This metric delves into understanding how an AI model discriminates or prioritizes certain input features over others when making predictions or classifications. The working principle of the selectivity metric involves quantifying the model's preference or emphasis on specific features by assessing the magnitude of their impact on the model's decision-making process. Mathematically, this metric could involve calculations related to feature importance scores, such as determining the weight or significance attributed to individual features within the model's architecture. The selectivity metric assesses how crucial or influential each feature is in influencing the model's output or predictions. It might involve methodologies like feature importance analysis, examining how alterations or removal of particular features affect the model's performance. Furthermore, this metric could encompass statistical measures or algorithms aimed at ranking or scoring features based on their impact on the model's predictions.

In the following, we provide a Python implementation for evaluating the selectivity metric for the explanation generated from the image classification problem.

```
import numpy as np
import matplotlib.pyplot as plt
import torch
import torchvision
import quantus
from typing import Union

# Load pre-trained MobileNet V3 model.
model = torchvision.models.models.mobilenet_v3_small(pretrained=True)
model = model.to(device)

# Save x_batch and y_batch as numpy arrays that will be used to call
metric instances.
x_batch, s_batch, y_batch = x_batch.cpu().numpy(), s_batch.cpu().numpy(),
y_batch.cpu().numpy()
# Make sure s_batch is of correct shape.
s_batch = s_batch.reshape(len(x_batch), 1, 224, 224)
# Quick assert.
assert [isinstance(obj, np.ndarray) for obj in [x_batch, y_batch, s_batch,
a_batch]]

# Create the selectivity experiment.
selectivity = quantus.Selectivity(
    patch_size=56,
    perturb_baseline="black",
)

# Call the metric instance to produce scores.
results = {method: selectivity(model=model,
                               x_batch=x_batch,
                               y_batch=y_batch,
                               a_batch=None,
                               device=device,
                               explain_func=quantus.explain,
                               explain_func_kwargs={"method": method}) for
method in ['Saliency']}

# Plot example!
selectivity.plot(results=results)
results
```

11.4.7 SensitivityN metric

The SensitivityN metric proposed by Ancona *et al.* in 2018 operates as a significant evaluation criterion within XAI, specifically aiming to measure the sensitivity of individual input features on a model's predictions in a nuanced manner. This metric's principle revolves around quantifying the sensitivity of the model's output concerning alterations or perturbations in individual input features. The working principle of the SensitivityN metric involves calculating the change in the model's predictions when specific input features are perturbed or modified while keeping the rest of the input unchanged. Mathematically, this metric employs partial derivatives or sensitivity analysis to assess how alterations in individual features impact the model's output. The SensitivityN metric evaluates the model's sensitivity to changes in input features, providing insights into which features significantly influence the model's predictions. This metric could encompass methodologies such as gradient-based approaches or sensitivity analysis techniques to quantify the impact of each feature on the model's output. Understanding feature sensitivity aids in interpreting the model's behavior and assists in feature selection or modification for improved model performance. This metric finds applications in various domains where pinpointing influential features affecting predictions is crucial, such as healthcare diagnostics, financial analysis, or image recognition tasks, enhancing the interpretability and trustworthiness of AI models.

In the following, we provide a Python implementation for evaluating the explanation generated from the image classifier using SensitivityN metric.

```
import numpy as np
import matplotlib.pyplot as plt
import torch
import torchvision
import quantus
from typing import Union

# Load pre-trained MobileNet V3 model.
model = torchvision.models.models.mobilenet_v3_small(pretrained=True)
model = model.to(device)

# Save x_batch and y_batch as numpy arrays that will be used to call
metric instances.
x_batch, s_batch, y_batch = x_batch.cpu().numpy(), s_batch.cpu().numpy(),
y_batch.cpu().numpy()
# Make sure s_batch is of correct shape.
s_batch = s_batch.reshape(len(x_batch), 1, 224, 224)
# Quick assert.
assert [isinstance(obj, np.ndarray) for obj in [x_batch, y_batch, s_batch,
a_batch]]

# Create the sensitivity-n experiment.
sensitivity_n = quantus.SensitivityN(
    features_in_step=224,
    n_max_percentage=0.8,
    similarity_func=quantus.similarity_func.correlation_pearson,
    perturb_func=quantus.perturb_func.baseline_replacement_by_indices,
    perturb_baseline='uniform',
    return_aggregate=False,
)
# Call the metric instance to produce scores.
results = sensitivity_n(model=model,
                        x_batch=x_batch,
                        y_batch=y_batch,
                        a_batch=None,
                        device=device,
                        explain_func=quantus.explain,
                        explain_func_kwargs={'method':
'IntegratedGradients'})


# Plot example!
sensitivity_n.plot(results=results)
```

11.4.8 Infidelity metric

The infidelity metric in the realm of XAI serves as a crucial benchmark to measure the fidelity or faithfulness of an AI model's explanations concerning its actual decision-making process. This metric operates on the principle of quantifying the disparity or inconsistency between the explanations provided by the model and the model's true underlying behavior or decision rationale. The working principle of the infidelity metric involves assessing the degree of discrepancy between the model's explanation and its actual decision function, aiming to quantify the level of fidelity or faithfulness in the interpretative output. Mathematically, this metric could involve calculations related to the divergence or dissimilarity between the model's explanations and its internal decision mechanisms, potentially utilizing metrics like cosine similarity, divergence measures, or loss functions. The infidelity metric aims to evaluate how well the explanations generated by an AI model truly represent its decision-making process. It involves methodologies focused on assessing the alignment or coherence between the model's explanations and the actual factors influencing its predictions.

In the following, we provide a Python implementation for evaluating the explanation generated from the image classifier using the infidelity metric.

```
import numpy as np
import matplotlib.pyplot as plt
import torch
import torchvision
import quantus
from typing import Union

# Load pre-trained MobileNet V3 model.
model = torchvision.models.models.mobilenet_v3_small(pretrained=True)
model = model.to(device)

# Save x_batch and y_batch as numpy arrays that will be used to call
metric instances.
x_batch, s_batch, y_batch = x_batch.cpu().numpy(), s_batch.cpu().numpy(),
y_batch.cpu().numpy()
# Make sure s_batch is of correct shape.
s_batch = s_batch.reshape(len(x_batch), 1, 224, 224)
# Quick assert.
assert [isinstance(obj, np.ndarray) for obj in [x_batch, y_batch, s_batch,
a_batch]]

# Return Infidelity scores in an one-liner - by calling the metric
instance.
quantus.Infidelity(
    perturb_baseline='uniform',
    perturb_func=quantus.perturb_func.baseline_replacement_by_indices,
    n_perturb_samples=5,
    perturb_patch_sizes=[56],
    display_progressbar=True,
)(model=model,
   x_batch=x_batch,
   y_batch=y_batch,
   a_batch=None,
   device=device,
   explain_func=quantus.explain,
   explain_func_kwargs={'method': 'Saliency'})
```

11.5 Robustness metrics

Assessing the robustness of XAI models against perturbations and adversarial attacks stands as a critical endeavor to ensure the reliability and stability of these models in real-world applications. Various metrics and methodologies have been developed to evaluate the resilience of XAI models when subjected to perturbations or maliciously crafted inputs. Robustness metrics aim to scrutinize the model's behavior when faced with alterations or perturbations in input data, ensuring that the model's predictions remain consistent and reliable. These metrics delve into understanding the model's susceptibility to adversarial attacks, where intentionally manipulated inputs aim to deceive or mislead the model's decision-making processes. Robustness assessments within XAI typically involve techniques such as sensitivity analysis, adversarial testing, perturbation analysis, and metrics that evaluate the model's performance under varying conditions or perturbed inputs [19].

11.5.1 Continuity test

The continuity test serves as a fundamental metric within the realm of XAI, focusing on evaluating the robustness and stability of AI models when subjected to slight variations or perturbations in input data. The working principle of the continuity test revolves around assessing the model's response continuity when input features undergo minor changes. This metric ensures that small variations or perturbations in input data do not cause drastic or unexpected alterations in the model's predictions or explanations. Mathematically, the continuity test might involve measuring the model's output changes or divergences in explanations due to introducing small modifications in input features. It typically employs mathematical formulations that quantify the deviation or variations between the model's responses before and after perturbations. For instance, continuity might be evaluated through metrics like Lipschitz continuity, which measures the model's sensitivity to input changes within a bounded region [20].

In the following, we provide a Python implementation for evaluating the explanation generated from the image classifier using the continuity test.

```
import numpy as np
import matplotlib.pyplot as plt
import torch
import torchvision
import quantus
from typing import Union

# Load pre-trained MobileNet V3 model.
model = torchvision.models.models.mobilenet_v3_small(pretrained=True)
model = model.to(device)

# Save x_batch and y_batch as numpy arrays that will be used to call
metric instances.
x_batch, s_batch, y_batch = x_batch.cpu().numpy(), s_batch.cpu().numpy(),
y_batch.cpu().numpy()
# Make sure s_batch is of correct shape.
s_batch = s_batch.reshape(len(x_batch), 1, 224, 224)
```

```
# Quick assert.
assert [isinstance(obj, np.ndarray) for obj in [x_batch, y_batch, s_batch,
a_batch]]

# Return continuity scores in an one-liner - by calling the metric
instance.
continuity_test = quantus.Continuity(
    patch_size=56,
    nr_steps=10,
    perturb_baseline='uniform',
    similarity_func=quantus.similarity_func.correlation_spearman,
)
scores = continuity_test(model=model,
                         x_batch=x_batch,
                         y_batch=y_batch,
                         a_batch=None,
                         explain_func=quantus.explain,
                         explain_func_kwargs={"method": "Saliency"},
                         device=device)
scores
```

11.5.2 Local Lipschitz estimate

The local Lipschitz estimate constitutes a pivotal metric within the domain of XAI, aiming to quantify the local smoothness or stability of AI models concerning variations in input data. This metric operates based on the foundational principle of estimating the Lipschitz constant locally, which denotes the rate of change or sensitivity of the model's predictions to perturbations in input features within a specific region. The working principle of the local Lipschitz estimate involves determining the Lipschitz constant for a particular region around a data point and assessing how small changes in input features correspond to alterations in the model's predictions or explanations [21]. Mathematically, the local Lipschitz estimate might involve computing gradients or derivatives of the model's output concerning input features and constraining these changes within a local region. It aims to provide an estimation of how sensitive the model is to input variations, offering insights into the model's stability and smoothness within that particular vicinity. By utilizing the local Lipschitz estimate, stakeholders gain insights into the model's local smoothness and stability, enhancing trust and reliability in AI-driven decisions in various real-world IoT applications, while also offering a measure of model interpretability and robustness [22].

In the following, we provide a Python implementation for evaluating the explanation generated from the image classifier using the local Lipschitz estimate.

```
import numpy as np

import matplotlib.pyplot as plt
import torch
import torchvision
```

```
import quantus
from typing import Union

# Load pre-trained MobileNet V3 model.
model = torchvision.models.models.mobilenet_v3_small(pretrained=True)
model = model.to(device)

# Save x_batch and y_batch as numpy arrays that will be used to call
metric instances.
x_batch, s_batch, y_batch = x_batch.cpu().numpy(), s_batch.cpu().numpy(),
y_batch.cpu().numpy()
# Make sure s_batch is of correct shape.
s_batch = s_batch.reshape(len(x_batch), 1, 224, 224)
# Quick assert.
assert [isinstance(obj, np.ndarray) for obj in [x_batch, y_batch, s_batch,
a_batch]]


# Return local lipschitzEstimate estimate scores in an one-liner - by
calling the metric instance.
quantus.LocalLipschitzEstimate(
    nr_samples=10,
    perturb_std=0.2,
    perturb_mean=0.0,
    norm_numerator=quantus.similarity_func.distance_euclidean,
    norm_denominator=quantus.similarity_func.distance_euclidean,
    perturb_func=quantus.perturb_func.gaussian_noise,
    similarity_func=quantus.similarity_func.lipschitz_constant,
)(model=model,
   x_batch=x_batch,
   y_batch=y_batch,
   a_batch=None,
   device=device,
   explain_func=quantus.explain,
   explain_func_kwargs={'method': 'Saliency'})
```

11.5.3 Max-sensitivity metric

The max-sensitivity metric is an important evaluation tool in XAI. It measures the maximum sensitivity of AI models to changes or perturbations in input features. This metric is based on the core premise of measuring the highest rate of change in the model's predictions with respect to changes in input data. The max-sensitivity metric operates by calculating the highest level of sensitivity exhibited by the model by assessing the largest potential alteration in output for a given modification in input attributes. Mathematically, this metric may include computing the highest derivative or gradient of the model's output concerning input features, with the goal of identifying the most significant changes in predictions when input features are modified. Its objective is to establish a maximum limit on the model's responsiveness, emphasizing the highest degree to which alterations in input characteristics impact the model's forecasts or justifications. Within safety-critical systems like autonomous vehicles or healthcare equipment, this measure can evaluate the highest level of sensitivity in the model's responses to changes in environmental or patient data. This ensures a comprehension of the model's behavior under extreme situations. By utilizing the max-sensitivity metric, stakeholders can obtain valuable information about the model's highest sensitivity to changes in input. This information is useful for assessing risks, making decisions, and ensuring the strength and dependability of AI models in various real-world scenarios. Ultimately, this contributes to building trust and transparency in AI-driven systems.

In the following, we provide a Python implementation for evaluating the explanation generated from the image classifier using the max-sensitivity metric.

```
import numpy as np
import matplotlib.pyplot as plt
import torch
import torchvision
import quantus
from typing import Union

# Load pre-trained MobileNet V3 model.
model = torchvision.models.models.mobilenet_v3_small(pretrained=True)
model = model.to(device)

# Save x_batch and y_batch as numpy arrays that will be used to call
metric instances.
x_batch, s_batch, y_batch = x_batch.cpu().numpy(), s_batch.cpu().numpy(),
y_batch.cpu().numpy()
# Make sure s_batch is of correct shape.
s_batch = s_batch.reshape(len(x_batch), 1, 224, 224)
# Quick assert.
assert [isinstance(obj, np.ndarray) for obj in [x_batch, y_batch, s_batch,
a_batch]]

# Return max sensitivity scores in an one-liner - by calling the metric
instance.
quantus.MaxSensitivity(
    nr_samples=10,
    lower_bound=0.2,
    norm_numerator=quantus.norm_func.fro_norm,
    norm_denominator=quantus.norm_func.fro_norm,
    perturb_func=quantus.perturb_func.uniform_noise,
    similarity_func=quantus.similarity_func.difference)(model=model,
   x_batch=x_batch,
   y_batch=y_batch,
   a_batch=None,
   device=device,
   explain_func=quantus.explain,
   explain_func_kwargs={"method": "Gradient"})
```

11.5.4 Average sensitivity (Avg Sensitivity) metric

The avg-sensitivity measure is a vital tool in XAI employed to evaluate the average sensitivity of AI models to changes in input data. This statistic is based on assessing the global rate of change in a model's predictions with respect to fluctuations throughout its whole input feature space. The avg-sensitivity measure operates by computing the mean sensitivity of the model, which quantifies the average rate of variation in the model's output in response to modifications in the input features. Mathematically, this measure usually entails calculating the mean derivative or gradient of the model's output with respect to input features, which represents the average influence of input fluctuations on the model's predictions. Calculating the mean sensitivity provides a thorough comprehension of how changes in the input space collectively impact the model's behavior, offering a measure of the model's overall responsiveness to variations in input [5].

In the following, we provide a Python implementation for evaluating the explanation generated from the image classifier using the avg-sensitivity metric.

```
import numpy as np
import matplotlib.pyplot as plt
import torch
import torchvision
import quantus
from typing import Union

# Load pre-trained MobileNet V3 model.
model = torchvision.models.models.mobilenet_v3_small(pretrained=True)
model = model.to(device)

# Save x_batch and y_batch as numpy arrays that will be used to call
metric instances.
x_batch, s_batch, y_batch = x_batch.cpu().numpy(), s_batch.cpu().numpy(),
y_batch.cpu().numpy()
# Make sure s_batch is of correct shape.
s_batch = s_batch.reshape(len(x_batch), 1, 224, 224)
# Quick assert.
assert [isinstance(obj, np.ndarray) for obj in [x_batch, y_batch, s_batch,
a_batch]]

# Return avg sensitivity scores in an one-liner - by calling the metric
instance.
quantus.AvgSensitivity(
    nr_samples=10,
    lower_bound=0.2,
    norm_numerator=quantus.norm_func.fro_norm,
    norm_denominator=quantus.norm_func.fro_norm,
    perturb_func=quantus.perturb_func.uniform_noise,
    similarity_func=quantus.similarity_func.difference,
)(model=model,
  x_batch=x_batch,
  y_batch=y_batch,
  a_batch=None,
  explain_func=quantus.explain,
  explain_func_kwargs={"method": "Saliency"},
  device=device)
```

11.5.5 Consistency metric

The consistency measure in XAI is a crucial parameter for evaluating the stability and coherence of an AI model's explanations or predictions across different data instances or subsets. The consistency metric is based on the notion of assessing the resilience and dependability of the model's output by quantifying the likeness or steadiness of its explanations when exposed to variations in input data. The metric seeks to measure the degree of consistency in the model's explanations or predictions by evaluating the level of alignment or invariance in the interpretative outputs across various samples or subsets.

The consistency metric assesses the model's reliability in delivering consistent explanations or predictions across different circumstances or subsets of the dataset. The process entails utilizing approaches such as computing statistical measurements, similarity scores, or other metrics to quantify the likeness or unlikeness of the model's interpretative outputs across various instances or subsets. This metric aids in comprehending the model's ability to generalize and its resilience to alterations in the input data by evaluating the consistency of the model's outputs. The consistency measure is based on mathematical formulations that include quantifying the similarity or stability of the interpretative outputs of the model across different data instances or subsets. The primary premise of measuring model consistency entails evaluating the level of similarity, variance, or stability in the model's explanations or predictions when exposed to different data circumstances. The specific mathematical representations may differ depending on the model and the evaluation criteria utilized.

In the following, we provide a Python implementation for evaluating the explanation generated from the image classifier using the consistency metric.

```
import numpy as np
import matplotlib.pyplot as plt
import torch
import torchvision
import quantus
from typing import Union

# Load pre-trained MobileNet V3 model.
model = torchvision.models.models.mobilenet_v3_small(pretrained=True)
model = model.to(device)

# Save x_batch and y_batch as numpy arrays that will be used to call
metric instances.
x_batch, s_batch, y_batch = x_batch.cpu().numpy(), s_batch.cpu().numpy(),
y_batch.cpu().numpy()
# Make sure s_batch is of correct shape.
s_batch = s_batch.reshape(len(x_batch), 1, 224, 224)
# Quick assert.
assert [isinstance(obj, np.ndarray) for obj in [x_batch, y_batch, s_batch,
a_batch]]
```

```
# Return Consistency scores in an one-liner - by calling the metric
instance.
quantus.Consistency(
    discretise_func=quantus.discretise_func.top_n_sign,
    return_aggregate=False,
)(model=model,
   x_batch=x_batch,
   y_batch=y_batch,
   a_batch=None,
   explain_func=quantus.explain,
   explain_func_kwargs={'method': 'Saliency'},
   device=device)
```

11.5.6 Relative input stability metric

The relative input stability metric is a crucial evaluation tool in XAI that aims to quantify the stability of an AI model's predictions when faced with changes or disturbances in input features, in relation to the size of those changes. This measure is based on the concept of assessing the impact of changes in input features on the model's predictions, with a focus on the consistency of predictions when the input data is altered. The relative input stability metric operates by assessing the model's response to fluctuations in input features and measuring the extent to which the model maintains stability in the face of these perturbations. Mathematically, this measure involves measuring the relationship between changes in the model's output and variations in the input. It indicates how stable the model's predictions are in relation to the magnitude of changes in the input features. Through the evaluation of relative stability, it provides valuable information regarding the model's coherence and dependability in generating predictions under different input circumstances.

In the following, we provide a Python implementation for evaluating the explanation generated from the image classifier using the relative input stability metric.

```
import numpy as np
import matplotlib.pyplot as plt
import torch
import torchvision
import quantus
from typing import Union

# Load pre-trained MobileNet V3 model.
model = torchvision.models.models.mobilenet_v3_small(pretrained=True)
model = model.to(device)

# Save x_batch and y_batch as numpy arrays that will be used to call
metric instances.
x_batch, s_batch, y_batch = x_batch.cpu().numpy(), s_batch.cpu().numpy(),
y_batch.cpu().numpy()
# Make sure s_batch is of correct shape.
s_batch = s_batch.reshape(len(x_batch), 1, 224, 224)
```

```
# Quick assert.
assert [isinstance(obj, np.ndarray) for obj in [x_batch, y_batch, s_batch,
a_batch]]

# Return Relative Input Stability scores in an one-liner - by calling the
metric instance.
quantus.RelativeInputStability(nr_samples=5)(
    model=model,
    x_batch=x_batch,
    y_batch=y_batch,
    explain_func=quantus.explain,
    explain_func_kwargs={"method": "Saliency"},
    device=device
)
```

11.5.7 Relative output stability metric

The relative output stability metric is a pivotal evaluation measure within XAI, focusing on assessing the stability of an AI model's predictions concerning variations in output or predictions resulting from changes in input features. This metric operates based on the fundamental principle of quantifying the consistency or stability of the model's predictions relative to changes in input data. The working principle of the relative output stability metric involves evaluating the model's output changes concerning variations in input features to determine how stable the predictions remain amidst varying input conditions. Mathematically, this metric might involve quantifying the ratio or proportion of changes in the model's predictions concerning variations in input, illustrating the stability of the model's output in relation to alterations in the input features. By assessing the relative output stability, it offers insights into the model's consistency and reliability in producing predictions when subjected to varying input conditions.

In the following, we provide a Python implementation for evaluating the explanation generated from the image classifier using the relative output stability metric.

```
import numpy as np
import matplotlib.pyplot as plt
import torch
import torchvision
import quantus
from typing import Union

# Load pre-trained MobileNet V3 model.
model = torchvision.models.models.mobilenet_v3_small(pretrained=True)
model = model.to(device)

# Save x_batch and y_batch as numpy arrays that will be used to call
metric instances.
x_batch, s_batch, y_batch = x_batch.cpu().numpy(), s_batch.cpu().numpy(),
y_batch.cpu().numpy()
# Make sure s_batch is of correct shape.
s_batch = s_batch.reshape(len(x_batch), 1, 224, 224)
```

```
# Quick assert.
assert [isinstance(obj, np.ndarray) for obj in [x_batch, y_batch, s_batch,
a_batch]]

# Return Relative Output Stability scores in an one-liner - by calling the
metric instance.
quantus.RelativeOutputStability(nr_samples=5)(
    model=model,
    x_batch=x_batch,
    y_batch=y_batch,
    explain_func=quantus.explain,
    explain_func_kwargs={"method": "Saliency"},
    device=device
)
```

11.6 Localization metrics

In the domain of XAI, a significant area of focus revolves around metrics aimed at localizing and attributing model predictions, contributing to a deeper understanding of how AI models arrive at specific decisions. These metrics emphasize the interpretability and transparency of AI systems by shedding light on the features or regions of input data that significantly influence the model's predictions. Metrics focused on localization and attribution aim to pinpoint the specific input features or regions that have the most substantial impact on the model's outputs, providing insights into the rationale behind the model's decision-making process. Through various localization techniques such as the pointing game, relevance mass accuracy, or attribution methods like feature attribution, these metrics seek to attribute significance to input features or regions, aiding in comprehending the model's decision process and facilitating domain experts' understanding of AI predictions, ultimately fostering trust and interpretability in AI-driven systems.

11.6.1 Pointing game metric

The pointing game metric is a pivotal evaluation method within the realm of XAI, designed to assess the model's capability to accurately pinpoint and highlight the relevant features in an input that significantly contribute to the model's predictions. This metric operates on the principle of measuring how well the model correctly identifies and highlights the crucial features in an input that are most influential in determining the output. The working principle of the pointing game metric involves comparing the model's identified important features or regions within the input data against a ground truth, typically provided by human experts or domain knowledge, to assess alignment or accuracy. Mathematically, this metric might involve calculating metrics such as precision, recall, or intersection-over-union (IoU) to quantify the overlap between the model's highlighted features and the ground truth, providing a measure of how accurately the model localizes significant aspects within the input data.

In the following, we provide a Python implementation for evaluating the explanation generated from the image classifier using the pointing game metric.

```
import numpy as np
import matplotlib.pyplot as plt
import torch
import torchvision
import quantus
from typing import Union

# Load pre-trained MobileNet V3 model.
model = torchvision.models.models.mobilenet_v3_small(pretrained=True)
model = model.to(device)

# Save x_batch and y_batch as numpy arrays that will be used to call
metric instances.
x_batch, s_batch, y_batch = x_batch.cpu().numpy(), s_batch.cpu().numpy(),
y_batch.cpu().numpy()
# Make sure s_batch is of correct shape.
s_batch = s_batch.reshape(len(x_batch), 1, 224, 224)
# Quick assert.
assert [isinstance(obj, np.ndarray) for obj in [x_batch, y_batch, s_batch,
a_batch]]

# Return effective complexity scores in an one-liner - by calling the
metric instance.
quantus.PointingGame(
)(model=model,
   x_batch=x_batch,
   y_batch=y_batch,
   a_batch=None,
   s_batch=s_batch,
   explain_func=quantus.explain,
   explain_func_kwargs={'method': 'Saliency'},
   device=device)
```

*11.6.2 Top-*K *intersection metric*

Among the many significant metrics used to evaluate the explainability of models in the field of XAI, the top-*K* intersection metric stands out. Its purpose is to determine how well the model's predictions match up with the ground truth, which is particularly important when several relevant features could impact the model's predictions. Here, *K* is the number of most important features or regions, and this measure is based on the idea of assessing how much the ground truth overlaps with the top-*K* features indicated by the model. Top-*K* intersection is based on the idea that you should compare the model's *K* most important features to the ground truth features, which are defined by human experts or domain expertise, and see how well they match. To calculate the overlap between the top-*K* features that the model has highlighted and the ground truth, this metric could use intersection-based metrics like the Jaccard index or the Dice coefficient. This would give a mathematical measure of how close the two sets of features are, which would be useful for feature localization. To help with precise object localization, this measure can be used in image identification tasks to evaluate how well an AI model finds and aligns with the *K* most important areas or patterns in a picture that correlate to particular objects or features.

In the following, we provide a Python implementation for evaluating the explanation generated from the image classifier using the top-K intersection metric.

```
import numpy as np

import matplotlib.pyplot as plt
import torch
import torchvision
import quantus
from typing import Union

# Load pre-trained MobileNet V3 model.
model = torchvision.models.models.mobilenet_v3_small(pretrained=True)
model = model.to(device)

# Save x_batch and y_batch as numpy arrays that will be used to call
metric instances.
x_batch, s_batch, y_batch = x_batch.cpu().numpy(), s_batch.cpu().numpy(),
y_batch.cpu().numpy()
# Make sure s_batch is of correct shape.
s_batch = s_batch.reshape(len(x_batch), 1, 224, 224)
# Quick assert.
assert [isinstance(obj, np.ndarray) for obj in [x_batch, y_batch, s_batch,
a_batch]]

# Return tki scores in an one-liner - by calling the metric instance.
quantus.TopKIntersection(
)(model=model,
   x_batch=x_batch,
   y_batch=y_batch,
   a_batch=None,
   s_batch=s_batch,
   explain_func=quantus.explain,
   explain_func_kwargs={"method": "Saliency"},
   device=device)
```

11.6.3 Relevance mass accuracy metric

The relevance mass accuracy metric proposed in the referenced paper focuses on evaluating the accuracy of an explanation method in terms of the distribution of relevance values within a defined ground truth mask without necessitating the binarization of the heatmap or precise alignment with the ground truth surface, a requirement often seen in metrics used for evaluations based on weak object localization. This metric aims to measure the accuracy of an explanation by assessing the proportion of relevance values attributed to pixels located within the ground truth region relative to the total relevance across the entire image. The working principle of the relevance mass accuracy metric involves computing the ratio of the sum of relevance values associated with pixels lying within the ground truth mask to the sum of all relevance values across the entire image. Mathematically, this metric is expressed as the summation of relevance values (r) at pixel locations (x, y) within the ground truth mask (G), divided by the total sum of relevance values across all pixels in the image (I). The formula for relevance mass accuracy is represented as:

$$\text{Relevance mass accuracy} = \frac{\sum_{(x,y)GG} r(x,y)}{\sum_{\text{all pixels in the image}} r(x,y)} \tag{11.1}$$

where $r(x,y)$ represents the relevance value at pixel location (x,y). G denotes the set of pixel locations within the ground truth mask. I represents the total number of pixels in the image.

This metric quantitatively assesses how much importance or relevance mass the explanation method assigns to pixels within the defined ground truth region compared to the entire image. A higher relevance mass accuracy value indicates a more accurate attribution of relevance to pixels within the ground truth mask, thereby signifying the effectiveness of the explanation method in capturing the significant areas in alignment with the ground truth.

In the following, we provide a Python implementation for evaluating the explanation generated from the image classifier using the relevance mass accuracy metric.

```
import numpy as np
import matplotlib.pyplot as plt
import torch
import torchvision
import quantus
from typing import Union

# Load pre-trained MobileNet V3 model.
model = torchvision.models.models.mobilenet_v3_small(pretrained=True)
model = model.to(device)

# Save x_batch and y_batch as numpy arrays that will be used to call
metric instances.
x_batch, s_batch, y_batch = x_batch.cpu().numpy(), s_batch.cpu().numpy(),
y_batch.cpu().numpy()
# Make sure s_batch is of correct shape.
s_batch = s_batch.reshape(len(x_batch), 1, 224, 224)
# Quick assert.
assert [isinstance(obj, np.ndarray) for obj in [x_batch, y_batch, s_batch,
a_batch]]


# Return relevane mass accuracy scores in an one-liner - by calling the
metric instance.
quantus.RelevanceMassAccuracy(
)(model=model,
   x_batch=x_batch,
   y_batch=y_batch,
   a_batch=None,
   s_batch=s_batch,
   explain_func=quantus.explain,
   explain_func_kwargs={'method': 'Saliency'},
   device=device)
```

11.6.4 Relevance rank accuracy metric

The relevance rank accuracy metric, outlined in the referenced paper, assesses the accuracy of an explanation method by evaluating the correspondence between the highest intensity relevance values identified by the method and their alignment with the predefined ground truth. This metric aims to quantify how effectively the most intense relevances, identified by the explanation method, correspond to the pixel locations specified within the ground truth mask. The working principle involves determining the ratio of high-intensity relevance values situated within the ground truth region to the total size of the ground truth mask. Mathematically, the relevance rank accuracy is calculated by first obtaining the highest intensity relevance values within the image. Subsequently, the metric counts the number of these high-intensity relevance values that coincide with pixel locations within the ground truth mask. This count is then divided by the total size of the ground truth mask. This count is then divided by the total size of the ground truth mask. The relevance rank accuracy metric can be expressed as the ratio of the count of high-intensity relevance values within the ground truth region to the total size of the ground truth mask:

$$P_{\text{top } K} = \{p_1, p_2, \ldots, p_K | R_{p_1} > R_{p_2} > \cdots > R_{p_K}\} \tag{11.2}$$

where $P_{\text{top } K}$ denotes the group of pixels with significance values $R_{p_1}, R_{p_2}, \ldots, R_{p_K}$ ordered in decreasing order till the Kth pixel. Next, the rank accuracy is calculated as:

$$\text{Rank accuracy} = \frac{|P_{top}K \cap GT|}{|GT|} \tag{11.3}$$

In the following, we provide a Python implementation for evaluating the explanation generated from the image classifier using the relevance rank accuracy metric.

```
import numpy as np
import matplotlib.pyplot as plt
import torch
import torchvision
import quantus
from typing import Union

# Load pre-trained MobileNet V3 model.
model = torchvision.models.models.mobilenet_v3_small(pretrained=True)
model = model.to(device)

# Save x_batch and y_batch as numpy arrays that will be used to call
metric instances.
x_batch, s_batch, y_batch = x_batch.cpu().numpy(), s_batch.cpu().numpy(),
y_batch.cpu().numpy()
# Make sure s_batch is of correct shape.
s_batch = s_batch.reshape(len(x_batch), 1, 224, 224)
```

```
# Quick assert.
assert [isinstance(obj, np.ndarray) for obj in [x_batch, y_batch, s_batch,
a_batch]]

# Return relevane rank accuracy scores in an one-liner - by calling the
metric instance.
quantus.RelevanceRankAccuracy(
)(model=model,
   x_batch=x_batch,
   y_batch=y_batch,
   a_batch=None,
   s_batch=s_batch,
   explain_func=quantus.explain,
   explain_func_kwargs={"method": "Saliency"},
   device=device)
```

11.7 Axiomatic metrics

Metrics based on axiomatic principles form a critical facet within the realm of XAI, setting benchmarks and standards that XAI models should ideally comply with. These metrics are rooted in fundamental axioms or principles, serving as evaluative tools to ascertain whether AI models align with crucial ethical, logical, or mathematical principles. Metrics adhering to axiomatic principles aim to ensure transparency, fairness, robustness, and reliability in AI systems, emphasizing traits such as consistency, non-discrimination, or stability, which are fundamental for trustworthy and ethical AI.

11.7.1 Completeness metric

The completeness metric within XAI serves as an evaluative measure designed to assess the extent to which an explanation generated by an AI model comprehensively captures all relevant information necessary for understanding its decisions or predictions. This metric revolves around the fundamental principle that an explanation provided by an AI system should encompass and elucidate all pertinent factors or features influencing the model's outputs. The working principle of the completeness metric involves quantifying the degree to which an explanation covers the critical factors contributing to the model's decision-making process. It assesses whether the provided explanation is comprehensive enough to include all influential features or elements within the input data that significantly impact the model's predictions. Mathematically, this metric might involve evaluating the coverage or inclusion of important features by comparing the explanation provided by the model against a ground truth or an established set of crucial factors identified by domain experts or through rigorous analysis. The completeness metric is pivotal in ensuring that explanations provided by AI models are thorough and encompass all relevant information, thereby enhancing transparency and trust in the decision-making process. Achieving completeness in explanations is vital, especially in critical domains such as healthcare, finance, and criminal justice, where decisions derived from AI systems must consider all pertinent information to avoid biases or oversight.

```
import numpy as np
import matplotlib.pyplot as plt
import torch
import torchvision
import quantus
from typing import Union

# Load pre-trained MobileNet V3 model.
model = torchvision.models.models.mobilenet_v3_small(pretrained=True)
model = model.to(device)

# Save x_batch and y_batch as numpy arrays that will be used to call
metric instances.
x_batch, s_batch, y_batch = x_batch.cpu().numpy(), s_batch.cpu().numpy(),
y_batch.cpu().numpy()
# Make sure s_batch is of correct shape.
s_batch = s_batch.reshape(len(x_batch), 1, 224, 224)
# Quick assert.
assert [isinstance(obj, np.ndarray) for obj in [x_batch, y_batch, s_batch,
a_batch]]

# Return completeness scores in an one-liner - by calling the metric
instance.
quantus.Completeness(
    abs=False,
    disable_warnings=True,
)(model=model,
   x_batch=x_batch,
   y_batch=y_batch,
   a_batch=a_batch,
   explain_func=quantus.explain,
   explain_func_kwargs={'method': 'Saliency'},
   device=device)
```

11.7.2 Non-sensitivity metric

The non-sensitivity metric, as described in the provided information from the source paper, aims to evaluate the stability of feature attributions or importance assigned by an AI model. It specifically focuses on identifying and assessing the features to which the model is not functionally dependent. Additionally, it considers the subset of indices denoting features with zero attribution, representing features with assigned zero importance. The computation of non-sensitivity involves finding the cardinality of the asymmetric difference between these two subsets. The rationale behind utilizing the non-sensitivity metric lies in ensuring that the attribution method correctly assigns zero importance only to those features to which the model is not functionally dependent on. This aligns with the proposed sensitivity axiom by Sundararajan *et al.*, aiming to maintain fidelity and accuracy in attributions. Furthermore, the number of non-zero attributions serves as a measure of complexity within the model, indicating the extent of attributed importance among features. However, it is important to note that metrics like monotonicity, non-sensitivity, and complexity primarily focus on individual features, prompting the proposal of effective complexity to encompass feature interactions, providing a more comprehensive evaluation of model complexities.

```
import numpy as np
import matplotlib.pyplot as plt
import torch
import torchvision
import quantus
from typing import Union

# Load pre-trained MobileNet V3 model.
model = torchvision.models.models.mobilenet_v3_small(pretrained=True)
model = model.to(device)

# Save x_batch and y_batch as numpy arrays that will be used to call
metric instances.
x_batch, s_batch, y_batch = x_batch.cpu().numpy(), s_batch.cpu().numpy(),
y_batch.cpu().numpy()
# Make sure s_batch is of correct shape.
s_batch = s_batch.reshape(len(x_batch), 1, 224, 224)
# Quick assert.
assert [isinstance(obj, np.ndarray) for obj in [x_batch, y_batch, s_batch,
a_batch]]

# Return non-sensitivity scores in an one-liner - by calling the metric
instance.
quantus.NonSensitivity(
    abs=True,
    eps=1e-5,
    n_samples=10,
    perturb_baseline="black",
    perturb_func=quantus.perturb_func.baseline_replacement_by_indices,
    features_in_step=6272,
)(model=model,
   x_batch=x_batch,
   y_batch=y_batch,
   a_batch=None,
   explain_func=quantus.explain,
   explain_func_kwargs={'method': 'Saliency'},
   device=device)
```

11.7.3 Input invariance metric

The input shift invariance metric aligns with the concept of assessing the reliability of saliency or attribution methods in XAI by testing their consistency in producing explanations when subjected to a semantically meaningless transformation of the input data across two neural networks. The goal outlined in the paper is to develop a methodology for measuring the reliability of explanations generated by XAI methods. This involves constructing a unit test where a semantically meaningless transformation, such as shifting the entire dataset by a constant vector, does not alter the model's functionality. The aim is to evaluate whether the explanations provided by the XAI method remain stable and consistent despite the model remaining unchanged. The bias of a neural network can compensate for the constant shift in the dataset, resulting in two neural networks with identical weights, intermediate activations, and predictions. Therefore, the expectation is that a reliable saliency or attribution method should produce identical explanations for both networks, as their underlying functionality and decision-making processes remain the same despite the constant shift in the input data.

```
import numpy as np
import matplotlib.pyplot as plt
import torch
import torchvision
import quantus
from typing import Union

# Load pre-trained MobileNet V3 model.
model = torchvision.models.models.mobilenet_v3_small(pretrained=True)
model = model.to(device)

# Save x_batch and y_batch as numpy arrays that will be used to call
metric instances.
x_batch, s_batch, y_batch = x_batch.cpu().numpy(), s_batch.cpu().numpy(),
y_batch.cpu().numpy()
# Make sure s_batch is of correct shape.
s_batch = s_batch.reshape(len(x_batch), 1, 224, 224)
# Quick assert.
assert [isinstance(obj, np.ndarray) for obj in [x_batch, y_batch, s_batch,
a_batch]]

# Return completeness scores in an one-liner - by calling the metric
instance.
quantus.InputInvariance(
    abs=False,
    disable_warnings=True,
)(model=model,
   x_batch=x_batch,
   y_batch=y_batch,
   a_batch=None,
   explain_func=quantus.explain,
   explain_func_kwargs={'method': 'Saliency'},
   device=device)
```

References

[1] Moustafa, N., Koroniotis, N., Keshk, M., Zomaya, A. Y., and Tari, Z. (2023). Explainable intrusion detection for cyber defences in the Internet of Things: Opportunities and solutions. *IEEE Communications Surveys & Tutorials*, *25*, 1775–1807.

[2] Zhang, J., and Tao, D. (2020). Empowering things with intelligence: A survey of the progress, challenges, and opportunities in artificial intelligence of things. *IEEE Internet of Things Journal*, *8*(10), 7789–7817.

[3] Jagatheesaperumal, S. K., Pham, Q. V., Ruby, R., Yang, Z., Xu, C., and Zhang, Z. (2022). Explainable AI over the Internet of Things (IoT): Overview, state-of-the-art and future directions. *IEEE Open Journal of the Communications Society*, *3*, 2106–2136.

[4] Pawlicki, M., Pawlicka, A., Uccello, F., *et al.* (2024). Evaluating the necessity of the multiple metrics for assessing explainable AI: A critical examination. *Neurocomputing*, *602*, 128282.

[5] Rosenfeld, A. (2021, May). Better metrics for evaluating explainable artificial intelligence. *In Proceedings of the 20th International Conference on Autonomous Agents and Multiagent Systems* (pp. 45–50).

[6] Sovrano, F., and Vitali, F. (2023). An objective metric for Explainable AI: How and why to estimate the degree of explainability. *Knowledge-Based Systems*, *278*, 110866.

[7] Colin, J., Fel, T., Cadène, R., and Serre, T. (2022). What I cannot predict, I do not understand: A human-centered evaluation framework for explainability methods. *Advances in Neural Information Processing Systems*, *35*, 2832–2845.

[8] Pessach, D., and Shmueli, E. (2022). A review on fairness in machine learning. *ACM Computing Surveys (CSUR)*, *55*(3), 1–44.

[9] Fabris, A., Esuli, A., Moreo, A., and Sebastiani, F. (2023). Measuring fairness under unawareness of sensitive attributes: A quantification-based approach. *Journal of Artificial Intelligence Research*, *76*, 1117–1180.

[10] Jiang, Z., Han, X., Fan, C., Yang, F., Mostafavi, A., and Hu, X. (2022, January). Generalized demographic parity for group fairness. In *International Conference on Learning Representations*.

[11] Giuliani, L., Misino, E., and Lombardi, M. (2023, July). Generalized disparate impact for configurable fairness solutions in ML. In *International Conference on Machine Learning* (pp. 11443–11458). PMLR.

[12] Garg, P., Villasenor, J., and Foggo, V. (2020, December). Fairness metrics: A comparative analysis. In *2020 IEEE International Conference on Big Data (Big Data)* (pp. 3662–3666). IEEE.

[13] Dunkelau, J., and Duong, M. K. (2022). Towards equalised odds as fairness metric in academic performance prediction. *arXiv preprint arXiv:2209.14670.*

[14] Romano, Y., Bates, S., and Candes, E. (2020). Achieving equalized odds by resampling sensitive attributes. *Advances in Neural Information Processing Systems*, *33*, 361–371.

[15] Jain, B., Huber, M., Elmasri, R., and Fegaras, L. (2020). Using bias parity score to find feature-rich models with least relative bias. *Technologies*, *8*(4), 68.

[16] Zhang, W., Aliannejadi, M., Pei, J., Yuan, Y., Huang, J. H., and Kanoulas, E. (2024). A Comparative Analysis of Faithfulness Metrics and Humans in Citation Evaluation. *arXiv preprint arXiv:2408.12398*.

[17] Chan, C. S., Kong, H., and Liang, G. (2022). A comparative study of faithfulness metrics for model interpretability methods. *arXiv preprint arXiv:2204.05514*.

[18] Pawlicki, M., Pawlicka, A., Uccello, F., *et al*. (2024). Evaluating the necessity of the multiple metrics for assessing explainable AI: A critical examination. *Neurocomputing*, *602*, 128282.

[19] Agarwal, C., Krishna, S., Saxena, E., *et al*. (2022). Openxai: Towards a transparent evaluation of model explanations. *Advances in Neural Information Processing Systems, 35*, 15784–15799.

[20] Huang, Q., Mezzi, E., Mutlu, O., *et al*. (2024, July). Beyond the Veil of Similarity: Quantifying Semantic Continuity in Explainable AI. In *World Conference on Explainable Artificial Intelligence* (pp. 308–331). Cham: Springer Nature Switzerland.

[21] Jordan, M., and Dimakis, A. G. (2020). Exactly computing the local lipschitz constant of relu networks. *Advances in Neural Information Processing Systems*, *33*, 7344–7353.

[22] Fazlyab, M., Robey, A., Hassani, H., Morari, M., and Pappas, G. (2019). Efficient and accurate estimation of lipschitz constants for deep neural networks. *Advances in Neural Information Processing Systems*, 32.

Chapter 12
Adversarial attacks and defense in explainable security intelligence

12.1 Introduction

In the landscape of contemporary IoT ecosystems, the integration of AI has propelled innovation but has concurrently engendered heightened vulnerabilities, prominently manifesting in adversarial attacks. This chapter navigates the intricate terrain of adversarial threats within the ambit of explainable XAI as applied to IoT security paradigms. It embarks upon a systematic exploration of adversarial incursions, elucidating their multifaceted nature, classification, and the inherent challenges they pose to the integrity and resilience of AI-driven IoT systems. The chapter dissects adversarial attacks through a nuanced lens, delving into their manifold variations including knowledge-based categorizations, the dichotomy between one-time and iterative attacks, distinctions between attacks during training versus inference phases, and the strategic divergence between targeted and non-targeted assaults. This comprehensive analysis aims to furnish a foundational understanding of adversarial threats, underlining their diverse modes of operation and emphasizing their pertinence in the context of XAI for bolstering IoT security.

Moreover, this chapter delineates the nuances of adversarial perturbation techniques, elucidating the intricate methods by which attackers manipulate input data to deceive AI models. It underscores the significance of evaluating adversarial attacks, expounding upon diverse metrics utilized to gauge attack success rates and perturbation magnitudes, thus facilitating a structured assessment of model robustness. Further, the chapter articulates an arsenal of defensive strategies encompassing preprocessing techniques, robust model architectures, and the deployment of gradient masking and obscuring methodologies. In this context, the chapter explicates the inherent tensions between explainability and robustness, emphasizing the delicate balance necessary for preserving XAI's interpretability while fortifying models against adversarial incursions. By integrating discussions on adversarial attacks specific to XAI systems and elucidating the challenges they pose to interpretability, this chapter endeavors to delineate defense mechanisms that fortify the integrity and trustworthiness of XAI in IoT applications, ensuring a resilient and XAI ecosystem amidst evolving adversarial landscapes.

In the dynamic realm of IoT-driven applications empowered by XAI, the imperative to comprehend and mitigate adversarial threats becomes vital for ensuring the robustness and security of these interconnected systems. Adversarial attacks, characterized by deliberate manipulations of input data to deceive AI models, pose an ominous specter capable of subverting the reliability and trustworthiness of IoT-integrated XAI systems. Understanding these attacks is pivotal as they transcend mere theoretical exploits, manifesting as real-world vulnerabilities that imperil the integrity of AI-driven decision-making processes within IoT infrastructures. The profound ramifications of adversarial attacks extend beyond compromising data confidentiality and system availability; they fundamentally challenge the interpretability and reliability of AI models, instigating dire consequences for critical applications like healthcare monitoring, smart transportation, and industrial automation. Hence, an in-depth comprehension of adversarial threats is foundational to fortify IoT systems against incursions, safeguarding them from nefarious manipulation and bolstering their resilience in the face of evolving attack vectors.

The imperative to address adversarial attacks in the domain of XAI for IoT systems transcends mere mitigation strategies; it embodies a proactive stance toward fortifying the underlying trustworthiness and reliability of AI algorithms. The intrinsic interconnectedness of IoT devices underscores the urgency of developing robust defenses against adversarial threats, given their potential to propagate across interconnected networks and compromise the entire ecosystem's security. Consequently, integrating defense mechanisms tailored to counter adversarial incursions becomes an indispensable facet of ensuring the long-term viability and efficacy of IoT-enabled XAI. Moreover, the nexus between explainability and security assumes paramount significance, where the interpretability of AI models serves as both a catalyst for comprehension and a defense mechanism against adversarial attacks. Therefore, the holistic understanding and proactive mitigation of adversarial threats in XAI not only bolster the security posture of IoT systems but also underpin the trustworthiness and transparency vital for fostering user confidence in these AI-enabled interconnected environments [1].

12.2 Adversarial attacks

Adversarial attacks represent a profound challenge in the realm of AI, constituting deliberate manipulations of input data aimed at deceiving machine-learning (ML) models. These attacks introduce imperceptible perturbations into input samples, strategically designed to mislead AI systems into making erroneous predictions or classifications. At the crux of adversarial attacks lies the exploitation of inherent vulnerabilities in AI algorithms, leveraging nuanced imperfections in the decision boundary to induce misclassifications or erroneous outputs without substantial alterations visible to human perception.

The principles underlying adversarial attacks emanate from the intricate workings of ML models, particularly neural networks, which exhibit high-dimensional decision boundaries susceptible to subtle alterations in input data. Leveraging the sensitivity of these models to minor variations in input, adversaries craft perturbations that exploit this vulnerability, causing misclassification by steering the model toward an incorrect output while appearing indistinguishable to human observers. This phenomenon challenges the conventional assumptions of model robustness, illuminating the intricacies and limitations of AI systems in perceiving and interpreting input data. The logic driving adversarial attacks is deeply rooted in the understanding that AI models, despite their high accuracy in various tasks, are susceptible to perturbations that evade human perception but profoundly impact model predictions. By probing and manipulating the decision-making process of AI models, adversaries exploit the brittleness of these systems, highlighting the discrepancy between human perception and the underlying computational processes. This discordance underscores the necessity for comprehensive defenses against adversarial attacks, aiming to fortify AI models by enhancing their resilience to subtle manipulations while maintaining high levels of accuracy and reliability in decision-making [2].

12.2.1 Knowledge of adversarial attacks

The taxonomy of adversarial attacks in the realm of AI delineates a nuanced landscape characterized by three primary categories: white-box, black-box, and gray-box attacks, each elucidating distinct strategies for manipulating ML models (refer to Figure 12.1). White-box attacks epitomize a scenario where adversaries possess complete knowledge of the target AI model's architecture, parameters, and training data. Leveraging this comprehensive understanding, adversaries meticulously craft perturbations to deceive the model, exploiting its vulnerabilities to induce misclassification or erroneous outputs. In contrast, black-box attacks epitomize a scenario where adversaries operate in complete ignorance of the model's internal architecture and parameters, relying solely on querying the model and observing its responses. Adversaries glean insights by submitting queries to the model and inferring its behavior from the output, strategically crafting adversarial inputs without explicit knowledge of the model's internal mechanisms. The inherent challenge of black-box attacks lies in their reliance on limited access to the model, necessitating innovative methods to manipulate the model effectively while circumventing the lack of detailed information.

Amidst these extremes lies the nuanced category of gray-box attacks, encompassing scenarios that amalgamate aspects of both white-box and black-box approaches. In gray-box attacks, adversaries possess partial knowledge of the target model, incorporating varying degrees of information about the model's architecture or parameters. This intermediate stance allows adversaries to exploit the available information to navigate the adversarial landscape more effectively, probing the model's vulnerabilities to craft perturbations that deceive the system while leveraging limited insights into its internal workings. The complexity of gray-box attacks stems from the intricate balance between exploiting partial knowledge and navigating the constraints posed by the lack of complete information, necessitating adaptive strategies to manipulate the model effectively. The taxonomy of

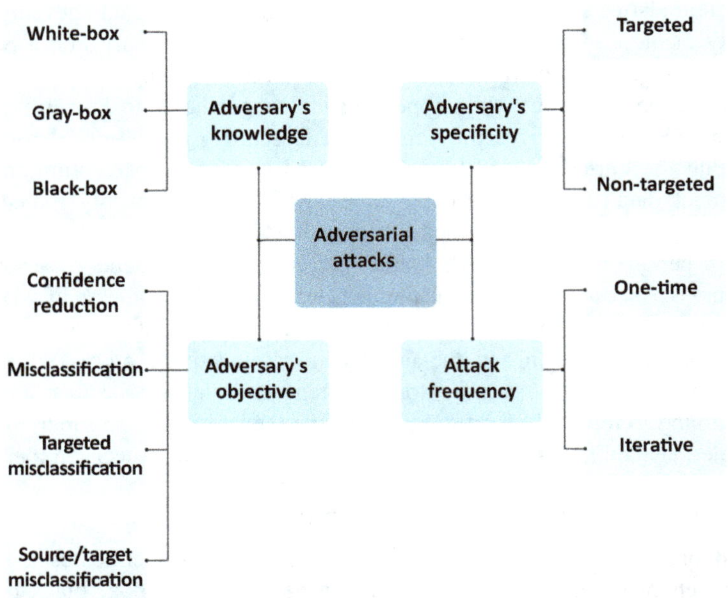

Figure 12.1 Visualization of the taxonomy of adversarial attacks according to different categorization criteria

adversarial attacks, comprising white-box, black-box, and gray-box categories, encapsulates a spectrum of strategies employed by adversaries to deceive AI models, highlighting the diverse challenges in defending against these incursions and fortifying the resilience of ML systems. Understanding these attack categories is instrumental in devising comprehensive defense mechanisms that mitigate vulnerabilities across varying degrees of adversary knowledge, fostering robust and resilient AI models resilient to diverse adversarial threats.

The transferability of adversarial attacks is a pivotal characteristic within the domain of ML security, illustrating the propensity of perturbations crafted to deceive one model to be efficacious across diverse models and domains. This phenomenon underscores the intricacies of adversarial vulnerabilities transcending individual models' idiosyncrasies, highlighting the broader susceptibility of ML algorithms to similar manipulation strategies. The transferability of attacks manifests when perturbations generated to deceive one model successfully induce misclassifications or erroneous outputs in other distinct models trained on similar or disparate datasets. Notably, transferability extends beyond models built on analogous architectures or trained on akin datasets, traversing diverse domains and architectures, which emphasizes the potential severity and pervasive nature of adversarial vulnerabilities across the ML landscape [3]. Understanding this transferability elucidates the pervasive nature of vulnerabilities within AI models, necessitating robust defense strategies that account for the potential propagation of attacks across models and domains [4].

The transferability of adversarial attacks across models and domains underscores the inherent challenges in fortifying ML systems against incursions, accentuating the need for comprehensive defense mechanisms that transcend model-specific idiosyncrasies. The phenomenon's implications are profound, compelling researchers and practitioners to explore generalized defense strategies capable of mitigating adversarial threats across diverse models and application domains. Moreover, the transferability of attacks unveils the intricate interplay between model architectures, training data, and the adversarial landscape, urging a deeper comprehension of the underlying vulnerabilities that render ML algorithms susceptible to similar exploitation. Thus, as the transferability of adversarial attacks signifies a broadened threat landscape that transcends individual models, it underscores the imperative for collaborative efforts in developing robust and universally applicable defense mechanisms, fostering resilience against adversarial manipulations that traverse models and domains within the ML ecosystem.

12.2.2 One-time versus iterative attacks

Adversarial attacks in the realm of ML encompass diverse strategies, among which single-step and iterative attacks represent distinct approaches, each characterized by unique methodologies and varying impacts on the robustness of AI models. Single-step attacks, also known as one-shot attacks, involve directly crafting perturbations in a single iteration to deceive the target model, aiming to produce misclassifications or erroneous outputs. These attacks typically optimize perturbations based on the gradient information derived from a single backward pass through the model. The simplicity and swiftness of single-step attacks make them computationally efficient; however, their impact often tends to be limited in inducing significant model misclassifications compared to their iterative counterparts. Single-step attacks tend to generate less potent adversarial examples, frequently causing slight alterations to input samples, which may result in lower success rates in evading model predictions when compared to iterative strategies. Conversely, iterative attacks embody a more nuanced approach, iteratively refining perturbations over multiple iterations to progressively deceive the target model. These attacks strategically accumulate imperceptible modifications to input samples, exploiting the model's vulnerabilities incrementally to induce misclassifications or erroneous predictions. Iterative attacks capitalize on the iterative optimization process to fine-tune perturbations, often resulting in more potent adversarial examples with higher success rates in evading model predictions. Despite being computationally more intensive than single-step attacks, iterative strategies tend to produce adversarial examples with higher effectiveness and impact, leading to more pronounced vulnerabilities in AI models' robustness and security.

The impacts of these attack strategies on AI model robustness are distinct, with single-step attacks often demonstrating lower potency in inducing misclassifications compared to iterative counterparts. While single-step attacks are computationally efficient, their generated adversarial examples might possess limited effectiveness in fooling models consistently across various scenarios. In contrast, iterative attacks exhibit higher success rates in deceiving models due to their progressive refinement of perturbations, leading to more potent adversarial examples capable of causing significant misclassifications and thereby challenging the reliability and trustworthiness of AI systems. Understanding the nuances between these strategies aids in devising comprehensive defense mechanisms capable of mitigating the impacts of both single-step and iterative attacks, thus fostering resilient AI models capable of withstanding diverse adversarial incursions within the ML landscape [5].

The evolution of adversarial attacks over multiple iterations represents a dynamic process wherein adversaries iteratively refine perturbations to progressively deceive ML models, profoundly impacting the robustness and reliability of these models. Iterative attacks embody a strategic refinement of adversarial examples through successive optimization steps, aiming to exploit the model's vulnerabilities incrementally. As adversaries iteratively refine perturbations, they craft imperceptible modifications to input samples, amplifying the potency of adversarial examples over multiple iterations. This incremental refinement process enables attackers to gradually navigate the decision boundaries of ML models, ultimately inducing significant misclassifications or erroneous predictions. The effects of iterative attacks on model robustness are multifaceted, highlighting the vulnerabilities inherent in AI systems when confronted with successive adversarial manipulations. Initially, iterative attacks may exhibit marginal impact in the early stages, generating adversarial examples that cause subtle deviations in model predictions. However, as the attack progresses through multiple iterations, the potency of adversarial examples intensifies, leading to more pronounced misclassifications and heightened model vulnerability. The gradual accumulation of imperceptible perturbations across iterations challenges the model's resilience, ultimately compromising its reliability and trustworthiness. Moreover, iterative attacks tend to produce adversarial examples with higher success rates in evading model predictions compared to their single-step counterparts, exacerbating the challenges in defending against these sophisticated manipulations.

Furthermore, the iterative refinement of attacks accentuates the need for robust defense mechanisms capable of withstanding successive adversarial manipulations. Defending against attacks evolving over multiple iterations necessitates adaptive strategies that account for the progressive nature of adversarial perturbations. Techniques such as adversarial training, where models are trained using adversarially crafted examples, or the integration of gradient masking methods to obscure sensitive information during training, represent potential avenues to fortify models against iterative attacks. Understanding the iterative evolution of attacks is pivotal in devising proactive defenses that mitigate the cumulative effects of successive manipulations, fostering robust and resilient ML models resilient to the evolving landscape of adversarial threats [3–5].

12.2.3 Training versus inference attacks

The distinction between adversarial attacks occurring during the training phase versus those manifesting during the inference phase delineates two distinct scenarios with disparate implications for the robustness and security of ML models (refer to Figure 12.2). Attacks during the training phase entail adversaries injecting perturbations into the training dataset or manipulating the training process itself. Such attacks aim to compromise the model's learning process, inducing biases, or distorting the decision boundaries by introducing adversarial examples during model training. By infiltrating the training data, adversaries aim to influence the model's learned representations, potentially resulting in compromised generalization and susceptibility to adversarial manipulations during inference. Conversely, attacks during the inference phase occur when adversaries exploit vulnerabilities in the trained model by presenting crafted adversarial examples at inference time. These attacks aim to deceive the model into making incorrect predictions or classifications by subtly altering input samples during the model's operational phase [5,6].

The implications of attacks during the training versus inference phases on model robustness differ significantly. Attacks during the training phase pose fundamental challenges to the learning process itself, potentially compromising the model's ability to learn accurate representations from the data. Adversarial manipulations during training can lead to the creation of inherently biased models, exhibit reduced generalization capabilities, and are more susceptible to adversarial examples during deployment. On the other hand, attacks during the inference phase target the operational integrity of the trained model. These attacks exploit vulnerabilities present in the already trained model, aiming to

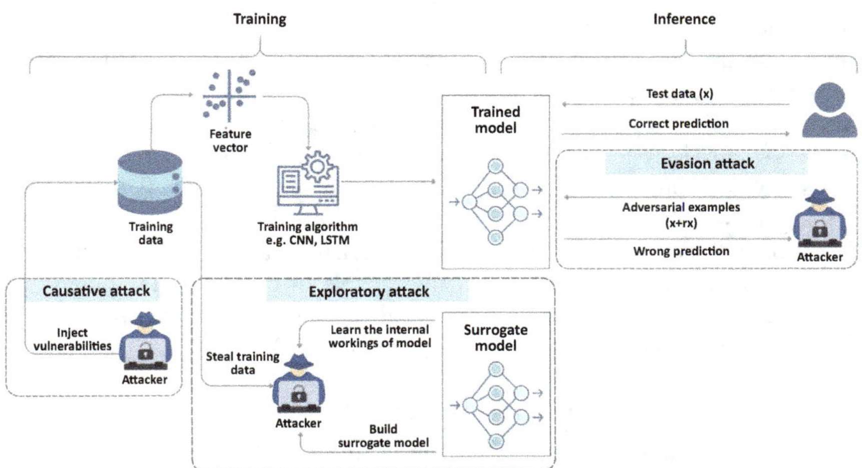

Figure 12.2 Visualization of adversarial attacks during training stage and inference stage

deceive it by presenting subtly modified inputs, thereby challenging the model's reliability and trustworthiness during its practical use. Understanding this dichotomy between attacks during the training and inference phases is pivotal in devising comprehensive defense mechanisms that fortify ML models at both stages, ensuring robustness and security across the entire lifecycle of AI systems.

Adversarial attacks manifest diverse implications on model behavior and security, with each type of attack engendering distinct challenges that undermine the reliability and robustness of ML models. Single-step attacks, characterized by their direct crafting of perturbations in a solitary iteration, often exhibit limited potency in inducing significant misclassifications or alterations in model behavior. While computationally efficient, these attacks tend to generate adversarial examples with slight modifications to input samples, impacting the model's behavior modestly by causing subtle deviations in predictions. Conversely, iterative attacks, evolving over multiple iterations to progressively refine perturbations, pose more pronounced implications on model behavior and security. These attacks incrementally manipulate input samples, resulting in more potent adversarial examples capable of inducing significant misclassifications or erroneous outputs, thereby challenging the model's reliability and undermining security by exploiting vulnerabilities in AI systems. The potency and cumulative impact of iterative attacks underscore their capacity to profoundly affect model behavior, amplifying security risks by compromising the model's trustworthiness and integrity during inference and decision-making processes.

The implications of these attacks on model security are multifaceted, encompassing challenges that transcend mere misclassification and extend to fundamental threats to the model's integrity. Single-step attacks, while less impactful in inducing substantial misclassifications, underscore the vulnerabilities present in the model's decision boundaries, thereby exposing potential weaknesses in model security. On the other hand, Iterative attacks, pose heightened security risks by leveraging progressively refined adversarial examples to compromise model behavior and reliability. These attacks not only jeopardize the accuracy and trustworthiness of AI systems but also accentuate broader security concerns by exploiting vulnerabilities that transcend individual models, potentially undermining the integrity of ML algorithms across various domains and applications. The implications of each type of attack on model behavior and security highlight the imperative for robust defense mechanisms capable of mitigating the diverse and evolving adversarial threats, ensuring the resilience and security of ML models in the face of sophisticated adversarial manipulations [6,7].

12.2.4 Targeted versus non-targeted attacks

Targeted and non-targeted attacks represent distinct paradigms in the landscape of adversarial ML, each driven by unique objectives and exhibiting characteristic strategies to deceive ML models (refer to Figure 12.3). Targeted attacks manifest as deliberate manipulations wherein adversaries aim to steer the model's prediction

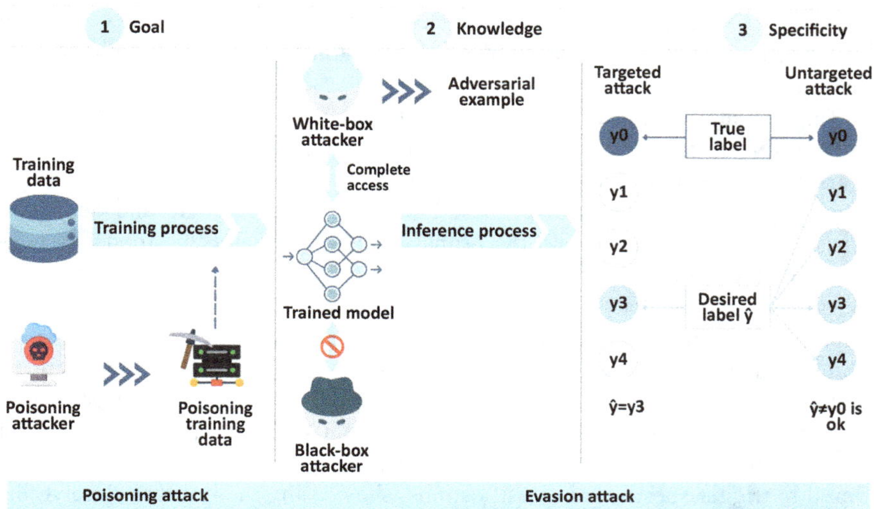

Figure 12.3 Illustration of different types of adversarial attacks across different stages of AI system

toward a specific incorrect output class. The primary objective of targeted attacks is to induce the model to confidently misclassify the input sample as a pre-defined target class chosen by the adversary. Adversaries intricately craft perturbations to alter the input features to maximize the model's confidence in predicting the desired target class. These attacks demand a sophisticated understanding of the model's decision-making process, often involving the exploration of adversarial examples tailored to trigger a specific misclassification, showcasing the strategic intent to undermine model accuracy for a particular predefined objective. Conversely, non-targeted attacks encompass a different approach, wherein adversaries aim to cause misclassifications without specifying a particular target output class. The objective of non-targeted attacks is to generate adversarial examples that induce any misclassification, regardless of the specific predicted output label. Adversaries iteratively craft perturbations to alter input samples to deviate from the correct prediction, aiming to cause the model to make errors without focusing on a specific desired output label.

Non-targeted attacks often focus on exploring vulnerabilities in the model's decision boundary without specifying a specific target class, showcasing a broader intent to undermine model reliability and trustworthiness by causing misclassifications across various output classes. The characteristics of non-targeted attacks lie in their exploration of weaknesses within the model's decision-making process, aiming to challenge its robustness by inducing errors across multiple classes rather than a specific predefined misclassification. Understanding the divergent objectives and characteristics between targeted and non-targeted attacks provides crucial insights into the motivations and strategies employed by adversaries to undermine the reliability and integrity of ML models, necessitating diverse defense strategies capable of effectively mitigating both types of adversarial threats.

Targeted attacks constitute a sophisticated class of adversarial manipulations meticulously crafted to subvert the decision-making process of ML models by compelling them to misclassify input samples into specific, pre-defined target classes chosen by the adversary. These attacks entail a strategic and deliberate alteration of input features with the primary goal of inducing the model to confidently predict the input as belonging to a particular, predefined target class. The intricacy of targeted attacks lies in their precision and directed nature, requiring a comprehensive understanding of the model's decision boundaries and vulnerabilities. Adversaries intricately manipulate input samples by introducing imperceptible perturbations, aiming to maximize the model's confidence in predicting the desired target class while subtly altering the input's features. Achieving the goal of targeted misclassification demands a deep exploration of adversarial examples tailored to trigger specific misclassifications, showcasing the intent to deceive the model into exhibiting highly confident yet erroneous predictions, further accentuating the strategic sophistication underlying these attacks [8].

Targeted attacks often involve complex optimization techniques such as iterative algorithms like the C&W attack or optimization-based approaches like the use of gradient information to craft adversarial examples that steer model predictions toward the predefined target class. These attacks require adversaries to possess an intricate understanding of the model's inner workings, encompassing its architecture, parameters, and decision boundaries. The objective of targeted attacks extends beyond mere manipulation of model outputs; it delves into the precision-driven quest to achieve a predefined misclassification outcome, showcasing the strategic intent to undermine model accuracy for specific, often malicious, objectives. The precision, sophistication, and strategic intent characterizing targeted attacks underscore the need for robust defenses capable of withstanding such highly directed and intentional manipulations, emphasizing the necessity for comprehensive strategies that fortify ML models against these intricate adversarial incursions.

Non-targeted attacks represent a significant facet within the domain of adversarial ML, embodying a strategic approach wherein adversaries aim to induce misclassifications in ML models without specifying a predefined target output class. Unlike targeted attacks that aim for a specific misclassification outcome, non-targeted attacks pursue the objective of causing the model to make errors without emphasizing a particular desired output label. These attacks involve the iterative manipulation of input samples to deviate the model's predictions from the correct output class, without focusing on steering the prediction toward any predetermined target class. Adversaries craft imperceptible perturbations intending to lead the model into making incorrect predictions across various output classes, showcasing a broader intent to challenge the model's reliability and trustworthiness without a narrow predefined goal. Non-targeted attacks explore vulnerabilities present in the model's decision boundary, aiming to undermine its robustness by inducing errors across multiple classes rather than steering toward a specific predefined misclassification, highlighting a broader and more exploratory nature compared to targeted attacks.

The characteristic feature of non-targeted attacks lies in their strategic intent to exploit weaknesses within the model's decision-making process, inducing errors across diverse output classes without specifically aiming for a predefined misclassification. Techniques employed in non-targeted attacks often revolve around iterative optimization methods like the fast gradient sign method (FGSM) or DeepFool, aiming to craft adversarial examples that cause the model to misclassify the input without a specific target label. By iteratively refining perturbations, adversaries aim to manipulate the model's decision boundary, subtly altering input samples to trigger misclassifications across multiple classes. Non-targeted attacks present a broader exploratory approach that seeks to challenge the overall reliability and robustness of ML models, emphasizing the need for comprehensive defense mechanisms capable of effectively mitigating these diverse and broad-ranging adversarial threats.

Different types of adversarial attacks have significant implications for the interpretability and trustworthiness of AI systems, profoundly impacting their reliability and transparency. Targeted attacks, with their precise manipulation of input samples to induce specific misclassifications, challenge the interpretability of AI systems by creating misleading explanations for model predictions. When successful, targeted attacks can generate adversarial examples that lead the model to confidently misclassify inputs into predefined target classes, thereby distorting the interpretability of the model's decision-making process. This distortion undermines the trustworthiness of explanations provided by the AI system, as these explanations are based on erroneous or manipulated predictions. As a result, the trust in the model's reliability and transparency diminishes, affecting the overall trustworthiness of the AI system's outputs and interpretations, thereby posing significant challenges in critical applications where transparency and interpretability are paramount.

Non-targeted attacks, aiming to induce misclassifications without specifying a particular output label, present challenges to the trustworthiness and interpretability of AI systems from a broader perspective. By inducing errors across multiple output classes, non-targeted attacks disrupt the interpretability of models by creating inconsistencies in the explanations provided for different predictions. These attacks challenge the coherence and consistency of the model's interpretability, impacting the trustworthiness of the explanations provided by the AI system. The misleading explanations generated due to adversarial manipulations undermine the users' trust in the model's reliability and transparency, hindering the system's overall trustworthiness. The impact of different types of attacks on interpretability and trustworthiness underscores the need for robust defense mechanisms and comprehensive explanations in AI systems to mitigate the effects of adversarial manipulations, ensuring transparent and reliable interpretations even in the face of sophisticated attacks.

12.3 Adversarial perturbation techniques

In this section, we delve into the intricate realm of adversarial perturbation techniques, illuminating the methodologies and nuances behind these manipulative strategies within the landscape of adversarial ML. Adversarial perturbation techniques represent a cornerstone in the arsenal of attacks deployed to deceive and compromise the integrity of ML models. These techniques involve the deliberate crafting of imperceptible alterations to input data, strategically designed to deceive models into making erroneous predictions or classifications. As adversaries continuously refine their approaches, understanding the mechanisms underpinning adversarial perturbation techniques becomes paramount for fortifying AI systems against such manipulations [9].

12.3.1 Fast gradient sign method

The FGSM stands as a seminal technique in the domain of adversarial ML, renowned for its simplicity and efficacy in crafting adversarial examples. FGSM operates on the principle of leveraging the gradients of the model's loss function with respect to the input features to generate perturbations. The method entails computing the gradient of the loss function concerning the input data and then perturbing the input in the direction that maximizes the loss, scaled by a small epsilon (ϵ) value. Specifically, FGSM crafts adversarial perturbations by perturbing each feature of the input data in the direction that maximizes the loss, essentially taking a step in the direction of the gradient of the loss function. The magnitude of this perturbation is constrained by the epsilon value, ensuring that the perturbations remain imperceptible while inducing significant changes in the model's output. Despite its simplicity, FGSM has demonstrated its efficacy in generating adversarial examples that successfully fool ML models, showcasing its utility as a foundational technique in understanding and fortifying models against adversarial vulnerabilities.

FGSM's effectiveness lies in its computational simplicity and efficiency, making it a prominent choice for generating adversarial examples. However, its simplicity also brings limitations, as FGSM tends to create perturbations that might be noticeable or easier to defend against compared to more sophisticated iterative methods. These perturbations might not be as potent in inducing misclassifications against models that are more robust or have been trained with adversarial training. Despite its limitations, FGSM serves as a fundamental benchmark in the study of adversarial attacks, enabling researchers to explore defense mechanisms and understand model vulnerabilities. Its ease of implementation and capacity to generate adversarial examples expediently have paved the way for further advancements in adversarial ML research, acting as a cornerstone technique shaping the discourse and evolution of defense strategies against adversarial threats.

12.3.2 Projected gradient descent

Projected gradient descent (PGD) represents a sophisticated and iterative adversarial attack technique that operates by iteratively refining adversarial perturbations to deceive ML models. Unlike single-step methods, PGD involves multiple iterations of gradient descent steps, gradually refining perturbations within a specified ε-bound to ensure that the generated adversarial examples remain imperceptible. PGD operates by iteratively applying gradient descent on the loss function with respect to the input data while projecting the perturbed data onto the ε-boundary, ensuring that the perturbations are within the desired range. Each iteration of PGD entails refining the perturbation in the direction that maximizes the loss, constraining the perturbations to remain within the specified ε-bound. This iterative refinement process allows PGD to generate potent adversarial examples, often exhibiting higher success rates in fooling ML models compared to simpler attack methods. While computationally more intensive than single-step methods, PGD's iterative nature enhances its effectiveness in crafting adversarial examples, making it a prominent technique in exploring model vulnerabilities and testing defense mechanisms against adversarial attacks.

PGD's iterative approach enables adversaries to craft more potent and robust adversarial examples compared to single-step methods by progressively refining perturbations over multiple iterations. The iterative nature of PGD provides adversaries with greater flexibility in exploring the model's vulnerabilities within the specified ε-bound. Furthermore, PGD's ability to generate adversarial examples that traverse the model's decision boundary more effectively than single-step methods underscores its significance in evaluating model robustness. However, the computational cost associated with multiple iterations and the necessity for a predetermined number of steps pose practical challenges in deploying PGD against large-scale models or real-time applications. Nevertheless, PGD's effectiveness in generating robust adversarial examples has made it a benchmark technique in adversarial ML research, playing a pivotal role in the development of defense strategies aimed at fortifying ML models against sophisticated adversarial threats.

12.3.3 Carlini–Wagner attack

The Carlini–Wagner (CW) attack stands as an advanced and versatile adversarial attack technique renowned for its ability to generate potent adversarial examples while simultaneously optimizing for stealth and imperceptibility. This attack method introduces a novel optimization-based approach that aims to generate adversarial perturbations by solving an optimization problem. The CW attack formulates the adversarial perturbation generation as an optimization problem, seeking to find minimal perturbations that induce misclassifications while adhering to specific constraints, such as imperceptibility or minimum distortion. Unlike simpler attacks, the CW attack incorporates a loss function that considers both the model's prediction and the magnitude of the perturbation, optimizing for minimal perturbation while ensuring the desired misclassification. This sophisticated approach allows CW to generate adversarial examples that are highly effective in fooling ML models while exhibiting imperceptible changes, making it a formidable technique for evaluating model robustness and security. The CW attack's versatility lies in its ability to generate adversarial examples under various threat models, accommodating different assumptions and constraints, and effectively bypassing defense mechanisms designed against simpler attack techniques.

The strength of the CW attack lies in its capability to produce potent adversarial examples while considering constraints on perturbation magnitude and imperceptibility. CW attacks are successful across various ML architectures and have demonstrated high success rates in evading detection by defense mechanisms, rendering it a crucial tool for evaluating and understanding model vulnerabilities. Despite its computational complexity, the CW attack's ability to generate adversarial examples that exhibit minimal distortion and high success rates in misclassification poses substantial challenges for defense mechanisms aiming to counter adversarial threats. Moreover, its adaptability to accommodate different threat scenarios and constraints amplifies its significance as a benchmark technique for assessing the robustness of ML models. The CW attack stands as a testament to the sophistication and ingenuity of adversarial attacks, offering insights into the intricate interplay between model vulnerabilities, optimization techniques, and imperceptible manipulations, thereby contributing significantly to the advancement of defense strategies against adversarial threats in ML [9].

12.4 Zeroth-order optimization attacks

Zeroth-order optimization (ZOO) attacks represent a category of adversarial attack methodologies that diverge from traditional gradient-based optimization techniques by operating without direct access to gradient information from the target model. Unlike methods such as FGSM or CW (CW) attacks that leverage gradient information to craft adversarial examples, ZOO attacks rely solely on querying the target model to obtain feedback on perturbed inputs, without directly accessing gradient information. This distinct approach poses challenges due to the lack of gradient knowledge, compelling ZOO attacks to employ heuristics and black-box optimization strategies to iteratively refine perturbations. By employing algorithms such as genetic algorithms, evolutionary strategies, or coordinate descent, ZOO attacks explore the input space through querying the model's outputs, seeking to minimize confidence in the correct prediction while ensuring perturbations remain imperceptible. The absence of direct gradient access in ZOO attacks adds complexity, often demanding a higher number of model queries and computational resources to generate effective adversarial examples, but it showcases an innovative approach that navigates the limitations of gradient-free optimization to deceive ML models efficiently.

The unique characteristic of ZOO attacks lies in their ability to manipulate models without direct access to gradient information, making them particularly relevant in scenarios where models' internal details are inaccessible or protected. However, the reliance on black-box queries and heuristic optimization methods renders ZOO attacks computationally demanding and less efficient compared to attacks utilizing gradient information. Yet, ZOO attacks present a formidable challenge in scenarios where gradient information is inaccessible or when models are highly guarded, illustrating their significance in evaluating the robustness of ML models under more constrained conditions. The innovative nature of ZOO attacks, which operate under limited information and heuristic optimization strategies, highlights their adaptability in challenging real-world settings where models' architectures or gradients are concealed or restricted. As adversaries continue to devise sophisticated methods to exploit ML models, understanding and addressing the vulnerabilities posed by ZOO attacks become integral in fortifying models against diverse adversarial threats in a broader spectrum of practical applications.

12.5 Evaluating adversarial attacks

In this section, we delve into the critical domain of evaluating adversarial attacks, exploring methodologies, metrics, and frameworks employed to assess the efficacy and impact of adversarial manipulations on ML models. The evaluation of adversarial attacks serves as a pivotal aspect in comprehending the vulnerabilities inherent in AI systems, elucidating the extent to which models are susceptible to adversarial manipulations. This section navigates through diverse evaluation techniques and metrics designed to quantify the effectiveness, robustness, and resilience of ML models against adversarial threats.

Metrics for assessing the effectiveness of adversarial attacks play a crucial role in quantifying the impact of manipulations on ML models. Success rate stands as a fundamental metric, measuring the percentage of adversarial examples that successfully deceive the model into making incorrect predictions. A high success rate indicates the potency of adversarial attacks in inducing misclassifications, highlighting the model's vulnerability to manipulations. Alongside success rate, distortion metrics quantify the magnitude of perturbations introduced to the input samples. Metrics like L2-norm or L-infinity norm measure the magnitude of perturbations, reflecting how much the input features have been altered to craft adversarial examples. Low distortion signifies imperceptible changes, emphasizing the stealth of adversarial manipulations, whereas higher distortion indicates more noticeable alterations. Furthermore, the transferability metric evaluates the generalization of adversarial examples across different models or architectures. This metric measures the success rate of adversarial examples crafted on one model when applied to another, highlighting the robustness of adversarial attacks and the transferability of vulnerabilities across diverse models. Metrics for assessing attack effectiveness enable a comprehensive understanding of the impact and robustness of adversarial manipulations, providing insights into the model's susceptibility to diverse attack strategies and aiding in the development of more resilient ML systems.

Moreover, metrics such as fooling rate and confidence reduction quantify the impact of adversarial attacks on the model's prediction confidence. Fooling rate measures the percentage of adversarial examples that induce high confidence in incorrect predictions, reflecting the degree of model confidence in erroneous classifications. Conversely, confidence reduction measures the average decrease in model confidence between correct and incorrect predictions, portraying the reduction in the model's certainty when facing adversarial inputs. These metrics shed light on how adversarial examples affect the model's decision-making process, emphasizing the model's confidence in erroneous predictions induced by manipulations. Additionally, metrics like targeted success rate measure the percentage of adversarial examples that successfully induce specific targeted misclassifications. Assessing the effectiveness of adversarial attacks through these diverse metrics elucidates the multifaceted impact of manipulations on model behavior, guiding the evaluation process and facilitating the development of robust defenses against adversarial threats in ML.

Assessing the robustness of ML models against adversarial attacks entails various evaluation techniques, each with its unique strengths and limitations. Adversarial training, a prominent technique, involves augmenting the training process with adversarially crafted examples, aiming to enhance the model's robustness against adversarial manipulations. While effective in bolstering resilience, adversarial training might not provide foolproof defenses, as adversaries constantly evolve their attack strategies, potentially bypassing defenses created using known adversarial examples. Another technique, robustness certification, aims to guarantee robustness by mathematically certifying that a model remains robust within a specified perturbation bound. However, this approach often relies on simplified assumptions or approximations, leading to conservative estimates of robustness that might not accurately reflect a model's actual resilience. Similarly, gradient masking methods obscure gradients during training to deter adversaries from accessing information useful for crafting adversarial examples.

These methods would not render models completely immune to attacks, as adversaries continuously devise strategies to bypass such defenses. Moreover, the computational overhead associated with some robustness evaluation techniques, such as adversarial training on large-scale models or conducting exhaustive search-based certification, poses practical limitations, hindering their widespread applicability in real-time or resource-constrained environments. Despite their significance, robustness evaluation techniques often face limitations and trade-offs. Adversarial examples crafted using current knowledge might not encompass the full spectrum of potential adversarial manipulations, leading to vulnerabilities against future attacks. Evaluating models under a limited set of adversarial scenarios might overlook other vulnerabilities present in different attack landscapes. Furthermore, robustness evaluation techniques are typically conducted under controlled settings and assumptions that might not fully represent real-world complexities, potentially limiting their generalizability. Additionally, the inherent trade-off between robustness and model performance poses a challenge, as enhancing robustness might lead to reduced accuracy on clean data or increased computational overhead. Addressing these limitations demands continual advancements in evaluation methodologies that consider diverse attack strategies, model architectures, and real-world scenarios, aiming to develop more comprehensive and practical techniques for assessing the robustness of ML models against adversarial threats [10].

Creating a comprehensive evaluation framework for assessing the robustness of ML models against adversarial attacks poses multifaceted challenges due to the evolving nature of adversarial threats and the complexity of AI systems. One significant challenge lies in the diverse spectrum of attack strategies employed by adversaries, ranging from simple to sophisticated methods. Designing an evaluation framework that encompasses the breadth of potential adversarial manipulations while maintaining scalability and efficiency presents a formidable task. Adversarial attacks constantly evolve, rendering any static evaluation framework susceptible to becoming obsolete over time. Moreover, the variability in model architectures, datasets, and application domains further complicates the creation of a one-size-fits-all evaluation framework. Addressing these challenges requires an adaptable and dynamic evaluation framework capable of accommodating a wide range of adversarial scenarios, fostering the development of defense mechanisms that transcend specific attack methodologies and adapt to emerging threats in real-time.

The interpretability and explainability of adversarial attacks present intricate challenges in devising a comprehensive evaluation framework. Understanding the motivations behind adversarial manipulations and their implications on model behavior remains a complex task. Adversarial attacks can deceive models in subtle ways, making it challenging to detect and interpret the rationale behind incorrect predictions induced by manipulations. Evaluating the impact of these attacks not only requires metrics for quantifying model vulnerabilities but also demands methodologies to interpret the underlying causes of model misclassifications. The integration of interpretability techniques within the evaluation framework becomes crucial for elucidating the mechanisms exploited by adversarial attacks and guiding the development of more robust defenses. Bridging the gap between the interpretability of adversarial manipulations and their effects on model behavior is imperative in creating a comprehensive evaluation framework that provides deeper insights into the vulnerabilities of AI systems against adversarial threats.

12.6 Adversarial defense

Herein, we navigate the landscape of adversarial defense mechanisms within XAI systems, exploring a plethora of strategies and methodologies devised to fortify AI systems against adversarial attacks while ensuring interpretability and transparency. As adversarial threats continue to pose challenges to the reliability and trustworthiness of ML models, the integration of robust defense mechanisms within XAI systems becomes imperative. Here, we aim to provide insights into the interplay between robustness and interpretability, paving the way for developing more resilient and transparent AI systems capable of withstanding the evolving landscape of adversarial threats.

12.6.1 Preprocessing techniques

12.6.1.1 Data augmentation

Data augmentation is a fundamental technique in bolstering the robustness of ML models against adversarial attacks while contributing to improved generalization and model performance. This methodology involves expanding the training dataset by applying various transformations or modifications to existing data samples, thereby augmenting the diversity and richness of the training set. Pertinent transformations encompass geometric alterations such as rotation, translation, scaling, flipping, or affine transformations, as well as color space manipulations like brightness adjustment, contrast variations, or color jittering. These augmentations introduce variations in the data distribution, enabling models to learn invariant features and patterns and fostering improved resilience against adversarial manipulations. By augmenting the training data with diverse transformations that simulate real-world variations, data augmentation helps in reduce overfitting, enhance model generalization, and fortify models against adversarial attacks by targeting specific features or patterns present in the original training set.

Augmenting datasets with diverse transformations facilitates the learning of more robust and invariant representations, contributing to the development of models that exhibit improved performance and resilience across a broader spectrum of input variations and adversarial perturbations. However, while data augmentation presents numerous benefits in enhancing model robustness, its efficacy in fortifying models against all forms of adversarial attacks remains limited. Adversaries might adapt their strategies to bypass defenses augmented with diverse training data, thereby necessitating more comprehensive defense strategies. Additionally, the effectiveness of data augmentation heavily relies on the chosen transformations and their relevance to the specific domain or task. In certain cases, over-reliance on specific augmentation techniques might inadvertently introduce vulnerabilities, where adversarial manipulations exploit these transformations to craft potent adversarial examples. Moreover, the computational overhead associated with generating augmented datasets and training models on an expanded dataset could pose practical challenges, particularly in resource-constrained environments. Thus, while data augmentation serves as a pivotal technique in enhancing model robustness against certain adversarial threats, its limitations necessitate the integration of additional defense mechanisms to create more resilient and interpretable AI systems capable of withstanding diverse adversarial challenges [11].

12.6.1.2 Input normalization

Input normalization is a foundational technique in fortifying ML models against adversarial attacks by standardizing input data, thereby reducing the model's susceptibility to subtle variations or perturbations. This preprocessing step involves scaling input features to a standardized range, often normalizing them to have zero mean and unit variance or scaling them within a predefined range. Normalization techniques such as *z*-score normalization or min-max scaling bring input features to a consistent scale, aiding models in learning robust and stable patterns by mitigating the impact of input variations. By normalizing input data, the model becomes less sensitive to minor fluctuations or perturbations in the input, reducing the model's vulnerability to adversarial manipulations that exploit these subtle variations. Furthermore, input normalization fosters improved convergence during model training, facilitates better weight initialization, and assists optimization algorithms in navigating the loss landscape more efficiently. This standardization process contributes to enhanced model generalization by ensuring that the model learns meaningful representations and features from the input data, thereby fortifying the model against certain types of adversarial perturbations that capitalize on input variations or inconsistencies.

While input normalization serves as a robust preprocessing technique, its effectiveness in mitigating adversarial threats remains constrained against more sophisticated attacks. Adversarial manipulations can still exploit higher-dimensional information or specific features that normalization techniques might not fully address. Moreover, adversarial attacks can adapt to standardized input distributions, requiring additional defense strategies beyond normalization to bolster model robustness. Additionally, the assumptions made during normalization, such as assuming Gaussian distributions or uniform feature importance, might not hold in complex real-world datasets, potentially limiting the efficacy of normalization techniques. Furthermore, the efficacy of input normalization in combating adversarial threats might vary across different domains or applications, demanding a nuanced approach tailored to the specific characteristics and vulnerabilities present in diverse datasets and models. Hence, while input normalization forms a crucial component in fortifying models against certain adversarial manipulations, it necessitates complementary defense mechanisms and a holistic approach in creating more resilient and interpretable AI systems capable of withstanding diverse adversarial challenges.

12.6.1.3 Randomization techniques

Randomization techniques offer a proactive defense strategy against adversarial attacks by introducing randomness into the model or training process, aiming to disrupt adversaries' attempts to craft effective adversarial examples. These techniques encompass diverse strategies, such as input or model randomization, injecting variability into the training or inference phases to enhance model robustness. Input randomization involves adding random noise or perturbations to input samples during training, making the model less reliant on precise feature representations and reducing its vulnerability to small, imperceptible perturbations. By augmenting the training data with randomized variations, the model learns to generalize better and becomes less sensitive to specific input variations exploited by adversaries. Similarly, model randomization techniques introduce randomness into the model's architecture or parameters, altering the model's internal behavior during training or inference. Methods like dropout, stochastic depth, or randomized layer connections introduce variability within the model, encouraging the learning of more diverse representations and preventing adversaries from exploiting specific model features or patterns. These randomized modifications to the model's architecture or inputs serve as a proactive defense mechanism, bolstering model robustness by disrupting adversarial attempts to craft targeted or precise attacks based on deterministic model behavior.

However, while randomization techniques provide an effective defense against certain adversarial threats, their applicability and effectiveness might vary across different attack scenarios and model architectures. Introducing randomness into models could potentially affect their interpretability and explainability, challenging efforts to understand and interpret model predictions accurately. Moreover, the efficacy of randomization techniques might depend on the degree and type of randomness introduced, as excessive randomization might hinder model learning or lead to performance degradation on clean data. Additionally, the computational overhead associated with randomization techniques, especially when incorporating significant randomness into model architectures or training processes, might pose challenges in deploying these defenses in resource-constrained environments or real-time applications. Thus, while randomization techniques represent a proactive defense strategy against adversarial attacks by enhancing model robustness, they necessitate a careful balance between introducing randomness and preserving model interpretability, demanding further exploration and refinement to create more resilient and interpretable AI systems capable of withstanding diverse adversarial challenges.

12.6.1.4 Feature squeezing

Feature squeezing serves as a promising defense mechanism against adversarial attacks by reducing the input space's granularity, thereby making it harder for adversaries to craft effective adversarial perturbations. This technique works by compressing or quantizing the input space, typically by reducing the precision of input features, to create a smoother and less expressive input space. Various methods, such as bit-depth reduction, spatial smoothing, or color discretization, are employed to squeeze the feature space, reducing the number of distinguishable values or features available to an adversary. By decreasing the granularity of input features, feature squeezing diminishes the potential variability an adversary can exploit to generate imperceptible perturbations, thereby hindering the construction of potent adversarial examples. Additionally, feature squeezing methods aim to maintain a balance between preserving necessary information for model accuracy and reducing the space in which adversarial perturbations can be effective. This reduction in the input space's expressiveness not only enhances model robustness by making it more resilient to adversarial manipulations targeting specific fine-grained details but also assists in regularizing models against overfitting, fostering improved generalization to unseen data.

Nonetheless, while feature squeezing demonstrates promise in mitigating adversarial threats, its effectiveness and generalizability across diverse datasets and models remain subject to certain limitations. The choice of squeezing parameters or compression techniques might impact the trade-off between model accuracy and robustness, potentially causing performance degradation or affecting the model's capacity to learn intricate details from the data. Moreover, the efficacy of feature squeezing could vary across different types of adversarial attacks, as adversaries might adapt their strategies to exploit reduced feature space in unforeseen ways. Furthermore, the degree of feature squeezing applied needs to be carefully calibrated to maintain a balance between reducing vulnerability to attacks and preserving necessary information for model accuracy, demanding a nuanced approach tailored to specific datasets and applications. Therefore, while feature squeezing presents a valuable defense mechanism against certain adversarial threats by constraining the input space's granularity, its practical implementation requires careful considerations and further exploration to ensure the creation of more robust and interpretable AI systems capable of withstanding diverse adversarial challenges.

12.6.2 Robust model architectures

12.6.2.1 Adversarial training

Adversarial training stands as a pivotal defense mechanism in fortifying ML models against adversarial attacks by augmenting the training process with adversarially crafted examples. This iterative approach involves exposing the model to adversarial examples during training, aiming to enhance its robustness against manipulations by iteratively refining its decision boundaries. Adversarial examples, generated by perturbing input data to mislead the model without changing the ground truth label, are incorporated into the training dataset. The model is then retrained on this augmented dataset, adapting its parameters to better accommodate and correctly classify these adversarial inputs. Through this process, adversarial training encourages the model to learn more resilient and generalizable features, enabling it to better distinguish between genuine data and adversarial manipulations. As the training progresses, the model gradually adjusts to adversarial input variations, becoming more adept at recognizing and mitigating potential adversarial threats. Adversarial training serves as an effective defense strategy, augmenting the model's ability to withstand diverse adversarial attacks and improving its robustness against sophisticated manipulations, thereby bolstering its reliability and trustworthiness.

However, while adversarial training demonstrates efficacy in enhancing model robustness, its practical implementation presents certain challenges. The process of adversarial training often demands increased computational resources and time due to the iterative generation of adversarial examples and subsequent model retraining. Moreover, adversarial training might lead to a phenomenon known as "overfitting to adversarial examples," where the model becomes overly focused on the specific adversarial manipulations present in the training set, potentially reducing its performance on clean data. Furthermore, the success of adversarial training heavily relies on the quality and diversity of generated adversarial examples, which might not cover the entire spectrum of potential adversarial manipulations. Adversaries may adapt their strategies, crafting more sophisticated attacks that bypass the defenses learned during adversarial training. Therefore, while adversarial training serves as a robust defense mechanism against many adversarial threats, ongoing research and refinement are essential to address its limitations and ensure the creation of more resilient and interpretable AI systems capable of withstanding evolving adversarial challenges.

12.6.2.2 Defensive distillation

Defensive distillation represents a defense strategy aimed at enhancing model robustness by training it to be more resistant to adversarial attacks through a two-step process involving temperature scaling. Initially introduced as a method to prevent model knowledge extraction, defensive distillation involves training a model on a softened version of the original training data, where the temperature parameter in the softmax function is increased during training. This augmented training process encourages the model to learn more abstract and smoother decision boundaries, reducing its sensitivity to small perturbations in the input space. Subsequently, a distilled model is created by training a new model on the softened logits, effectively creating a smoother and less confident model. The objective is to decrease the gradients of the loss function concerning input features, thereby limiting the effectiveness of small perturbations used in adversarial attacks. Defensive distillation aims to impart more generalizable knowledge to the model, making it less prone to overfitting on specific features or patterns present in the training data. This approach was initially proposed as a defense mechanism against knowledge-based attacks, but it has shown potential in augmenting model resilience against a subset of adversarial manipulations by encouraging models to learn more robust and generalizable representations.

While defensive distillation offers an innovative approach to enhance model robustness, its effectiveness in countering diverse adversarial threats remains limited. Adversaries may adapt their strategies to bypass defenses learned during the distillation process, crafting sophisticated attacks that exploit specific weaknesses not adequately addressed by this defense mechanism. Additionally, the overhead associated with defensive distillation, including increased computational resources and longer training times due to the augmentation of the training process, poses practical challenges, particularly in resource-constrained environments or real-time applications. Moreover, defensive distillation might not provide comprehensive protection against adversarial attacks, as it primarily focuses on specific types of manipulations and might not generalize well across different attack scenarios or model architectures. Therefore, while defensive distillation presents a novel approach to bolstering model robustness against certain adversarial threats, further research and exploration are necessary to address its limitations and to develop more versatile and effective defense mechanisms against the evolving landscape of adversarial attacks.

12.6.2.3 Ensemble methods

Ensemble methods stand out as a robust defense strategy against adversarial attacks by leveraging the diversity and collective decision-making of multiple models to enhance overall robustness. These methods involve training and combining multiple individual models, often with diverse architectures or trained on different subsets of data, to create an ensemble. Each model within the ensemble contributes its unique perspective and strengths in understanding the data distribution, reducing the risk of all models being susceptible to the same adversarial perturbations. By aggregating predictions from multiple models, ensemble methods aim to improve model generalization and robustness by mitigating the impact of potential adversarial manipulations on individual models. The ensemble's collective decision-making tends to be more reliable and resistant to adversarial attacks, as adversarial examples that deceive one model might not have the same effect across the entire ensemble. Furthermore, ensemble methods have shown efficacy in enhancing model interpretability by enabling the assessment of consensus among models or identifying instances where models diverge in predictions, providing valuable insights into potential adversarial inputs or uncertainties in model decision-making. However, while ensemble methods demonstrate promise in fortifying models against adversarial threats, they are not immune to certain limitations and challenges. Constructing and maintaining diverse and robust ensembles might require considerable computational resources and training time, particularly when dealing with large-scale models or complex architectures. The scalability of ensemble methods might be limited in resource-constrained environments due to the computational overhead associated with maintaining multiple models. Additionally, the effectiveness of ensemble methods heavily relies on the diversity and independence among constituent models; if the models within the ensemble are too similar or vulnerable to the same types of attacks, the ensemble's robustness might be compromised. Moreover, ensembles might encounter challenges in real-time applications or deployment scenarios where computational efficiency is crucial, as ensemble predictions might introduce latency or computational burdens. Therefore, while ensemble methods offer a potent defense strategy against adversarial threats by leveraging model diversity, addressing challenges related to scalability, diversity, and computational efficiency is essential to harness the full potential of ensemble-based defenses in creating more resilient and interpretable AI systems.

12.6.2.4 Certified defense methods

Certified defense methods represent a paradigm shift in adversarial robustness by employing rigorous mathematical guarantees to ensure model robustness against specified perturbations within a certain perturbation radius. These methods aim to provide certifiable guarantees that a model will produce the correct prediction within a defined radius around each input sample, regardless of potential adversarial manipulations within that radius. Certified defenses operate by formulating a formal verification process, typically using techniques like interval bound propagation or convex relaxation, to compute a certified robustness bound around each input sample. This bound guarantees that any perturbation within the specified radius will not cause the model's prediction to deviate from the ground truth label. These methods offer a level of assurance in model predictions by guaranteeing robustness against bounded adversarial manipulations, enhancing model trustworthiness and reliability. Certified defenses provide a more principled approach to adversarial robustness, offering mathematical guarantees and formal proofs of a model's resilience within a specific range of adversarial perturbations, thereby instilling confidence in the model's behavior under potential attacks. However, while certified defense methods offer compelling formal guarantees of robustness, their practical applicability and scalability to complex models or high-dimensional data remain challenging. Computing certified robustness bounds for high-dimensional input spaces or large-scale models can be computationally intensive and might require substantial computational resources, limiting their feasibility in real-time or resource-constrained environments. Moreover, certified defenses often rely on simplifying assumptions or approximations, leading to conservative estimates of robustness that might not fully reflect a model's actual resilience in complex real-world scenarios. Additionally, the guarantees provided by certified defenses are typically limited to specific predefined adversarial perturbations within the certified radius, potentially overlooking vulnerabilities to different types of attacks or variations outside the certified bounds. Addressing these challenges demands advancements in scalable certification methods and the exploration of techniques that offer more comprehensive and adaptable certified defenses, ensuring the creation of AI systems with robustness assurances against diverse adversarial threats.

12.6.3 Gradient masking and obscuring

Gradient masking techniques, employed as a defense strategy against adversarial attacks, aim to obscure or misrepresent gradient information accessible to adversaries during the model training or inference processes. These methods often involve intentionally suppressing or obfuscating gradient signals by manipulating the model architecture or training dynamics, hindering adversaries from accurately estimating or computing gradients to craft effective adversarial perturbations. One prevalent gradient masking technique involves applying non-differentiable or non-linear operations within the model architecture, such as using activation functions like ReLU (rectified linear unit) or max-pooling layers, to introduce non-smoothness in the model's decision surface. By introducing non-linearity or discontinuities in the gradients, these techniques impede adversaries' attempts to compute precise and reliable gradients required for crafting adversarial examples. Moreover, employing defensive distillation, where the model is trained on softened logits, is another gradient masking technique that suppresses gradient information, reducing the effectiveness of adversarial attacks. These masking strategies aim to mitigate adversaries' access to accurate gradient information, thereby impeding their ability to craft potent adversarial perturbations, enhancing model robustness, and fortifying AI systems against certain types of attacks. However, while gradient masking techniques offer a defense mechanism against adversarial attacks by limiting gradient-based adversarial methods, they present inherent trade-offs and limitations. Obscuring gradients within the model architecture might hinder model interpretability and transparency, as it becomes challenging to comprehend the reasoning behind model predictions or understand the decision-making process. Additionally, gradient masking methods are not foolproof and might not provide comprehensive protection against all forms of attacks, as adversaries continuously evolve strategies to bypass these defenses. Furthermore, the impact of gradient masking techniques on model generalization or performance on clean data requires careful consideration, as these defensive measures might inadvertently affect the model's learning dynamics or its ability to capture meaningful features. Therefore, while gradient masking techniques offer a means to mitigate adversarial threats by obscuring gradient information, balancing their effectiveness with the preservation of model interpretability and generalization remains an ongoing challenge in the quest to develop robust and interpretable AI systems resilient to adversarial manipulations.

Strategies aimed at obscuring model information from adversaries represent a crucial aspect of defending against adversarial attacks in ML. These strategies primarily focus on limiting adversaries' access to crucial information, such as gradients, model architecture, or learned representations, to hinder their ability to craft effective adversarial perturbations. One prevalent strategy involves gradient masking, where intentional obfuscation or suppression of gradient signals is introduced within the model architecture. This approach disrupts adversaries' attempts to compute accurate gradients necessary for crafting adversarial examples. Techniques like non-linear activations (e.g., ReLU), max-pooling layers, or defensive distillation employing softened logits obscure gradient information by introducing non-smoothness or non-differentiability in the model's decision boundaries. Another strategy revolves around limiting access to model parameters or architecture details, making it challenging for adversaries to infer critical information about the model's inner workings. Methods such as model compression, where only a compressed or distilled version of the model is publicly available, or model parameter quantization, which reduces parameter precision, restrict adversaries' insights into the model's specifics, thwarting their attempts to exploit vulnerabilities.

Additionally, differential privacy techniques offer another avenue to obscure model information by adding noise to the training process, thereby preventing adversaries from extracting sensitive information about individual data samples or model parameters. These techniques inject controlled randomness or noise into the training data or model updates, preserving privacy and obscuring fine-grained information that could aid adversaries in crafting targeted attacks. Moreover, employing ensemble methods, where predictions are aggregated from multiple diverse models, obscures specific model vulnerabilities by leveraging collective decision-making and increasing uncertainty for adversaries attempting to extract model details. While these strategies play a crucial role in obscuring model information, there exist trade-offs between model robustness, interpretability, and performance. Balancing the effectiveness of these strategies while ensuring model transparency and generalization remains a challenge, demanding continual research to fortify ML models against adversarial threats while maintaining their interpretability and utility in real-world applications.

12.7 Adversarial attacks on XAI systems

In this section, our exploration dives into the domain of adversarial attacks specifically targeted at XAI systems. These attacks, with their nuanced focus on undermining the interpretability and transparency of AI models, pose significant challenges in ensuring the reliability and trustworthiness of these systems, particularly in critical applications such as IoT security. This section endeavors to dissect the intricate landscape of adversarial assaults tailored toward XAI, aiming to shed light on the vulnerabilities inherent in XAI frameworks. By examining adversarial attacks that exploit the interpretability features designed to foster transparency in AI decision-making processes, this section aims to unravel the strategies adversaries employ to manipulate or deceive XAI systems. Moreover, the section delves into the implications of such attacks, emphasizing the critical need to fortify these systems against adversarial manipulations while preserving their interpretability, explicability, and reliability in the face of evolving threats. Through an in-depth analysis and elucidation of adversarial incursions targeting XAI systems, this section seeks to provide comprehensive insights into the vulnerabilities, impacts, and potential defense strategies essential in safeguarding the integrity and trustworthiness of these crucial AI frameworks.

Adversarial attacks tailored for XAI systems represent a sophisticated and nuanced form of manipulation specifically aimed at undermining the interpretability and transparency mechanisms within these systems. XAI systems are designed to provide comprehensible explanations for their decisions, making them susceptible to targeted adversarial assaults seeking to distort or manipulate these explanatory mechanisms. Adversaries may exploit vulnerabilities in the interpretability components, such as feature importance indicators or attention mechanisms, aiming to manipulate these explanations without significantly altering the model's predictions. By subtly modifying or perturbing certain input features, adversaries can deceive XAI systems into generating misleading or false explanations, potentially leading to incorrect or biased inferences. These attacks not only jeopardize the reliability and trustworthiness of the AI system's outputs but also obscure the interpretability and transparency crucial for users to comprehend and trust the AI's decisions. Adversaries may also target the post-hoc interpretability methods employed in XAI systems, such as generating adversarial perturbations that mislead post-hoc interpretability models, thereby distorting the explanations provided to users without affecting the primary model's predictions. This tailored form of adversarial attack undermines the fundamental goal of XAI systems, which is to provide transparent and understandable justifications for their decisions, making it imperative to fortify these systems against such manipulations while preserving their interpretability and trustworthiness [12].

Moreover, adversarial attacks targeting XAI systems often exploit the intricate relationship between model interpretability and model performance. Adversaries seek to compromise the interpretability components of XAI systems without significantly degrading the overall predictive accuracy, making these attacks more challenging to detect and counteract. Manipulating feature importance scores, altering saliency maps, or perturbing attention mechanisms while keeping the model's predictive performance intact poses a significant threat. These attacks undermine the reliability of interpretability tools by providing misleading explanations, causing users to potentially trust erroneous or biased justifications for AI decisions. Furthermore, the impact of such attacks extends beyond the immediate misinterpretation of model decisions, potentially leading to biased outcomes or compromised decisions in critical applications, including healthcare diagnostics or autonomous systems. Addressing these targeted adversarial attacks on XAI systems necessitates a comprehensive understanding of the interplay among model interpretability, performance, and adversarial vulnerabilities. Developing robust defenses that ensure both the interpretability and the reliability of AI explanations while mitigating the impact of potential manipulations becomes crucial in safeguarding the integrity and trustworthiness of XAI systems in the face of sophisticated adversarial threats.

Adversarial attacks on AI models significantly impact their interpretability and transparency, undermining the fundamental principles crucial for user trust and comprehension of AI-driven decisions. These attacks aim to distort the explanations provided by the models, leading to misinterpretations or misleading insights into the decision-making process. For instance, in XAI systems, attacks manipulating saliency maps or altering feature importance scores can misguide users by highlighting irrelevant or deceptive features as crucial contributors to the model's decisions. This distortion of interpretability jeopardizes users' ability to understand the reasoning behind AI predictions, hindering their trust in the system and potentially leading to erroneous decisions. Additionally, attacks targeting post-hoc interpretability methods, such as generating adversarial examples to deceive explanation models while preserving the primary model's predictions, further undermine the reliability of explanations. This manipulation creates discrepancies between the actual decision-making process and the explanations provided, obscuring the transparency essential for users to comprehend and accept AI-generated insights. Furthermore, adversarial attacks not only distort interpretability but also compromise the overall transparency of AI models by concealing vulnerabilities or biases in decision-making. Attacks exploiting model vulnerabilities without altering predictions can mask systemic biases or weaknesses within the model, leading to potentially biased or unjust decisions without providing adequate explanations for such outcomes. This lack of transparency not only diminishes users' confidence in AI systems but also raises ethical concerns regarding fairness, accountability, and potential societal impacts. Adversarial attacks that obfuscate the reasons behind AI decisions while maintaining seemingly accurate predictions pose a critical challenge in ensuring transparent and trustworthy AI applications. Addressing these impacts requires a concerted effort to fortify AI systems against adversarial threats while preserving and enhancing their interpretability and transparency, fostering user trust and understanding in AI-driven decision-making processes.

12.8 Explainability and robustness: a balancing act

This section scrutinizes the inherent tensions and dependencies between AI models' interpretability and their resilience against adversarial manipulations. Our endeavor is to unravel the intricate relationship where the quest for transparent decision-making meets the imperative need for fortifying AI systems against sophisticated adversarial assaults. Through an in-depth analysis and comprehensive examination, this section aims to shed light on the challenges, trade-offs, and potential synergies between interpretability and robustness [13].

The intricate relationship between model explainability and robustness stands at the forefront of discussions within the realm of AI, delineating a delicate trade-off crucial in shaping AI systems' reliability and trustworthiness. XAI frameworks aim to render AI models more transparent and understandable, enabling users to comprehend the rationale behind model decisions. However, the pursuit of heightened explainability often encounters a trade-off with model robustness against adversarial attacks. The mechanisms implemented to enhance interpretability, such as attention mechanisms or feature importance scores, inadvertently create vulnerabilities exploitable by adversaries to craft adversarial manipulations. Measures that bolster interpretability, such as simpler or more transparent models, might lack the complexity necessary to capture intricate decision boundaries, rendering them more susceptible to adversarial perturbations. Conversely, enhancing robustness through defenses like gradient masking or adversarial training could obfuscate explanations by introducing non-linearities or perturbations that undermine the interpretability of model decisions. This intricate trade-off necessitates a delicate balancing act: reinforcing model robustness against adversarial threats while preserving an adequate level of transparency and comprehensibility in AI systems, crucial for user trust and acceptance.

The trade-off between model explainability and robustness poses multifaceted challenges across various domains of AI applications. In safety-critical domains like healthcare or autonomous vehicles, where model decisions significantly impact human lives, the need for transparent and interpretable AI models is paramount. Yet, these very domains also demand heightened robustness against adversarial manipulations. Balancing these conflicting requirements becomes even more complex when considering regulatory compliance or ethical implications. While striving for increased model interpretability to ensure accountability and fairness, AI systems need to withstand potential adversarial attacks that may exploit these interpretability features. Moreover, achieving a harmonious equilibrium between explainability and robustness necessitates a deeper understanding of the vulnerabilities and trade-offs inherent in different interpretability methods and defense strategies. Exploring innovative approaches that reconcile these contrasting objectives becomes pivotal in advancing AI systems capable of providing understandable and reliable decision-making while fortifying themselves against adversarial pressures, ultimately shaping trustworthy and resilient AI solutions for real-world applications [12].

Maintaining a delicate equilibrium between model explainability and security represents a pivotal challenge in developing robust and interpretable AI systems. Strategies aimed at striking this balance often involve a multidimensional approach addressing both interpretability and security concerns concurrently. One approach involves incorporating robustness considerations into the design of XAI systems from their inception. This strategy entails developing novel interpretability methods that inherently integrate defense mechanisms against adversarial attacks. Designing XAI frameworks with embedded defenses, such as adversarial training during the interpretability learning phase or integrating robustness-aware interpretability tools, aims to create systems that are inherently more resistant to adversarial manipulations without compromising their interpretability. Additionally, employing diverse ensemble methods that amalgamate multiple interpretable models and simultaneously fortifying them against adversarial threats offers a synergistic approach. Ensemble models, comprising diverse interpretable models, not only enhance transparency in decision-making but also leverage collective decision-making to enhance robustness by averting individual model vulnerabilities, thereby maintaining a balance between explainability and security.

Furthermore, exploring adaptive and context-aware strategies presents another avenue for achieving a harmonious equilibrium between explainability and security in AI systems. These strategies involve dynamically adjusting the level of interpretability based on contextual factors such as the application domain, risk tolerance, or the presence of potential adversarial threats. Implementing adaptable interpretability levels, where explanations are provided in varying degrees of detail contingent upon the perceived risk or criticality of the decision, enables a flexible balance between transparency and security. Additionally, deploying reinforcement learning approaches that continuously adapt model interpretability based on evolving threats or user requirements offers a proactive defense mechanism. These adaptive strategies aim to tailor the level of interpretability to specific scenarios, striking a balance between comprehensibility and resilience against adversarial manipulations, thereby fostering trustworthy and robust AI systems adaptable to diverse real-world applications.

12.9 Explainable adversarial attacks

In this section, we embark on a comprehensive exploration into the evolving landscape of defense mechanisms tailored explicitly for XAI systems against adversarial attacks. In this section, we delve into the innovative domain of defense strategies designed to fortify the interpretability and transparency aspects of AI models in the face of sophisticated adversarial manipulations. The focus here is to dissect and elucidate the cutting-edge defenses specifically tailored to safeguard the interpretability components of AI systems, aiming to bolster their resilience against adversarial assaults while preserving their crucial trait of providing understandable and trustworthy explanations for their decisions.

Reverse engineering of adversarial attacks stands as a pivotal strategy in gaining profound insights into the vulnerabilities and intricacies of ML models, providing crucial knowledge for devising effective defense mechanisms. This process involves meticulous analysis and deconstruction of adversarial attacks crafted to manipulate AI models, aiming to unravel the underlying techniques and methodologies utilized by adversaries. By scrutinizing the adversarial perturbations and the methodologies used to generate them, researchers gain valuable insights into the weaknesses of AI models, uncovering the specific features or decision boundaries vulnerable to manipulation. Through this reverse engineering process, intricate details about the model's sensitivity to perturbations, the impact of manipulated features, and the underlying vulnerabilities are unearthed, offering a deeper understanding of the mechanisms exploited by adversaries to deceive AI systems. Additionally, reverse engineering adversarial attacks facilitate the identification of patterns or common traits across different attack strategies, enabling researchers to derive general principles and guidelines for fortifying models against various adversarial manipulations. This in-depth analysis not only aids in fortifying model defenses but also fosters a profound comprehension of the interplay between model complexity, interpretability, and robustness, essential in the continual evolution of more resilient and trustworthy AI systems.

Moreover, reverse engineering adversarial attacks play a pivotal role in fostering advancements in XAI, where transparency and interpretability are paramount. By dissecting these attacks targeted at XAI systems, researchers gain insights into the vulnerabilities within interpretability methods or explanation models, understanding how adversarial manipulations distort or misrepresent explanations provided by AI systems. This analysis aids in fortifying XAI frameworks against manipulations specifically designed to undermine interpretability, preserving the reliability and trustworthiness of the explanations offered by these systems. Additionally, reverse engineering adversarial attacks targeting XAI elucidate the intricate trade-offs between model transparency and robustness, offering guidance in balancing these conflicting objectives. Furthermore, this process contributes to the creation of more resilient and interpretable XAI systems by uncovering the intricate vulnerabilities exploited by adversaries and guiding the development of defense mechanisms tailored to safeguard against these sophisticated adversarial manipulations, ensuring the reliability and trustworthiness crucial in critical real-world applications [13].

Adversarial attacks serve as a potent tool for unveiling latent weaknesses and vulnerabilities within ML models, offering profound insights into their limitations and susceptibilities to manipulations. These attacks, meticulously crafted to deceive AI systems, exploit subtle vulnerabilities in the model's decision boundaries or feature representations, revealing the brittleness of these systems under specific conditions. By perturbing input data in imperceptible ways, adversarial attacks highlight specific areas of the input space where the model exhibits high sensitivity to slight variations, uncovering regions susceptible to misclassification or erroneous predictions. This exploration not only showcases the model's shortcomings but also reveals the inadequacies in capturing robust and generalizable patterns from the data. Furthermore, adversarial attacks highlight instances where the model's decision-making process contradicts human intuition, shedding light on potential blind spots or biases encoded within the model's learned representations. Through these attacks, weaknesses in the model's ability to generalize across diverse data distributions or its failure to account for subtle variations in input features are elucidated, providing invaluable insights into the limitations that might compromise the model's reliability in real-world scenarios.

Moreover, adversarial attacks serve as a diagnostic tool, uncovering the model's lack of robustness against variations that lie beyond the training data distribution. These attacks simulate scenarios where the model fails to generalize or extrapolate its learned patterns, exposing vulnerabilities that adversaries can exploit to deceive the system. The revealed weaknesses offer guidance for fortifying the model against potential exploits, driving advancements in defense mechanisms that mitigate these vulnerabilities. Additionally, adversarial attacks pinpoint specific areas within the model's decision boundaries that exhibit inconsistencies or regions where the model's predictions are particularly susceptible to manipulation. This granular insight aids in understanding the complex interplay among model complexity, feature representations, and decision-making, providing a roadmap for improving model robustness and resilience against adversarial manipulations. By leveraging adversarial attacks as a probing tool, researchers gain a deeper understanding of the intricacies and limitations inherent in ML models, fostering advancements toward more robust and trustworthy AI systems capable of withstanding diverse real-world challenges [15].

Balancing transparency and security within adversarial settings represents a formidable challenge in the development of robust and interpretable AI systems. Achieving transparency, essential for users to comprehend AI-driven decisions, often involves divulging detailed explanations of model predictions. However, this very transparency can be exploited by adversaries to launch sophisticated attacks aimed at manipulating or undermining AI systems. Striking a delicate equilibrium between transparency and security demands a multifaceted approach. One strategy involves implementing transparent AI frameworks fortified with robustness mechanisms that safeguard against adversarial manipulations. By integrating interpretable models with built-in defense mechanisms, such as adversarial training or feature-based explanations resilient to manipulations, researchers aim to provide understandable insights while fortifying the system against potential attacks. Additionally, adaptive transparency levels that dynamically adjust the depth of explanations based on the perceived risk or presence of adversarial threats offer a flexible approach. This adaptive strategy tailors the level of transparency to specific contexts, balancing the need for comprehension with the imperative of safeguarding against potential adversarial incursions, thereby fostering resilient and adaptable AI systems [16,17].

Furthermore, addressing the conundrum between transparency and security requires a comprehensive assessment of the trade-offs inherent in different defense mechanisms and interpretability strategies. Techniques that enhance model robustness, such as gradient masking or adversarial training, might inadvertently obfuscate explanations or compromise interpretability. Conversely, methods aimed solely at bolstering transparency, like simplified or more interpretable models, might lack the complexity required to capture intricate decision boundaries, rendering them more susceptible to adversarial manipulations. Thus, the integration of defenses that preserve model transparency while fortifying against adversarial threats becomes crucial. Additionally, fostering collaborations between multidisciplinary teams comprising experts in AI, cybersecurity, and ethics is essential.

This collaborative approach allows for a holistic evaluation of AI systems, considering both their interpretability and resilience against potential adversarial attacks. By navigating this complex landscape, researchers strive to develop AI systems that strike an optimal balance between transparency and security, fostering trust and reliability in AI-driven decision-making while mitigating vulnerabilities to adversarial threats [18,19].

12.10 Explainable defense mechanisms

This section marks an incisive exploration into the cutting-edge realm of defense mechanisms specifically tailored for XAI systems amidst adversarial landscapes. We focus squarely set on elucidating and scrutinizing innovative defense strategies meticulously designed to fortify the interpretability and resilience aspects of AI models against sophisticated adversarial assaults. Our endeavor here is to navigate through the intricacies of defense methodologies, advancements, and paradigms tailored explicitly to safeguard the interpretability components of AI systems.

Defense mechanisms tailored for XAI models represent a critical frontier in safeguarding the interpretability and resilience of AI systems against adversarial attacks. These defense strategies, specifically engineered to preserve the transparency and comprehensibility of AI models, aim to fortify the vulnerabilities exposed by adversarial manipulations. One notable approach involves integrating interpretability methods with robustness mechanisms during the model's design phase. By fusing interpretability features with adversarial training or incorporating defenses within the explanation generation process, these methodologies strive to imbue XAI models with intrinsic resilience against adversarial threats. Additionally, advancements in defense mechanisms focus on devising novel interpretability tools robust to manipulations. Such tools leverage diverse interpretability methods resilient to adversarial attacks, ensuring that the explanations provided by AI models remain trustworthy and accurate even in the presence of sophisticated adversarial perturbations. These defense mechanisms not only fortify the interpretability aspects but also contribute to the overall robustness and reliability of XAI systems, crucial for their adoption across various high-stakes domains.

Moreover, an in-depth exploration of defense mechanisms tailored for XAI models involves the development of hybrid strategies that reconcile interpretability and robustness. Hybrid defenses amalgamate interpretable models with ensemble methods, adversarial training, or uncertainty quantification techniques to reinforce the model's resilience while retaining its transparent decision-making process. Furthermore, exploring methods that adapt the level of interpretability based on the risk posed by potential adversarial threats offers a dynamic approach. Adaptive defenses adjust the depth or granularity of explanations contingent upon the perceived risk, balancing transparency with resilience against manipulations. These strategies not only mitigate adversarial vulnerabilities but also promote an understanding of the intricate interplay between interpretability and security. By scrutinizing and developing defense mechanisms tailored explicitly for XAI systems, researchers aim to foster the advancement of AI models capable of providing trustworthy and understandable explanations while ensuring resilience against sophisticated adversarial incursions [20].

The development of techniques that simultaneously enhance the robustness of AI models while preserving their interpretability stands as a critical endeavor in the advancement of XAI systems. These techniques aim to fortify AI models against adversarial attacks without compromising the clarity and transparency of their decision-making processes. One notable strategy involves the integration of interpretability methods with adversarial training during the model's learning phase. By combining interpretable models with robustness-enhancing mechanisms, such as adversarial training, researchers strive to create AI systems that are inherently resilient to adversarial manipulations while offering understandable explanations for their decisions. Additionally, advancements in defense techniques focus on developing interpretable models that leverage intrinsic robustness characteristics. These models incorporate interpretability features resilient to manipulations, ensuring that the explanations provided by the AI systems remain reliable and accurate even in the presence of sophisticated adversarial perturbations. These approaches foster AI models that are not only interpretable but also possess a heightened level of resilience against adversarial threats, crucial for their deployment in critical domains [21].

Furthermore, the development of techniques that bolster both interpretability and robustness encompasses the exploration of hybrid defense strategies. Hybrid models merge interpretable architectures with ensemble methods or uncertainty quantification techniques, reinforcing the model's resilience against adversarial attacks while preserving its transparent decision-making process. Moreover, investigating adaptive interpretability levels tailored to the risk posed by potential adversarial threats offers a dynamic approach. These adaptive defense strategies adjust the depth or granularity of explanations based on the perceived risk, ensuring a balance between transparency and resilience against manipulations. Such approaches not only fortify AI systems against adversarial vulnerabilities but also contribute to a deeper understanding of the intricate relationship between interpretability and security.

References

[1] Baniecki, H. and Biecek, P. (2023). Adversarial attacks and defenses in explainable artificial intelligence: a survey. *arXiv preprint arXiv:2306.06123*.

[2] Liang, H., He, E., Zhao, Y., Jia, Z., and Li, H. (2022). Adversarial attack and defense: a survey. *Electronics*, *11*(8), 1283.

[3] Michel, A., Jha, S. K., and Ewetz, R. (2022). A survey on the vulnerability of deep neural networks against adversarial attacks. *Progress in Artificial Intelligence*, *11*, 131.

[4] Suryotrisongko, H., Musashi, Y., Tsuneda, A., and Sugitani, K. (2022). Robust botnet DGA detection: blending XAI and OSINT for cyber threat intelligence sharing. *IEEE Access*, *10*, 34613–34624.

[5] Kong, Z., Xue, J., Wang, Y., Huang, L., Niu, Z., and Li, F. (2021). A survey on adversarial attack in the age of artificial intelligence. *Wireless Communications and Mobile Computing*, *2021*, 1–22.

[6] Tian, Q., Zhang, S., Mao, S., and Lin, Y. (2022). Adversarial attacks and defenses for digital communication signals identification. *Digital Communications and Networks*, *10*, 756–764.

[7] Capuano, N., Fenza, G., Loia, V., and Stanzione, C. (2022). Explainable artificial intelligence in cybersecurity: a survey. *IEEE Access*, *10*, 93575–93600.

[8] Ilahi, I., Usama, M., Qadir, J., *et al.* (2021). Challenges and countermeasures for adversarial attacks on deep reinforcement learning. *IEEE Transactions on Artificial Intelligence*, *3*(2), 90–109.

[9] Tang, B., Wang, J., Yu, Z., *et al.* (2022). Advanced persistent threat intelligent profiling technique: a survey. *Computers and Electrical Engineering*, *103*, 108261.

[10] Akhtar, N. and Mian, A. (2018). Threat of adversarial attacks on deep learning in computer vision: a survey. *IEEE Access*, *6*, 14410–14430.

[11] Mavroeidis, V. and Bromander, S. (2017). Cyber threat intelligence model: an evaluation of taxonomies, sharing standards, and ontologies within cyber threat intelligence. In *2017 European Intelligence and Security Informatics Conference (EISIC)* (pp. 91–98). Piscataway, NJ: IEEE.

[12] Akhtar, N., Mian, A., Kardan, N., and Shah, M. (2021). Advances in adversarial attacks and defenses in computer vision: a survey. *IEEE Access*, *9*, 155161–155196.

[13] Charmet, F., Tanuwidjaja, H. C., Ayoubi, S., *et al.* (2022). Explainable artificial intelligence for cybersecurity: a literature survey. *Annals of Telecommunications*, *77*(11–12), 789–812.

[14] Ibrahim, A., Thiruvady, D., Schneider, J. G., and Abdelrazek, M. (2020). The challenges of leveraging threat intelligence to stop data breaches. *Frontiers in Computer Science*, *2*, 36.

[15] Zhang, Z., Al Hamadi, H., Damiani, E., Yeun, C. Y., and Taher, F. (2022). Explainable artificial intelligence applications in cyber security: state-of-the-art in research. *IEEE Access*, *10*, 93104–93139, doi: 10.1109/ACCESS.2022.3204051.

[16] Sun, L., Dou, Y., Yang, C., *et al.* (2022). Adversarial attack and defense on graph data: a survey. *IEEE Transactions on Knowledge and Data Engineering*, *35*(8), 7693–7711, doi: 10.1109/TKDE.2022.3201243.

[17] Ge, W., Wang, J., Lin, T., Tang, B., and Li, X. (2023). Explainable cyber threat behavior identification based on self-adversarial topic generation. *Computers & Security*, *132*, 103369.

[18] Preuveneers, D. and Joosen, W. (2021). Sharing machine learning models as indicators of compromise for cyber threat intelligence. *Journal of Cybersecurity and Privacy*, *1*(1), 140–163.

[19] Tcydenova, E., Kim, T. W., Lee, C., and Park, J. H. (2021). Detection of adversarial attacks in AI-based intrusion detection systems using explainable AI. *Human-centric Computing and Information Sciences*, *107*, 102303.

[20] Klawikowska, Z., Mikołajczyk, A., and Grochowski, M. (2020). Explainable AI for inspecting adversarial attacks on deep neural networks. In *Artificial Intelligence and Soft Computing: 19th International Conference, ICAISC 2020, Zakopane, Poland, October 12–14, 2020, Proceedings, Part I 19* (pp. 134–146). Springer International Publishing.

[21] Alenizi, J. and Alrashdi, I. (2023). SFMR-SH: secure framework for mitigating ransomware attacks in smart healthcare using blockchain technology. *Sustainable Machine Intelligence Journal*, 2. doi: 10.61185/SMIJ.2023.22104.

Chapter 13
Federated learning meets explainable AI at the edge of things

13.1 Introduction

Federated learning (FL) and explainable artificial intelligence (XAI) stand as pivotal advancements in the realm of modern Internet of Things (IoT) applications, offering innovative paradigms to address critical challenges in decentralized learning and model interpretability. FL, a paradigmatic shift in machine learning (ML), has gained prominence due to its efficacy in training models across a network of edge devices while preserving data privacy. Unlike traditional centralized approaches, FL operates by distributing the learning process among a multitude of local devices, enabling model training on data generated at the edge without the need for raw data aggregation in a central repository. This decentralized learning framework not only safeguards sensitive user information but also empowers IoT systems by harnessing the collective intelligence of distributed devices, thereby enhancing model performance and adaptability in dynamic edge environments. On the other hand, XAI has emerged as a critical attribute in contemporary IoT ecosystems, aiming to demystify the decision-making process of complex ML models. In essence, XAI endeavors to elucidate the reasoning behind AI-driven decisions, enabling stakeholders to comprehend and trust model outputs. With the surge of AI-powered applications in IoT, interpretability becomes paramount, particularly in critical domains where transparent and accountable decision-making mechanisms are essential [1].

The integration of FL and XAI in modern IoT applications represents a symbiotic alliance poised to revolutionize the landscape of decentralized learning and model transparency at the edge. The fusion of FL with XAI heralds a new era of collaborative learning where edge devices autonomously contribute to model refinement without compromising data privacy or security. By distributing model training across a network of interconnected devices, FL capitalizes on the computational power of edge nodes while preserving sensitive data within local silos, thereby mitigating privacy risks associated with centralized data storage. Simultaneously, XAI augments this paradigm by enhancing the interpretability and transparency of AI-driven decisions within IoT systems. The amalgamation of FL and XAI not only facilitates the creation of robust and privacy-preserving models but also ensures that the decisions made by these models are explicable and comprehensible to stakeholders, fostering trust and accountability in AI-powered IoT applications. This holistic approach paves the way for developing resilient, adaptive, and ethically sound IoT ecosystems capable of addressing diverse challenges across industries while maintaining integrity, privacy, and reliability in decision-making processes at the edge [2].

The notion of the "edge of things" epitomizes a pivotal paradigm in contemporary IoT architectures, delineating the periphery where data processing, computation, and decision-making occur proximate to data sources or end-user devices. In distributed and resource-constrained IoT environments, the concept of the edge embodies a strategic vantage point where computational tasks are executed in close proximity to where data is generated, negating the need for continuous data transmission to distant cloud infrastructures. This proximity-driven computational approach holds profound relevance in resource-constrained settings characterized by limitations in bandwidth, latency, and energy resources. By leveraging edge computing, IoT devices offload computational burdens from centralized data centers, orchestrating localized data processing and analysis. This strategy not only ameliorates network congestion but also expedites real-time responses crucial for time-sensitive applications such as autonomous vehicles, healthcare monitoring, and industrial automation. Moreover, the edge paradigm facilitates autonomous decision-making in distributed IoT ecosystems spanning diverse geographical locations or remote areas, enabling devices to function effectively even in scenarios with intermittent network connectivity or constrained computational capabilities [3].

Within resource-constrained IoT environments, the concept of the edge is significant due to its ability to optimize computational tasks and alleviate network constraints while catering to diverse IoT applications. The "edge of things" paradigm empowers IoT devices with localized intelligence, allowing them to autonomously execute data processing and inferential tasks. In scenarios where network bandwidth is limited or unreliable, the edge's proximity to data sources ensures timely data analysis and decision-making without excessive reliance on continuous connectivity to remote servers. This decentralized approach not only reduces data transmission overhead but also enhances system resilience by mitigating the impact of network failures or latency issues. Furthermore, in resource-constrained environments marked by constrained energy resources or computational capabilities, edge computing optimizes the utilization of available resources by performing computations closer to the data source. Consequently, this strategic placement of computational capabilities at the edge of IoT architectures caters to the unique challenges posed by distributed and resource-constrained settings, thereby enabling efficient, responsive, and adaptive IoT systems capable of meeting diverse application requirements across various domains.

13.2 Fundamentals FL

In this section, we delve into the fundamental tenets and mechanisms underlying FL, elucidating its pivotal role in addressing challenges posed by data heterogeneity, privacy concerns, and resource limitations inherent in decentralized IoT environments. By exploring its core principles and working mechanisms, this section aims to provide a comprehensive understanding of FL, highlighting its significance in enabling collaborative model training without centralized data aggregation. By unraveling the intricate architecture and methodologies central to FL, we aim to delineate its applicability and potential in revolutionizing ML paradigms within distributed and resource-constrained IoT ecosystems.

13.2.1 Definition and motivation

FL represents a decentralized ML paradigm designed to train AI models across a network of edge devices while preserving data privacy. This approach allows devices to collaboratively learn a shared model without sharing raw data centrally. In FL, each edge device independently computes model updates based on its local data and then transmits only the updates' parameters or gradients to a central server or aggregator. Subsequently, the server aggregates these updates to generate a global model, which is then refined and redistributed to participating devices for further refinement. This decentralized learning paradigm enables collaborative model training while mitigating privacy risks, as individual user data remains local and is never shared or centralized, ensuring confidentiality and addressing privacy concerns associated with traditional centralized ML methods.

The motivation behind FL stems from the growing emphasis on data privacy and the proliferation of edge devices, such as smartphones, IoT sensors, and other connected devices. While foundational in ML, traditional centralized training methods, confront inherent limitations that impede their efficacy in diverse real-world scenarios, particularly in distributed and resource-constrained environments typified by the IoT. One of the primary challenges stems from the necessity of aggregating data from distributed sources into a centralized repository for model training. In scenarios where data privacy and regulatory constraints are paramount, this centralized approach poses significant risks, as sensitive information from various sources is amalgamated, potentially compromising individual privacy rights. Moreover, transferring massive volumes of raw data from edge devices to centralized servers engenders substantial bandwidth consumption. It exacerbates latency issues, especially in IoT ecosystems characterized by intermittent or constrained network connectivity. The impracticality of transferring extensive datasets to a central location burdens the network infrastructure and poses challenges in handling diverse data formats and structures inherent in distributed IoT environments. Thus, the rigidity of traditional centralized training methods becomes increasingly unsuitable and unviable in scenarios where data privacy, network constraints, and data heterogeneity are critical considerations [4].

Furthermore, the reliance on centralized training methodologies becomes untenable in scenarios characterized by stringent resource constraints and the need for real-time decision-making. In resource-constrained IoT environments, where edge devices possess limited computational capabilities and energy resources, the computational burden imposed by centralized training becomes a significant impediment. The constant necessity of transmitting data to centralized servers for model training and subsequent reiteration of model updates back to edge devices strains computational resources and amplifies latency issues, inhibiting timely responses essential for critical applications. Additionally, the dependence on continuous network connectivity for centralized training exacerbates vulnerabilities to network failures or security breaches, rendering IoT ecosystems susceptible to disruptions and compromises in sensitive operations. Consequently, the unsuitability of traditional centralized training methods becomes apparent in scenarios necessitating efficient resource utilization, reduced latency, and enhanced resilience—essential attributes in contemporary IoT deployments across diverse domains [5].

13.2.2 Key concepts in FL

In distributed ML frameworks like FL, several essential terms play pivotal roles in orchestrating collaborative model training across a network of devices. Clients, within the context of FL, denote individual or local devices participating in the learning process, typically situated at the edge of the network. These clients, often IoT devices or mobile phones, possess local datasets and execute local computations to update model parameters based on their respective data. The server, on the other hand, acts as the central orchestrator in FL, responsible for coordinating the learning process across diverse clients. It facilitates the aggregation of local updates from multiple clients and oversees the convergence toward a global model while preserving data privacy by avoiding direct access to raw client data.

Local updates represent the incremental changes to model parameters computed by individual clients using their local data. These updates encapsulate the learning derived from local datasets while ensuring data privacy by retaining data locally without transmission to a central server. Subsequently, the aggregation process plays a crucial role in FL, where the server consolidates these diverse local updates into a unified global model. This aggregation mechanism varies, employing techniques like federated averaging (FedAvg) or other aggregation methods that amalgamate local updates to refine the global model collaboratively. The resultant global model reflects the collective intelligence gleaned from distributed client updates, embodying a centralized representation of knowledge while preserving the privacy of individual client data.

This intricate interplay of clients, server, local updates, aggregation, and the global model constitutes the core architecture of FL. Clients contribute local knowledge without compromising sensitive data, the server orchestrates the collaboration while maintaining data privacy, local updates capture individual learning, aggregation consolidates insights, and the global model embodies the collective intelligence of the distributed network—creating a sophisticated framework to enable collaborative ML while respecting privacy and preserving efficiency in resource-constrained, distributed environments.

Federated training, a fundamental process within the realm of FL, encompasses a series of intricate steps orchestrated to collaboratively train ML models across a network of decentralized devices while preserving data privacy and security. The main steps in federated training encompass a choreographed sequence crucial for achieving convergence toward a global model representation (see Figure 13.1).

Step 1: Client sampling and initialization

The federated training process commences with the selection of participating clients from the network of edge devices. These clients are sampled either randomly or strategically, considering factors such as device availability, computational capabilities, and data diversity. Upon selection, each client initializes a local copy of the global model provided by the central server. This initialization primes the clients to begin their individual learning processes based on their local datasets.

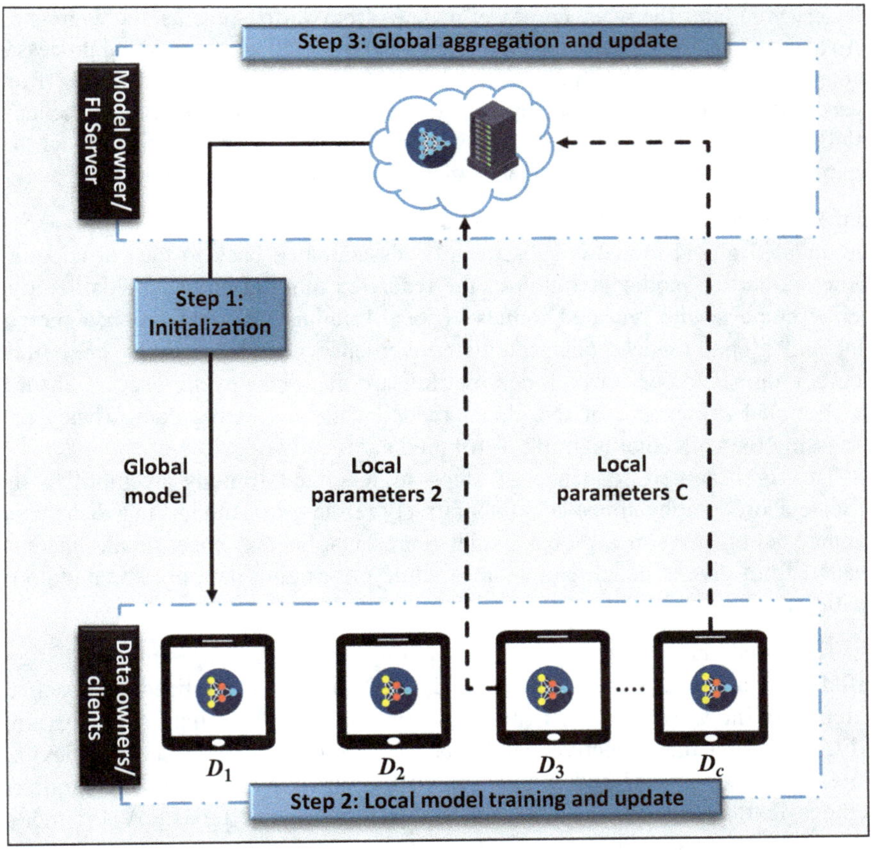

Figure 13.1 Showcases of the primary stages involved in FL within an IoT network

Step 2: Local model training (local updates)
Each selected client independently executes model training using its local data without transmitting raw data to the central server. Leveraging its local dataset, the client performs iterative computations to update the model parameters. These local updates encapsulate the learning derived from the client's specific data distribution, enabling personalized model refinement while upholding data privacy by keeping sensitive information local.

Step 3: Communication with the server (model updates)
Following local model training, the updated model parameters or gradients are communicated back to the central server. This phase involves the transmission of model updates, encapsulating the alterations made to the local model during training. The server aggregates these diverse model updates received from multiple clients, typically employing FedAvg or other aggregation techniques to reconcile variations in model updates and synthesize a collective understanding used to refine the global model.

Step 4: Aggregation and global model update
The server collates the received model updates from various clients and aggregates them to compute an updated version of the global model. This aggregation process is pivotal, as it involves reconciling the heterogeneity of updates stemming from diverse local datasets and computational environments. By amalgamating these local insights, the server orchestrates the convergence toward a refined global model that encapsulates the collective intelligence gleaned from distributed client updates.

Step 5: Model distribution and iteration
The updated global model is subsequently disseminated back to the participating clients, initiating another iteration of the federated training process. This iterative cycle continues with repeated rounds of local training, communication, aggregation, and global model updates until convergence criteria, such as predefined accuracy thresholds or convergence metrics, are met. This cyclic process ensures the continual refinement of the global model while respecting data privacy and harnessing the collective learning distributed across edge devices.

This orchestrated sequence of steps in federated training exemplifies the intricate choreography aimed at collaboratively refining ML models in a distributed manner, culminating in a global model representation that encapsulates insights gleaned from diverse local data sources while prioritizing data privacy and computational efficiency [6].

The collaborative learning process, epitomized by FL, represents a paradigm shift from traditional centralized methods, offering a decentralized approach to training ML models across a network of distributed devices. Collaborative learning in FL hinges on the collective intelligence harnessed from diverse edge devices while preserving data privacy, which is in stark contrast to conventional centralized methods that necessitate aggregating data into a singular repository for model training. In this collaborative approach, each participating device—referred to as a client—engages in local model training using its unique dataset without sharing raw data with a central server. The essence of collaborative learning lies in orchestrating multiple clients autonomously contributing incremental insights derived from their local data distributions, culminating in a unified global model.

The divergence between collaborative learning and traditional centralized methods lies primarily in the orchestration of model training and data aggregation. Unlike centralized approaches that entail the amalgamation of data into a central repository for model training, collaborative learning in FL operates in a federated manner, distributing the learning process across multiple edge devices. This decentralized strategy preserves data privacy by ensuring that sensitive information remains local, obviating the need for raw data transmission to a central server. In contrast, traditional methods necessitate the transmission of comprehensive datasets to a centralized location, raising concerns about data privacy, security, and the potential exposure of sensitive information. Furthermore, the collaborative learning process in FL integrates a federated aggregation mechanism to harmonize the diverse model updates generated by individual clients. This aggregation step, conducted by the central server, reconciles the heterogeneity inherent in local updates stemming from different devices with varied data distributions and computational capabilities. The amalgamation of these localized insights leads to refining a global model that encapsulates the collective intelligence derived from distributed devices without compromising individual data privacy. Conversely, traditional methods lack this distributed aggregation mechanism, often entailing a singular, centralized training process devoid of individual client contributions. This results in limited adaptability to diverse data distributions and potential privacy concerns associated with centralized data storage and processing.

13.2.3 Advantages and challenges

FL offers many substantial benefits, including enhanced data privacy and reduced communication overhead, distinguishing it from traditional centralized ML approaches.

Data privacy preservation: FL is a robust framework for maintaining data privacy, particularly in scenarios involving sensitive information. Unlike conventional methods that necessitate the aggregation of raw data into a centralized repository, FL operates on a decentralized premise, allowing individual devices or clients to retain their data locally. This distributed learning paradigm ensures that raw data remains on the edge devices, preventing the exposure of sensitive information during the training process. By minimizing the need to transmit sensitive data to a central server, FL inherently mitigates risks associated with potential data breaches, unauthorized access, or privacy infringements. This approach aligns with privacy regulations and user confidentiality concerns, making FL an attractive solution in domains such as healthcare, finance, and personal devices where data privacy is critical [7].

Reduced communication overhead: Another significant advantage of FL is its ability to curtail communication overhead in distributed environments, particularly in scenarios characterized by limited network bandwidth or intermittent connectivity. Traditional ML models often necessitate the continuous transmission of large volumes of data between edge devices and central servers for model training and updates, leading to substantial communication costs and latency issues. However, FL minimizes this communication burden by enabling local model training on individual devices without transmitting raw data. Only model updates or gradients derived from local learning are communicated to the central server for aggregation. This reduction in data transmission significantly diminishes network traffic, conserves bandwidth, and alleviates latency concerns. Consequently, FL is well-suited for resource-constrained environments, such as IoT ecosystems, edge computing settings, or mobile devices, where efficient utilization of communication resources is imperative for optimal performance.

While offering significant advantages, FL, also confronts several challenges that impede its seamless implementation and effectiveness in distributed environments. These challenges encompass non-IID (non-independently and identically distributed) data, communication efficiency, black-box models, and model divergence, each posing unique obstacles in the FL paradigm.

Non-IID data distribution: One of the primary challenges in FL arises from the heterogeneity of data distributions across edge devices or clients. In real-world scenarios, data collected from various devices may exhibit substantial disparities in terms of data type, quantity, quality, and statistical characteristics. This non-IID nature of data can significantly impact the learning process, as the assumption of identical data distributions across devices—central to many ML algorithms—might not hold. Such disparities hinder model convergence, resulting in challenges when aggregating updates from diverse clients. Coping with non-IID data distribution remains a critical challenge in FL, necessitating advanced techniques to adaptively handle and mitigate the effects of disparate data distributions.

Communication efficiency: FL relies on efficient communication between edge devices and a central server for model updates. However, in scenarios where network bandwidth is limited, communication overhead becomes a significant bottleneck. Transmitting model updates or gradients from numerous devices to the central server incurs substantial communication costs and latency. Addressing communication efficiency is pivotal to mitigating these challenges, requiring innovative strategies such as compression techniques, selective client participation, or differential privacy measures to optimize communication without compromising model accuracy or data privacy.

Black-box models and interpretability: The inherent complexity of ML models employed in FL, often termed "black-box" models, poses a challenge in terms of model interpretability. Understanding the decision-making process of these intricate models becomes challenging, raising concerns about transparency and explainability. In sensitive domains like healthcare or finance, where interpretability is crucial for trust and compliance, the opacity of black-box models presents a significant hurdle. Overcoming this challenge necessitates the development of explainable AI techniques tailored to FL settings, enabling stakeholders to comprehend and trust the decisions made by these complex models.

Model divergence and convergence: Model divergence and convergence issues emerge due to the collaborative nature of FL. Diverse edge devices contribute updates to a central model, potentially leading to inconsistencies or divergence in model behavior. This phenomenon arises from variations in local datasets, computational capabilities, or learning dynamics across devices. Ensuring model convergence while mitigating divergence poses a substantial challenge in FL, demanding sophisticated aggregation techniques, regularization methods, or adaptive learning algorithms to harmonize diverse local updates and foster convergence toward a globally coherent model [8,9].

13.3 Taxonomy of FL methods

In the landscape of FL, a diverse taxonomy of methods has emerged, each addressing specific challenges and nuances inherent in decentralized learning paradigms across distributed devices. This section delves into the taxonomy of FL methods, offering a systematic exploration and categorization of diverse approaches employed to facilitate collaborative model training while respecting data privacy and accommodating the intricacies of edge environments. The taxonomy encompasses a spectrum of methodologies, encompassing techniques to handle non-IID data distributions, optimize communication efficiency, enhance model interpretability, and ensure convergence amidst diverse local updates.

13.3.1 Network topology

FL methods can be taxonomized based on network topology, encompassing distinct approaches that cater to the varying organizational structures and complexities present in decentralized learning scenarios across distributed networks (see Figure 13.2). This taxonomy classifications based on network topology showcase diverse paradigms within FL, each tailored to accommodate varying network architectures, communication capabilities, and privacy requirements, offering a spectrum of methodologies to adapt to the complexities and dynamics of decentralized learning in distributed environments. Understanding these categorizations provides a foundation for selecting suitable FL methodologies aligned with specific network structures and operational constraints prevalent in diverse edge computing ecosystems [10].

Centralized FL represents a fundamental topology where all edge devices or clients directly communicate with a centralized server during the collaborative model training process. In this framework, edge devices initiate local model training using their respective datasets and subsequently transmit model updates or gradients to a central server for aggregation and refinement of the global model. This topology centralizes the coordination of the learning process, allowing the server to orchestrate model updates and maintain a unified representation of the global model. However, despite its simplicity and ease of management, centralized FL faces challenges concerning privacy and communication overhead. The direct transmission of model updates from each client to the central server raises privacy

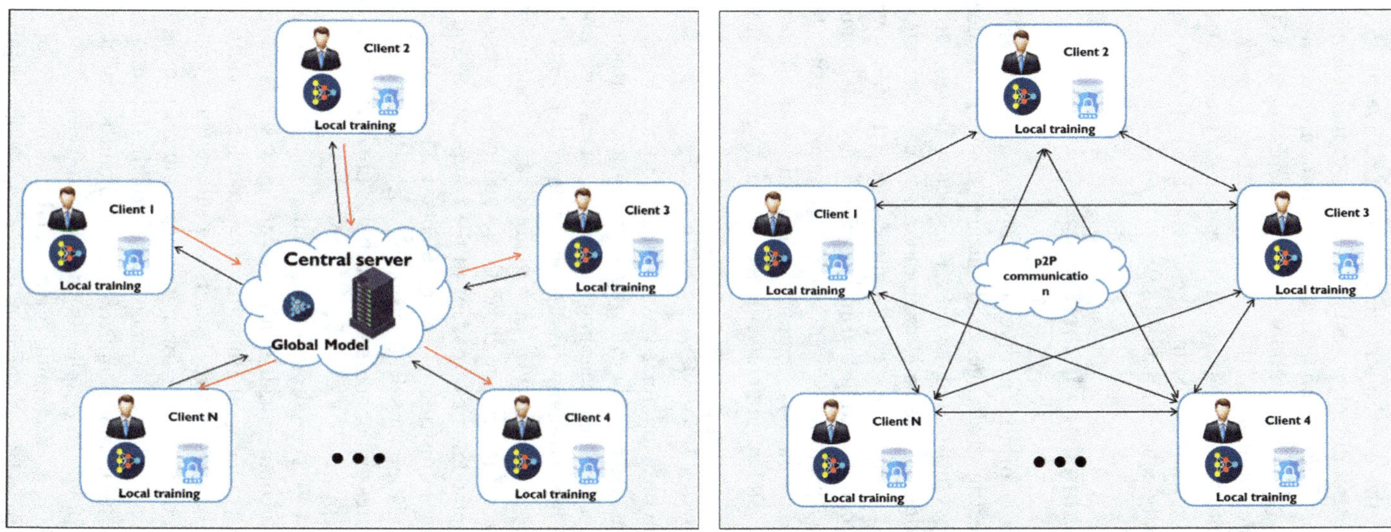

Figure 13.2 Visual comparison between centralized and decentralized FL

concerns, as sensitive information might be exposed during data transmission. Moreover, the centralized communication architecture entails substantial communication costs and latency, especially in scenarios with a large number of participating devices or limited network bandwidth, hindering scalability and efficiency in resource-constrained edge environments [11].

Centralized FL remains a foundational approach but encounters limitations in scenarios where data privacy and communication efficiency are paramount. While the centralized coordination streamlines the aggregation process and ensures a unified global model representation, its shortcomings in addressing data privacy and communication overhead necessitate advanced techniques and adaptations. Strategies such as differential privacy measures or encryption techniques are being explored to mitigate privacy risks, while optimization methods, such as model compression or selective participation, aim to alleviate communication overhead. Despite its challenges, centralized FL serves as a cornerstone topology, paving the way for more sophisticated and privacy-preserving approaches within FL, offering a baseline for comparison and evolution toward more decentralized and privacy-centric methodologies tailored to diverse edge computing environments.

Decentralized FL signifies a paradigm shift from centralized architectures, where edge devices engage in collaborative model training without reliance on a central coordinating entity. In this topology, devices communicate directly with each other in a peer-to-peer manner, eliminating the necessity for a central server orchestrating the learning process. This framework decentralizes the coordination of model updates, allowing devices to collaborate autonomously by exchanging model parameters or gradients among themselves. Unlike centralized approaches, decentralized FL mitigates privacy concerns associated with the direct transmission of sensitive data to a central server, as devices interact directly with each other, preserving the confidentiality of local datasets during communication. However, despite its merits in enhancing privacy, decentralized FL poses challenges in synchronization and coordination among devices. The absence of a central entity overseeing the learning process may lead to difficulties in harmonizing diverse local updates, potentially causing model divergence or inefficiencies in convergence. Additionally, in scenarios with many devices or disparate computational capabilities, ensuring synchronization and effective collaboration among heterogeneous devices becomes a considerable challenge, impacting the efficiency and efficacy of the learning process [12].

Decentralized FL represents an evolution toward a more distributed and privacy-centric FL paradigm. Its emphasis on direct device-to-device communication fosters improved privacy preservation by avoiding centralized data aggregation points. However, to address synchronization and coordination challenges inherent in this topology, various strategies are being explored. These include FedAvg algorithms with adaptive learning rates to manage heterogeneity, and gossip-based communication protocols to facilitate the efficient exchange of model updates among devices. Moreover, techniques such as local model ensembling or federated distillation aim to enhance convergence and cooperation among decentralized devices. Decentralized FL stands as a promising avenue, offering increased data privacy while concurrently demanding innovative solutions to navigate synchronization complexities and effectively harness the collective intelligence distributed across diverse edge devices in FL ecosystems.

Distributed FL emerges as an advanced topology in the landscape of decentralized ML, fostering a peer-to-peer communication framework where edge devices engage in collaborative model training while interacting with multiple peers. This topology extends the decentralization paradigm of FL by allowing devices to communicate and exchange model updates with multiple other devices, enabling a distributed network where each device collaborates with several peers. Unlike centralized or decentralized approaches, distributed FL introduces a multiparty interaction model, empowering devices to engage in bidirectional communication with various peers, thereby diversifying the collaborative learning process. However, the complexity of managing interactions among multiple devices poses significant challenges in terms of synchronization, coordination, and scalability. The intricacies of orchestrating communication and aggregating diverse updates from numerous peers require sophisticated algorithms and protocols to ensure efficient collaboration while addressing heterogeneity in local datasets and computational capabilities across distributed devices [13].

Distributed FL represents a pivotal evolution in FL, aiming to harness the collective intelligence dispersed across a network of diverse edge devices. The distributed nature of interactions fosters enhanced collaboration, enabling devices to engage in bidirectional information exchange beyond the constraints of traditional centralized or decentralized architectures. To address challenges inherent in this topology, various strategies are being explored. These include adaptive aggregation algorithms that adaptively weight contributions from different peers based on their reliability or expertise, and federated reinforcement learning techniques tailored for efficient interaction among distributed devices. Additionally, privacy-preserving mechanisms such as secure multi-party computation and homomorphic encryption are being researched to ensure confidentiality and security while facilitating collaborative learning among multiple distributed peers. Distributed FL stands at the forefront of innovative research endeavors, aspiring to leverage the collective intelligence of distributed networks while navigating the complexities of multi-party interactions in FL settings.

Hierarchical FL presents a sophisticated topology where edge devices are organized into hierarchical structures, fostering localized collaborations within defined clusters or tiers. This hierarchical arrangement introduces a multi-level architecture where devices interact and collaborate within their respective clusters before transmitting aggregated updates to higher-level nodes for further aggregation. Such a hierarchical framework facilitates localized model training and aggregation at different levels of the hierarchy, enabling a tiered approach to collaborative learning. This topology provides a scalable structure that streamlines communication and aggregation processes by conducting local aggregations within clusters before propagating the refined updates hierarchically. However, managing hierarchical relationships and orchestrating efficient collaborations among diverse clusters while ensuring coherence in the global model representation pose

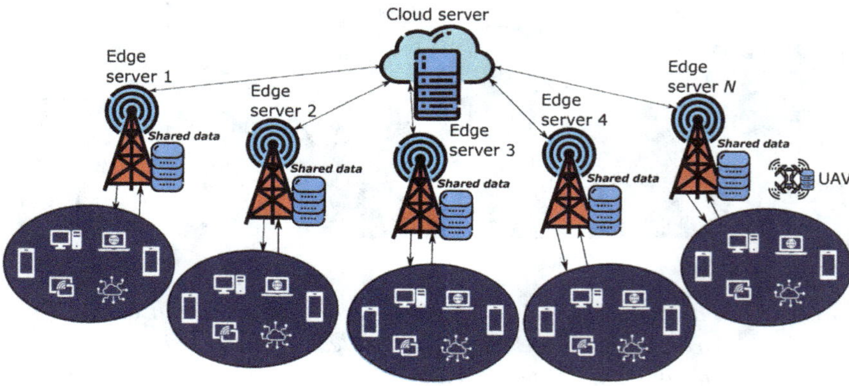

Figure 13.3 Visualization of hierarchical FL in IoT

significant challenges. The complexity of synchronization and coordination across hierarchical levels necessitates sophisticated algorithms and coordination mechanisms to balance local autonomy within clusters and global convergence across the hierarchy [14].

Hierarchical FL stands as an innovative approach that offers a balance between local autonomy and global convergence, catering to the complexities of diverse edge computing environments (see Figure 13.3). Strategies to address challenges in this topology involve hierarchical aggregation schemes that consolidate local updates at different tiers while ensuring consistency in the global model representation. Additionally, adaptive hierarchical learning algorithms and cross-tier communication protocols aim to facilitate efficient interactions among clusters at various levels of the hierarchy. Moreover, privacy-preserving techniques compatible with hierarchical architectures, such as differential privacy mechanisms or FL with secure aggregation, are being explored to ensure data confidentiality while enabling collaborative learning across hierarchical levels. Hierarchical FL represents a promising avenue for FL, leveraging structured hierarchical relationships to optimize collaboration and convergence while accommodating the diverse dynamics inherent in complex distributed systems.

13.3.2 Communication mechanism

FL exhibits diverse taxonomies based on communication mechanisms, encompassing horizontal FL, vertical FL, and federated transfer learning, each tailored to address specific challenges inherent in decentralized collaborative model training (see Figure 13.4). Horizontal FL involves devices collaborating on similar features or data samples while maintaining distinct data. This mechanism facilitates model training across devices with similar data distributions, optimizing model performance without transmitting raw data. Vertical FL, on the other hand, enables collaboration across devices possessing complementary but non-overlapping feature sets, allowing model training by jointly analyzing different aspects of data without

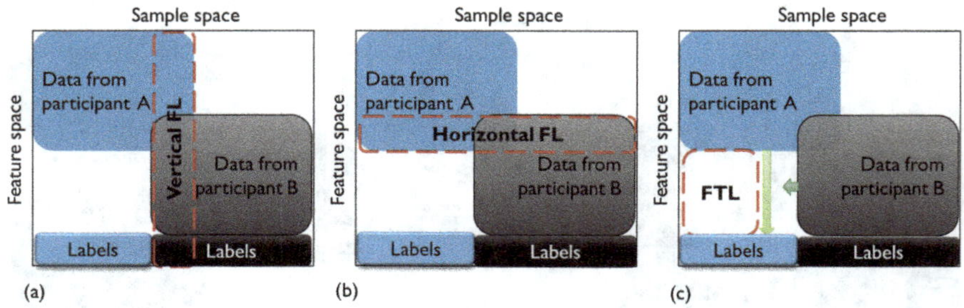

Figure 13.4 A breakdown of FL categorized according to the utilized data partitioning: (a) Vertical federated learning, (b) horizontal federated learning, and (c) federated transfer learning

exchanging raw data. This mechanism proves valuable in scenarios where merging disparate datasets can enhance model robustness while preserving data privacy. Furthermore, federated transfer learning extends FL by leveraging knowledge transfer between tasks or domains, enabling devices to benefit from pre-trained models or learned representations from other devices, thereby facilitating faster convergence and improved performance in scenarios with limited local data. These taxonomies based on communication mechanisms underscore FL's adaptability to diverse data distribution patterns, privacy constraints, and optimization objectives, providing a comprehensive framework to navigate the complexities of collaborative learning in decentralized environments [15].

Horizontal FL represents a communication mechanism where edge devices possess similar feature sets but maintain distinct data samples, enabling collaborative model training without exchanging raw data. In this topology, devices share common features relevant to the learning task but have diverse data instances, allowing for model aggregation and refinement without compromising data privacy. Horizontal FL emphasizes collaboration among devices with comparable data distributions, facilitating model convergence and performance improvement by aggregating insights gleaned from diverse but similar datasets. However, challenges arise in synchronizing diverse local updates while maintaining the integrity of the global model, particularly in scenarios where data heterogeneity among devices results in variations in learning dynamics or statistical distributions. Ensuring effective aggregation while managing diverse local updates without compromising model accuracy remains a significant challenge in horizontal FL, necessitating advanced aggregation techniques and synchronization strategies to reconcile disparities in local datasets and facilitate coherent convergence toward a refined global model representation.

The horizontal FL mechanism offers notable advantages by fostering collaboration among devices possessing analogous feature spaces, mitigating concerns related to data privacy and raw data transmission (see Figure 13.5). By enabling model training without sharing raw data, horizontal FL preserves the confidentiality of local datasets, complying with stringent privacy regulations and assuaging

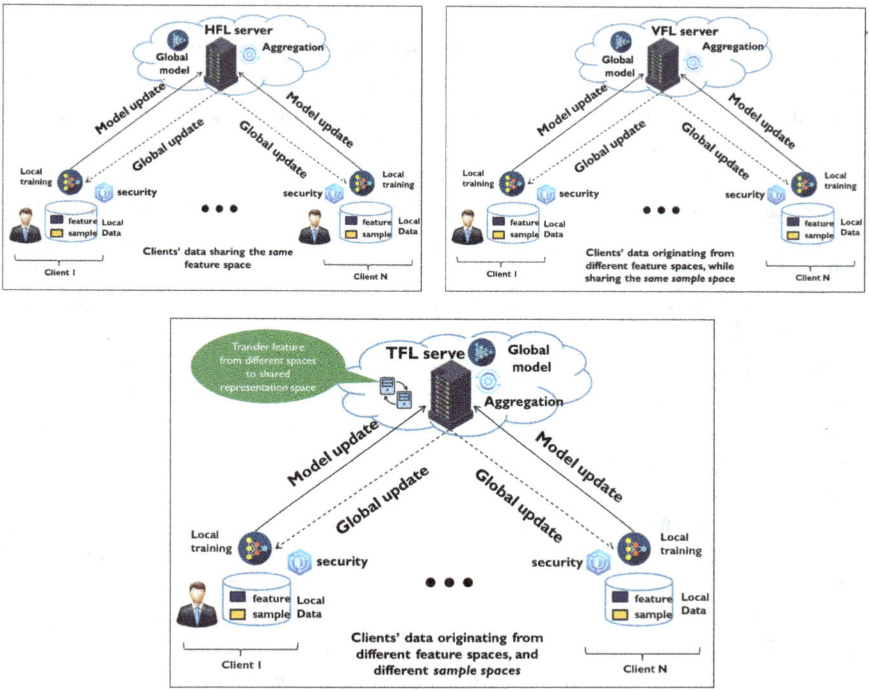

Figure 13.5 Visual comparison among horizontal FL, vertical FL, and federated transfer learning

privacy concerns associated with data transmission. Nevertheless, addressing the challenges of data heterogeneity and synchronization complexities poses critical research avenues within horizontal FL. Advanced techniques such as FedAvg with adaptive learning rates, model personalization strategies, or differential privacy measures tailored for feature-level collaboration aim to enhance model convergence and privacy preservation while accommodating diverse feature distributions among collaborating devices. Horizontal FL stands as a promising approach in FL, leveraging shared feature spaces to harness collective insights while safeguarding data privacy, albeit demanding innovative solutions to navigate data heterogeneity and synchronization intricacies across diverse but similar datasets.

Vertical FL embodies a communication mechanism wherein edge devices possess identical samples or instances but exhibit distinct feature sets, allowing collaborative model training without exchanging raw data. In this topology, devices maintain data instances that align but differ in the types of features available, enabling joint analysis of complementary but non-overlapping feature spaces across devices. Vertical FL facilitates collaborative learning by aggregating insights derived from diverse feature sets while preserving data privacy, as the raw data remains localized on each device. However, challenges arise in reconciling disparate feature spaces and effectively aggregating local updates to ensure coherent convergence toward a refined global model. Managing feature mismatches and integrating insights across complementary but non-overlapping feature sets present significant hurdles, demanding innovative techniques and alignment strategies to harmonize diverse features and optimize model performance without compromising data confidentiality.

The vertical FL mechanism offers distinctive advantages by fostering collaboration among devices possessing matching data instances while maintaining feature diversity, thus enabling joint learning across complementary feature spaces. By allowing collaborative model training without the need for raw data exchange, vertical FL preserves data privacy and confidentiality, aligning with privacy regulations and assuring stakeholders of data security. Nevertheless, addressing challenges associated with feature mismatch and effective aggregation remains pivotal within vertical FL. Strategies such as federated transfer learning techniques, feature alignment methods, or federated meta-learning approaches aim to bridge feature gaps and enhance aggregation across diverse feature spaces while preserving data privacy, contributing to improved model convergence and performance in scenarios with complementary but non-overlapping feature sets. Vertical FL stands as an innovative approach in FL, leveraging matching data instances with diverse feature spaces to harness collaborative insights while ensuring data confidentiality, albeit requiring nuanced techniques to navigate feature heterogeneity and aggregation complexities across diverse but complementary features [16].

Federated transfer learning emerges as an extension of FL that leverages the transfer of knowledge, representations, or pre-trained models from one device to aid in the learning process of another. This communication mechanism allows edge devices to benefit from shared knowledge or pre-existing model representations without directly exchanging raw data. By transferring learned insights or model parameters, devices can enhance their local models, facilitating faster convergence, improved performance, and generalization in scenarios with limited local data or computational resources. Federated transfer learning emphasizes knowledge transfer across devices, enabling devices with fewer samples or resources to benefit from the expertise gleaned by other devices, thereby fostering a more efficient and collaborative learning ecosystem. However, challenges persist in determining effective strategies for knowledge transfer, adapting transferred knowledge to suit local contexts, and ensuring the privacy and security of shared knowledge during the transfer process, warranting sophisticated techniques and protocols to facilitate seamless knowledge exchange while preserving data confidentiality.

Federated transfer learning presents a promising avenue in FL by capitalizing on knowledge sharing and transfer among distributed devices to enhance model performance and convergence. By enabling devices to leverage shared knowledge or pre-trained models, federated transfer learning addresses limitations in scenarios with sparse local data or computational constraints. Yet, addressing challenges related to effective knowledge transfer, domain adaptation, and privacy preservation remains pivotal within this mechanism. Advanced techniques such as model distillation, domain adaptation strategies, or privacy-preserving transfer learning protocols aim to refine knowledge transfer processes, adapt transferred knowledge to local contexts, and uphold data privacy during knowledge exchange. Federated transfer learning stands at the forefront of innovative research endeavors, aiming to optimize collaborative learning by harnessing shared knowledge and expertise across distributed devices while navigating challenges associated with knowledge transfer and adaptation across diverse edge computing environments [17].

13.3.3 Aggregation methods

In the realm of FL, taxonomy based on aggregation methods delineates distinct approaches employed to amalgamate and refine local updates from distributed devices into a unified global model. This taxonomy encompasses diverse aggregation methods such as FedAvg, secure aggregation, weighted aggregation, and compression-based aggregation, each tailored to address specific challenges and requirements inherent in collaborative model training across decentralized networks.

FedAvg stands as a foundational aggregation method within the realm of FL, instrumental in orchestrating collaborative model training across distributed devices while preserving data privacy. The working principle of FedAvg involves an iterative process orchestrated by a central server, aiming to refine a global model by aggregating model parameter updates derived from participating edge devices. Initially, each device independently conducts local model training using its respective dataset, generating local updates based on its learning experience. These local updates, often in the form of gradients or parameter differentials, are then transmitted to the central server for aggregation. During aggregation, the central server aggregates the received updates from multiple devices by computing the weighted average of these updates, synthesizing a collective understanding while maintaining a unified global model. Subsequently, the aggregated update is disseminated back to the participating devices, initiating another iteration of local model training. This cyclic process continues for multiple rounds, allowing the global model to progressively refine itself by aggregating diverse insights derived from the distributed devices' local datasets. The effectiveness of FedAvg lies in its simplicity and scalability, enabling model convergence across distributed devices without necessitating raw data exchange. However, challenges persist in ensuring convergence and mitigating heterogeneity among local updates stemming from devices with diverse data distributions or computational capabilities. Moreover, FedAvg's reliance on averaging local updates assumes identical data distributions among devices, potentially posing challenges in scenarios with non-IID data or feature disparities. To address these challenges, adaptations of FedAvg, such as weighted aggregation schemes or adaptive learning rates, aim to manage heterogeneity and optimize model convergence across diverse edge devices while preserving data privacy—an essential principle in FL methodologies [18].

Secure aggregation stands as a pivotal aggregation method within FL, designed to ensure the privacy and confidentiality of model updates during the aggregation process across distributed devices. The working principle of secure aggregation revolves around preserving data confidentiality by enabling devices to contribute their updates in an encrypted form, allowing the central server to aggregate these encrypted updates without accessing raw data or compromising sensitive information. In this mechanism, participating devices encrypt their local model updates using cryptographic techniques, such as secure multi-party computation or homomorphic encryption, before transmitting them to the central server. This encryption ensures that the central server cannot decipher the individual updates contributed by devices but can perform computations on the encrypted data. The central server aggregates the encrypted updates, generating an encrypted aggregate, which is then decrypted to obtain the final aggregated update. This process ensures that the central server never has access to the raw updates or intermediate states, preserving data privacy and confidentiality throughout the aggregation procedure. The effectiveness of secure aggregation lies in its ability to facilitate collaborative model training while safeguarding the privacy of individual contributions from distributed devices. By leveraging cryptographic protocols, secure aggregation ensures that sensitive information remains encrypted throughout the aggregation process, addressing concerns related to data exposure or privacy breaches during model refinement. However, challenges persist in terms of computational overhead associated with encryption and decryption operations, which can significantly impact the efficiency of secure aggregation, particularly in resource-constrained edge environments. Advanced cryptographic techniques and optimization strategies aim to alleviate these computational challenges, striving to enhance the efficiency and scalability of secure aggregation while ensuring robust privacy preservation—a fundamental tenet in the domain of FL methodologies.

Quantization-based aggregation represents an innovative method within FL, focusing on reducing communication overhead during the aggregation of model updates from distributed devices. The working principle of quantization-based aggregation involves compressing the transmitted model updates before aggregation, thereby minimizing the amount of data transmitted between devices and the central server. This method operates by quantizing or reducing the precision of model parameters or gradients before transmission. Participating devices quantize their updates by converting floating-point values to lower precision representations, such as fixed-point or integer representations, reducing the size of transmitted data. Upon receiving these quantized updates, the central server aggregates the compressed updates, performing the aggregation process on the reduced precision representations. Subsequently, the aggregated update is disseminated back to the participating devices, and the process iterates for subsequent rounds of model refinement. By reducing the precision of updates before transmission and aggregation, quantization-based aggregation significantly diminishes communication overhead, conserving network bandwidth and mitigating latency issues, particularly in scenarios with limited network resources or large-scale distributed systems.

The effectiveness of quantization-based aggregation lies in its ability to optimize communication efficiency without compromising the accuracy or integrity of the global model. However, challenges persist in managing the trade-off between compression levels and model accuracy, as aggressive quantization might result in information loss or degradation in model performance. Balancing compression ratios to minimize communication costs while preserving model accuracy remains a critical consideration within quantization-based aggregation. Advanced techniques such as adaptive quantization schemes or error compensation strategies aim to mitigate information loss and maintain model fidelity while optimizing communication efficiency. Quantization-based aggregation stands as a promising approach within FL methodologies, offering a pathway to alleviate communication bottlenecks and enhance scalability in large-scale distributed learning scenarios, provided that the compression methods employed strike a balance between communication efficiency and model accuracy.

Gradient compression represents a pivotal technique in FL, focusing on minimizing communication overhead by compressing the gradients exchanged between edge devices and the central server during the model training process. The working principle of gradient compression involves reducing the size of gradient updates before transmission, optimizing the amount of information transmitted while maintaining model accuracy. Participating devices compute gradients during local model training, which are then compressed before transmission to the central server. This compression can take various forms, such as quantization, sparsification, or encoding techniques. Quantization involves reducing the precision of gradient values, typically by converting floating-point values to lower precision representations, thereby reducing the amount of data transmitted. Sparsification techniques identify and transmit only significant gradient values while discarding or representing less critical values with minimal information. Encoding mechanisms leverage algorithms to encode gradients in a more compact format, further reducing the size of transmitted data. Upon receiving these compressed gradients, the central server aggregates and updates the global model, iterating the process for subsequent rounds of model refinement. By compressing gradients before transmission, gradient compression significantly minimizes communication costs, conserves bandwidth, and alleviates latency concerns, particularly in scenarios with limited network resources or high communication overhead.

The effectiveness of gradient compression lies in its ability to balance communication efficiency with model accuracy, addressing challenges associated with substantial communication overhead in FL setups. However, optimizing compression methods without compromising model convergence and performance remains a key challenge. Aggressive compression techniques might lead to information loss or degradation in model accuracy, necessitating a trade-off between compression levels and model fidelity. Advanced techniques such as adaptive compression algorithms, error compensation mechanisms, or personalized compression strategies tailored to individual device characteristics aim to mitigate information loss and maintain model accuracy while optimizing communication efficiency. Gradient compression stands as a promising approach in FL methodologies, offering a pathway to alleviate communication bottlenecks and enhance scalability in large-scale distributed learning settings, provided that the compression methods employed strike a balance between communication efficiency and model accuracy tailored to specific edge computing environments.

Personalized or adaptive aggregation stands as a nuanced approach within FL, aiming to tailor the aggregation process to individual edge devices based on their characteristics, data distributions, or model performance. The working principle of personalized or adaptive aggregation involves adapting the aggregation mechanism to suit the unique attributes of participating devices, optimizing model updates' aggregation to enhance convergence and performance. In this approach, participating devices conduct local model training using their respective datasets, generating individual updates based on their learning experiences. These updates are then transmitted to the central server, which adapts the aggregation mechanism based on device-specific factors such as data distribution, learning dynamics, or computational capabilities. Devices exhibiting faster convergence or greater accuracy may receive higher aggregation weights or prioritization during the aggregation process, allowing their updates to contribute more substantially to the global model. Conversely, devices with slower convergence or less reliable updates might receive lower weights or adaptive learning rates, ensuring that their contributions are accounted for without significantly impacting the global model's performance. This adaptive approach to aggregation aims to optimize model convergence and performance across diverse edge devices, effectively utilizing their collective intelligence while accommodating variations in learning dynamics and data distributions.

The effectiveness of personalized or adaptive aggregation lies in its ability to adaptively tailor the aggregation process to suit the characteristics and performance of individual devices, addressing challenges associated with heterogeneity in distributed learning environments. However, determining suitable adaptation strategies and balancing personalized updates' contributions without compromising the global model's integrity remains a significant challenge. Advanced techniques such as adaptive federated optimization algorithms, dynamic aggregation schemes, or federated meta-learning approaches aim to adaptively adjust aggregation mechanisms to individual device characteristics while maintaining global model coherence. Personalized or adaptive aggregation stands as a promising avenue within FL methodologies, offering a pathway to optimize model convergence and performance by leveraging the strengths of individual devices while navigating challenges associated with heterogeneity and variations in learning dynamics across distributed edge environments [19].

Hierarchical aggregation represents an innovative approach within FL, aiming to orchestrate the aggregation process in a hierarchical structure across multiple tiers or clusters of edge devices. The working principle of hierarchical aggregation involves organizing participating devices into hierarchical levels or clusters based on specific criteria, such as geographical proximity, computational capabilities, or expertise. Each tier conducts local model training and generates individual updates based on its dataset, initiating local aggregations within the cluster. These locally aggregated updates are then transmitted hierarchically to higher-level nodes or clusters for further aggregation at each hierarchical level. The hierarchical structure facilitates localized collaboration and aggregation within clusters before propagating refined updates to higher tiers, culminating in the generation of a globally aggregated update. This hierarchical aggregation process continues iteratively, allowing for the refinement of the global model by integrating insights and updates across diverse hierarchical levels, catering to localized characteristics and collective intelligence within each tier.

The effectiveness of hierarchical aggregation lies in its ability to manage and orchestrate collaborative model training in a structured hierarchical framework, accommodating varying levels of granularity and scalability across distributed devices. However, challenges persist in synchronizing and harmonizing updates across hierarchical levels, addressing heterogeneity in learning dynamics, and ensuring consistent model convergence throughout the hierarchy. Advanced techniques such as adaptive hierarchical aggregation algorithms, cross-tier communication protocols, or localized adaptation strategies aim to reconcile disparities across hierarchical levels while maintaining coherence in the global model representation. Hierarchical aggregation stands as a promising approach within FL methodologies, offering a structured framework to optimize collaboration and aggregation across hierarchical tiers, thereby leveraging localized insights and expertise within distributed networks while addressing challenges associated with managing hierarchical relationships in diverse edge computing environments.

13.3.4 Model architecture

Within the taxonomy of FL, categorization based on model architecture delineates two distinct paradigms homogeneous FL and heterogeneous FL, each addressing specific characteristics and challenges pertaining to the uniformity or diversity of model architectures across distributed devices.

Homogeneous FL encapsulates a paradigm where all participating edge devices employ identical or uniform model architectures during local model training and collaborative learning processes. In this context, uniformity in model architectures ensures that each device utilizes the same type of model structure, comprising identical neural network architectures, layers, and model parameters. The fundamental premise of homogeneous FL revolves around maintaining consistency and uniformity in the model representation across distributed devices. This uniformity streamlines the aggregation process, allowing straightforward consolidation of model updates from diverse devices into a unified global model. Homogeneous FL simplifies the coordination and synchronization of model updates among devices by assuming consistency in model structures, enabling efficient aggregation and convergence while minimizing complexities associated with reconciling disparate architectures or parameters during the collaborative learning process. However, this paradigm might face limitations in scenarios where diverse edge devices possess varying computational capabilities or data distributions, potentially affecting the efficacy of a uniform model representation across all participating devices. The effectiveness of homogeneous FL lies in its ability to facilitate seamless collaboration and aggregation by employing identical model architectures across distributed devices, simplifying the aggregation and convergence processes. Nevertheless, challenges arise when considering variations in computational capacities or data heterogeneity among devices, as the assumption of uniformity may not always align with the diverse characteristics inherent in decentralized edge environments. Addressing these challenges might necessitate adaptations in the training process or model architectures to account for variations in computational resources or data distributions, ensuring that the uniform model representation remains effective across heterogeneous devices. Homogeneous FL represents a foundational approach, emphasizing uniformity in model architectures across distributed devices while potentially requiring adaptations to accommodate the dynamic nature of edge computing environments for optimal collaborative learning outcomes.

Heterogeneous FL delineates a paradigm where participating edge devices employ diverse or varying model architectures, configurations, or parameters during local model training, fostering model diversity across the distributed network. In this context, heterogeneous FL embraces the versatility of different devices utilizing distinct neural network architectures, configurations, or variations in model parameters tailored to their specific tasks, data characteristics, or computational constraints. This paradigm acknowledges and accommodates the inherent diversity among edge devices, allowing for flexibility in model architectures that align with individual device capabilities or domain-specific requirements. The key premise of heterogeneous FL revolves around leveraging the strengths of model diversity, enabling devices to utilize customized or specialized architectures that suit their local datasets, computational resources, or task-specific constraints. However, this diversity in model architectures introduces complexities in the aggregation process, necessitating mechanisms to reconcile and amalgamate disparate model updates from devices with varying structures or parameters into a coherent global model representation [20].

The effectiveness of heterogeneous FL lies in its adaptability to accommodate the diverse characteristics of edge devices, allowing for customized model architectures and configurations tailored to individual requirements. This paradigm embraces the heterogeneity prevalent in decentralized networks, empowering devices to utilize different neural architectures or parameter variations while contributing collectively to the global learning process. Nonetheless, reconciling diverse model architectures or configurations during the aggregation process poses challenges, requiring sophisticated aggregation algorithms or adaptation strategies to merge updates effectively while maintaining the integrity and performance of the global model. Heterogeneous FL stands as an innovative approach within FL methodologies, capitalizing on model diversity across distributed devices while demanding nuanced techniques to harmonize and aggregate diverse model updates for optimal collaborative learning outcomes in heterogeneous edge computing environments.

13.3.5 Participant selection

In the taxonomy of FL, the categorization based on participant selection distinguishes between two prominent methodologies: random selection FL and strategic selection FL, each addressing distinct participant selection strategies within distributed learning frameworks. Random selection FL adopts an approach where devices are chosen arbitrarily or randomly from the pool of available edge devices to participate in the collaborative learning process. This methodology emphasizes equal opportunity for all devices to contribute to the global model, without considering specific criteria or characteristics during participant selection. Conversely, strategic selection FL operates on the premise of deliberately choosing participating devices based on predefined criteria or strategic considerations. This methodology involves the strategic selection of devices considering various factors such as device reliability, computational capabilities, data quality, or expertise in specific domains. Strategic selection FL aims to optimize the collective intelligence of participating devices by selecting devices strategically, potentially improving the convergence speed, model performance, or efficiency of the collaborative learning process by leveraging specific device attributes. Categorizing FL methodologies based on participant selection offers insights into the diverse participant selection strategies, illuminating methodologies that prioritize random selection for inclusivity and those emphasizing strategic selection for tailored contributions and optimized collaborative learning outcomes within distributed learning ecosystems [21].

Random selection represents a participant selection strategy within FL, where devices are chosen arbitrarily or without specific criteria from the available pool of edge devices to partake in the collaborative learning process. In this methodology, participant selection occurs without considering specific device characteristics, data quality, or expertise, emphasizing equal opportunity for all devices to contribute to the global model. The fundamental premise of random selection revolves around ensuring inclusivity and fairness in engaging devices, allowing each device in the network to have an equal chance of participating in the collaborative model training. However, the random selection strategy may lead to the involvement of devices with varying computational capabilities, data quality, or expertise levels, potentially impacting the convergence speed or overall performance of the global model. While random selection ensures a democratic approach to participation, it might encounter challenges in optimizing model convergence or accommodating devices with heterogeneous characteristics, necessitating mechanisms to manage disparities in device contributions and maximize collaborative learning efficiency despite varied device capabilities.

The effectiveness of random selection lies in its emphasis on inclusivity and equal opportunity, allowing all devices within the network to participate in the collaborative learning process without bias or specific criteria. This approach ensures broad engagement and involvement across diverse devices in FL setups. Nonetheless, challenges may arise in scenarios where diverse devices with varying data distributions or computational capabilities participate, potentially impacting the overall convergence or efficiency of the global model. Strategies such as adaptive aggregation mechanisms, federated optimization techniques, or personalized learning algorithms aim to mitigate disparities arising from random selection, optimizing the collaboration among heterogeneous devices to achieve more efficient and effective collaborative learning outcomes. Random selection stands as a fundamental approach within FL methodologies, providing a democratic engagement model while requiring adaptations to optimize learning efficiency across devices with varying characteristics and capabilities.

Strategic selection in FL represents a participant selection strategy where devices are intentionally chosen based on predefined criteria, considering factors such as data quality, device capabilities, expertise in specific domains, or other strategic considerations. This methodology aims to optimize the collaborative learning process by strategically selecting participants that align with specific criteria, thereby enhancing the efficiency, convergence speed, or overall performance of the global model. Strategic selection involves a deliberate selection process that prioritizes devices exhibiting desirable attributes, ensuring that participants contribute meaningfully to the collaborative model training. By employing specific criteria or strategic considerations, such as selecting devices with higher data quality or superior computational capabilities, strategic selection FL aims to harness the collective intelligence of strategically chosen devices to accelerate convergence, improve model accuracy, or address domain-specific requirements more effectively. However, while strategic selection optimizes participant engagement for tailored contributions, it may inadvertently exclude certain devices or introduce biases if the selection criteria are not comprehensive or inclusive, requiring careful consideration and adjustments to ensure fairness and inclusivity in participant engagement.

The effectiveness of strategic selection lies in its deliberate and targeted approach to participant engagement, optimizing the collaborative learning process by deliberately selecting devices based on predefined criteria or strategic considerations. This method allows for a more tailored engagement, potentially leading to enhanced model convergence, improved accuracy, or domain-specific optimization by leveraging devices with specific attributes or expertise. Nevertheless, challenges may arise concerning fairness, inclusivity, or bias in participant engagement if the selection criteria are not comprehensive or do not encompass the diversity of devices within the network. Strategies focusing on fairness-aware selection criteria, inclusive engagement policies, or adaptive participant selection mechanisms aim to mitigate biases and ensure equal opportunity for meaningful contributions across diverse devices while optimizing the efficiency and performance of strategic selection FL methodologies. Strategic selection is an advanced approach within FL, emphasizing intentional and tailored engagement to harness specific device attributes or capabilities for more effective collaborative learning outcomes [19–21].

13.3.6 Convergence techniques

Within the taxonomy of FL, classification based on synchronization delineates between synchronous FL and asynchronous FL methodologies, each offering distinct approaches to managing the coordination and synchronization of model updates across distributed devices (see Figure 13.6).

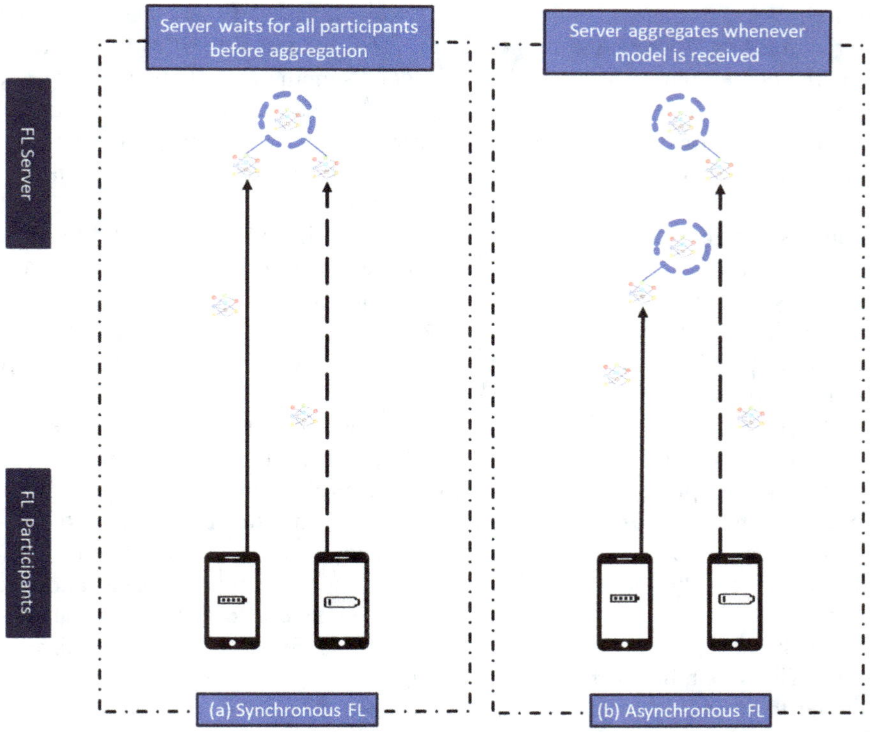

Figure 13.6 Visual comparison between synchronous FL and asynchronous FL

Synchronous FL denotes a methodology where all participating devices coordinate and synchronize their local model updates with a central server or among themselves at predefined synchronization points or intervals. In this approach, simultaneously initiate their local model training and transmit their updates for aggregation and incorporation into the global model at specific synchronized epochs. The fundamental principle of synchronous FL is ensuring uniformity and consistency in the collaborative learning process across distributed devices by enforcing synchronized updates. This synchronous coordination facilitates uniform model convergence and aggregation, aligning the learning trajectories among participating devices, and ensuring that updates are integrated into the global model coherently. However, the strict synchronization requirements may lead to potential bottlenecks or delays if devices with varying computational capabilities or data availability struggle to synchronize updates within the predefined intervals, potentially hindering the overall efficiency of the learning process.

The effectiveness of synchronous FL lies in its ability to maintain uniformity and consistency in the model updates across participating devices by enforcing synchronized update transmission at predefined intervals. This approach ensures coherence in the learning process and simplifies the aggregation of updates into a unified global model. Nonetheless, challenges arise in accommodating devices with diverse computational capacities or fluctuating network conditions, as strict synchronization requirements may lead to inefficiencies or delays in update transmission. Strategies such as adaptive synchronization intervals, dynamic participation criteria, or asynchronous fallback mechanisms aim to mitigate these challenges, ensuring effective synchronization while accommodating the heterogeneity prevalent in distributed edge environments. Synchronous FL stands as a foundational approach, emphasizing strict synchronization for uniformity in collaborative learning, albeit requiring adaptations to optimize synchronization efficiency across devices with varying capabilities and constraints.

Asynchronous FL represents a methodology where participating devices update the global model independently without adhering to strict synchronization requirements or predefined intervals. In this approach, devices engage in local model training and transmit their updates asynchronously, independent of synchronized epochs or coordination with other devices. The core principle of asynchronous FL revolves around offering flexibility in the learning process, allowing devices to contribute updates whenever they complete a local training round or whenever data becomes available without waiting for specific synchronization points. This flexibility permits devices to update the global model autonomously, catering to varying computational capacities, data availability, or network conditions among distributed devices. However, this decentralized and asynchronous update transmission might introduce disparities in the learning trajectories among devices, potentially leading to challenges in aggregating updates coherently into the global model or managing consistency in convergence speed across the distributed network.

The effectiveness of asynchronous FL lies in its flexibility and adaptability, allowing devices to update the global model independently based on their specific schedules or availability of data. This approach accommodates diverse computational capabilities and network conditions among devices, potentially enhancing the overall efficiency of collaborative learning by asynchronously leveraging data availability. Nevertheless, challenges arise in maintaining consistency and coherence in model convergence across devices due to the lack of strict synchronization, potentially leading to delays in aggregating updates or managing convergence trajectories. Strategies such as adaptive aggregation techniques, convergence regularization mechanisms, or asynchronous-aware optimization algorithms aim to mitigate these challenges, ensuring effective model aggregation while accommodating the autonomous and asynchronous update transmission characteristic of asynchronous FL methodologies. Asynchronous FL stands as a flexible and adaptive approach, allowing for decentralized and autonomous update transmission across distributed devices, albeit requiring strategies to harmonize updates and optimize convergence across heterogeneous edge environments.

13.3.7 Failure tolerance

Within the taxonomy of FL, categorization based on failure tolerance delineates between fault–tolerant FL and resilient FL methodologies, each addressing distinct approaches in managing and responding to failures or disruptions within the distributed learning framework. Fault–tolerant FL encompasses methodologies designed to withstand and mitigate failures or disruptions during the collaborative learning process. This approach focuses on preemptive measures and strategies to detect, isolate, and recover from faults, ensuring the continuous operation and integrity of the learning process even in the presence of failures or disruptions. Contrarily, resilient FL emphasizes adaptability and responsiveness to failures or disruptions by dynamically adjusting and reconfiguring the learning process to continue operation despite encountering faults. Resilient FL methodologies aim to quickly adapt and recover from failures, ensuring the continuity of the learning process by dynamically adjusting to changes or disruptions within the distributed network. Categorizing FL methodologies based on failure tolerance offers insights into methodologies emphasizing fault tolerance for continuous operation and those focusing on adaptability and resilience to overcome disruptions, illuminating strategies that prioritize fault anticipation and recovery and those emphasizing adaptability and responsiveness to ensure continuity within distributed learning ecosystems.

Resilient FL embodies a methodology designed to ensure the continuity of the collaborative learning process, even in scenarios where some participating devices encounter unreliable connections, failures, or disruptions. This approach prioritizes adaptability and responsiveness, aiming to dynamically adjust and reconfigure the learning process to maintain operation and mitigate the impact of faults or disruptions within the distributed network. Resilient FL methodologies incorporate adaptive mechanisms or algorithms capable of identifying and responding to unreliable connections or failures among devices. These methodologies focus on strategies such as dynamic re-scheduling of training rounds, prioritizing robustness in communication protocols, or employing fault-tolerant algorithms to detect and accommodate devices with intermittent or unstable connections. By dynamically adapting to the varying reliability of connections or device availability, resilient FL ensures the continuity of the learning process, mitigating the impact of disruptions and enabling the collaborative training to proceed effectively despite encountered faults.

The effectiveness of resilient FL lies in its ability to adapt and respond swiftly to failures, disruptions, or unreliable connections, ensuring the continuity and progress of the collaborative learning process within distributed environments. This approach allows for the accommodation of devices with varying levels of reliability or intermittent connectivity, minimizing the impact of faults on the overall learning process. However, challenges persist in effectively managing and adapting to diverse failure scenarios or unreliable connections, necessitating sophisticated adaptive mechanisms or fault-tolerant strategies capable of promptly responding and recovering from encountered disruptions. Strategies focusing on adaptive learning schedules, fault-tolerant optimization algorithms, or resilient communication protocols aim to mitigate the impact of unreliable connections or disruptions, ensuring the resilience and continuity of the collaborative learning process within resilient FL methodologies operating in distributed and dynamic edge computing environments.

13.3.8 Resource allocation

Within the taxonomy of FL, classification based on resource allocation discerns between distinct methodologies: bandwidth-efficient FL, computation-efficient FL, and energy-efficient FL, each tailored to optimize specific resources within distributed learning frameworks. Bandwidth-efficient FL methodologies focus on minimizing data transmission and communication overhead among distributed devices during the collaborative learning process. Computation-efficient FL methodologies aim to optimize computational resources across devices by employing algorithms or strategies that minimize computational complexities during local model training or aggregation processes. Energy-efficient FL methodologies prioritize minimizing energy consumption among distributed devices during collaborative learning processes.

Computation-efficient FL encompasses methodologies to minimize the computational burdens imposed on client devices during local model training and aggregation processes within distributed learning frameworks. This approach focuses on strategies and algorithms designed to alleviate the computational load on edge devices, thereby enhancing scalability and efficiency in FL setups. Computation-efficient FL methodologies often incorporate techniques such as federated optimization algorithms, which aim to reduce the computational complexities associated with local model updates and aggregation procedures. These algorithms optimize the aggregation process, diminishing the computational requirements for individual devices by employing more efficient aggregation mechanisms or by limiting the computation-intensive operations performed on client devices. Additionally, model quantization or compression techniques may be employed to reduce the computational demands during model training, enabling devices to process and transmit smaller-sized updates without compromising the overall accuracy or effectiveness of the collaborative learning process.

The effectiveness of computation-efficient FL lies in its ability to streamline and minimize computational requirements on client devices, thereby enhancing the scalability and feasibility of FL in resource-constrained environments. By employing algorithms and strategies tailored to reduce computational complexities during model updates and aggregation, computation-efficient FL methodologies alleviate the burden on edge devices, making FL more accessible and adaptable to diverse computational capabilities across the network. However, challenges persist in optimizing computational efficiency without sacrificing the quality or accuracy of the learned models. Strategies focusing on federated optimization algorithms, adaptive computation schedules, or model compression methods aim to balance computational efficiency and model accuracy, ensuring the effective operation of computation-efficient FL methodologies in distributed learning ecosystems while addressing resource constraints and computational limitations on client devices.

Energy-efficient FL constitutes a methodology specifically designed to minimize energy consumption across distributed devices during the collaborative learning process. This approach emphasizes strategies to reduce the energy footprint associated with model updates, transmission, and computations, thus optimizing the overall energy efficiency of FL frameworks. Energy-efficient FL methodologies encompass a range of techniques, including energy-aware model scheduling, low-power optimization algorithms, or adaptive learning strategies tailored to minimize energy consumption without compromising the quality or effectiveness of the collaborative learning process. These methodologies prioritize the development of algorithms and protocols that enable devices to participate in model training and updates while conserving energy resources. Techniques such as smart scheduling of model updates to align with periods of lower energy consumption, utilization of low-power communication protocols, or adaptive learning rate strategies that minimize unnecessary computations contribute to the overarching goal of reducing energy consumption in FL setups.

The effectiveness of energy-efficient FL lies in its ability to minimize the energy footprint across distributed devices, making FL more sustainable and adaptable to energy-constrained environments. By employing energy-efficient algorithms and protocols, this methodology enables devices to participate in collaborative learning processes while conserving energy resources. Nevertheless, challenges persist in balancing energy efficiency with maintaining learning effectiveness and model accuracy. Strategies focusing on adaptive energy-aware scheduling, dynamic resource allocation, or low-power optimization algorithms aim to optimize energy consumption while ensuring that the collaborative learning process remains efficient and effective within energy-efficient FL methodologies, thus addressing the imperative of sustainable and energy-conscious FL frameworks.

13.4 Intersection of FL and XAI

This section explores the intriguing convergence of FL and XAI methodologies, addressing the significance, implications, and intricacies of combining collaborative learning across decentralized devices with the imperative need for transparency and interpretability in ML models. It scrutinizes how FL's collaborative learning distributed across edge devices intersects with XAI's quest for transparency, interpretability, and accountability in the decision-making of ML models, shedding light on the challenges and opportunities presented by their amalgamation within contemporary AI ecosystems.

13.4.1 Challenges in achieving explainability in FL

Making FL models explainable poses notable challenges owing to the distributed nature of the learning process across diverse and often uncoordinated edge devices.

One of the inherent complexities lies in the decentralized nature of FL, where model training occurs locally on individual devices using diverse datasets, limiting direct access to comprehensive data housed across the network. Consequently, the lack of centralized data aggregation impedes traditional explainability approaches reliant on full access to a unified dataset, rendering it challenging to generate holistic explanations for model decisions. Moreover, the privacy-preserving nature of FL, ensuring data confidentiality by not transmitting raw data to a central server, presents an additional layer of complexity. This privacy-enhancing aspect creates a barrier to comprehensively scrutinizing the decision-making processes of FL models, hindering the extraction of interpretable insights without compromising individual data privacy. Therefore, the decentralized and privacy-preserving characteristics inherent in FL pose significant hurdles in creating comprehensive and interpretable explanations for model predictions and decisions.

Furthermore, the heterogeneity prevalent among edge devices participating in FL exacerbates the challenge of making FL models explainable. The diversity in data distributions, computational capabilities, and local models across devices introduces variations in the learning process, hindering the generation of uniform and consistent explanations for model behavior. Discrepancies in local datasets or model architectures among devices contribute to variations in model updates and decision-making, complicating the extraction of generalized and coherent explanations across the distributed learning network. Additionally, the lack of transparency in the aggregation process, where model updates merge into a global model, further complicates the interpretability of FL models. The opaque nature of the aggregation mechanism obscures the understanding of how individual updates influence the global model's decisions, impeding the creation of transparent explanations for model predictions. Overall, the heterogeneity among devices and the opacity in the aggregation process within FL systems present formidable challenges in achieving explainability, necessitating innovative approaches to elucidate model behavior across diverse and decentralized learning environments.

13.4.2 Techniques for explainable FL

Model-agnostic approaches offer promising avenues to enhance the interpretability of FL models within decentralized and privacy-constrained environments. These approaches prioritize techniques that aim to extract interpretability without relying on specific model architectures or access to raw data, accommodating the inherent challenges posed by FL's distributed nature. One such method involves leveraging post-hoc interpretability techniques that analyze model behavior after training completion. These techniques encompass various approaches like feature importance analysis, surrogate models, or local explanation methods. Feature importance analysis identifies the relative significance of input features in influencing model predictions across distributed devices, providing insights into the contribution of different features without necessitating access to raw data. Surrogate models create simplified yet interpretable versions of the complex FL model, aiding in understanding its decision-making processes. Local explanation methods, such as

LIME (local interpretable model-agnostic explanations), generate interpretable explanations for individual predictions by approximating the behavior of the FL model around specific data points. These model-agnostic post-hoc approaches do not rely on model-specific characteristics, making them adaptable to diverse FL setups and preserving privacy by avoiding direct access to sensitive individual data while enhancing interpretability.

Additionally, SHAP (SHapley Additive exPlanations) and integrated gradients represent other model-agnostic techniques conducive to enhancing FL model interpretability. SHAP values quantify the impact of each feature on model predictions, providing a global perspective on feature importance while ensuring model-agnostic interpretability. These values assign credit to each feature based on its contribution to the prediction, enabling the assessment of each feature's influence on the model's decision-making process across diverse edge devices. Integrated gradients, another model-agnostic approach, quantifies feature importance by assessing how changes in feature values affect model predictions across the entire input space. By considering feature values from the baseline to the input data point, integrated gradients provide a nuanced understanding of feature contributions without necessitating access to individual device data. These model-agnostic techniques offer promising avenues to unravel FL model behavior without compromising privacy, ensuring interpretability and transparency across decentralized learning networks.

Extracting explanations from distributed models within FL settings necessitates innovative methodologies to overcome the challenges posed by decentralized learning across diverse edge devices while ensuring data privacy. Collaborative learning across distributed devices in FL mandates techniques that can generate explanations without aggregating sensitive individual data centrally. One methodology involves federated model explanations (FME), aiming to derive global and local explanations while respecting the privacy constraints of FL. FME leverages techniques that generate model explanations collaboratively across devices without exchanging raw data. These methodologies may include federated feature attribution methods that assess the impact of features on model predictions without accessing individual data points directly. Collaborative meta-learning approaches within FME allow devices to exchange knowledge in the form of model parameters or summaries, facilitating the generation of global model explanations while preserving individual data privacy. Additionally, federated model-agnostic explanations employ post-hoc interpretability techniques adaptable to various FL setups. By generating explanations independently on each device and aggregating interpretable insights without sharing raw data, these approaches enhance transparency and interpretability across the distributed FL model.

Moreover, decentralized explanation methods that embrace differential privacy principles emerge as an innovative methodology for extracting explanations from distributed FL models while preserving individual data confidentiality. Differential privacy-centric techniques ensure that model explanations do not compromise the privacy of individual data points. Federated versions of differential privacy, such as federated differential privacy or local differential privacy, integrate privacy preservation mechanisms into the explanation extraction process across distributed devices. These methods inject controlled noise or perturbations into local explanations before aggregation, ensuring that no single device's data can be inferred from the generated explanations. By adhering to privacy-preserving principles, decentralized explanation methodologies in FL bolster transparency and interpretability without compromising the confidentiality of individual data, addressing the challenges inherent in distributed learning environments while offering comprehensive and privacy-respecting explanations for FL models.

Transparency techniques compatible with FL algorithms play a pivotal role in addressing the challenges of explaining model decisions while preserving data privacy across distributed edge devices. One approach involves federated transparency techniques designed to elucidate model behavior and decision-making processes in a privacy-preserving manner. Federated SHAP techniques adapt the SHAP framework to FL setups, enabling the computation of feature importance scores across distributed devices without exchanging raw data. By leveraging the FedAvg mechanism in FL, federated SHAP aggregates local SHAP values obtained from individual devices, providing a global perspective on feature importance while maintaining individual data privacy. Additionally, federated surrogate models or explanations facilitate transparency by generating simplified and interpretable versions of FL models across decentralized devices. These surrogate models approximate the behavior of complex FL models, enabling insights into decision-making processes while avoiding the direct exchange of sensitive data. Such federated transparency techniques ensure interpretability in FL algorithms without compromising the confidentiality of individual device data.

Furthermore, decentralized and privacy-preserving techniques like federated and localized privacy-preserving explanations accommodate the constraints of FL algorithms while enhancing transparency. Federated differential privacy integrates privacy preservation mechanisms into the explanation extraction process across distributed devices. This technique ensures that individual data privacy remains intact while providing comprehensive and privacy-respecting explanations for FL models. Localized differential privacy-centric methods inject controlled noise or perturbations into local explanations before aggregation, ensuring that individual device data cannot be inferred from the generated explanations. These approaches strike a balance between transparency and data privacy, allowing for the extraction of interpretable insights from FL algorithms while safeguarding sensitive information. By employing these transparency techniques compatible with FL algorithms, researchers aim to elucidate model decisions and enhance interpretability across distributed learning networks, addressing the challenges of transparency in privacy-constrained environments.

13.5 Accountable FL

This section explores the multifaceted dimensions of accountable FL, addressing the mechanisms, standards, and ethical considerations essential for establishing accountability frameworks in FL setups. It scrutinizes the challenges, implications, and strategies pertinent to fostering accountability within FL, examining how accountability principles intersect with privacy, fairness, transparency, and regulatory compliance. Also, this section navigates through the emergent approaches and best practices designed to uphold accountability, offering insights into the evolving landscape of responsible AI in FL and its pivotal significance in ensuring ethical and trustworthy AI deployments across distributed environments.

13.5.1 Accountability in FL

Accountability and responsibility in FL models underscore essential principles that contribute to ethical, transparent, and trustworthy AI deployments across distributed edge devices. These principles serve as linchpins for ensuring that the decisions and outcomes derived from FL models align with ethical norms, regulatory standards, and societal expectations. Accountability in FL models involves elucidating decision-making processes to ensure stakeholders comprehend how and why the models make particular predictions or decisions. Responsibility within FL mandates that all entities involved in the collaborative learning process—including model developers, device owners, and organizations overseeing the FL framework—uphold ethical standards and bear the consequences of their actions, ensuring fairness, non-discrimination, and adherence to privacy-preserving measures. Both accountability and responsibility are pivotal for establishing trust among stakeholders and end-users, fostering confidence in FL models' reliability, fairness, and compliance with ethical guidelines, thus paving the way for ethically sound and socially beneficial AI applications in diverse settings.

Moreover, in the context of FL models, accountability and responsibility are crucial in addressing societal, ethical, and legal implications arising from model decisions made across decentralized devices. Ensuring accountability in FL models mitigates the risks associated with biased or erroneous predictions, providing avenues for recourse and redressal in case of adverse impacts caused by the models. Responsible practices within FL environments mandate a comprehensive understanding of the ethical ramifications of model deployments, necessitating ongoing monitoring, auditing, and adherence to ethical frameworks and regulatory standards. By embracing accountability and responsibility, FL stakeholders not only uphold ethical and legal norms but also cultivate a culture of transparency, fairness, and integrity, fostering public trust and confidence in deploying AI technologies within sensitive and privacy-conscious domains. Hence, prioritizing accountability and responsibility within FL models becomes integral to promoting ethical AI development, encouraging responsible innovation, and aligning AI applications with societal values and expectations.

Ethical considerations and legal implications stand as fundamental pillars within the subject of accountable FL, emphasizing the need to navigate complex ethical landscapes and regulatory frameworks governing AI applications in decentralized learning ecosystems. Ethical considerations in accountable FL encompass a spectrum of concerns, including fairness, transparency, privacy, and bias mitigation. Ensuring fairness entails safeguarding against discriminatory outcomes and biases that may emerge due to uneven distributions of data or model updates across diverse edge devices. Transparency within FL models becomes crucial, necessitating explanations for decision-making processes to ensure interpretability and trustworthiness. Moreover, upholding individual privacy remains paramount, mandating stringent measures to preserve data confidentiality and protect sensitive information while engaging in collaborative model training across decentralized devices. Addressing biases and striving for equitable outcomes underpin ethical imperatives, ensuring that FL models uphold fairness, justice, and societal well-being. Simultaneously, legal implications in accountable FL models intertwine with ethical concerns, navigating through a labyrinth of regulatory frameworks and compliance requirements. Adherence to legal standards encompasses considerations regarding data protection laws, privacy regulations, and sector-specific compliance requirements. FL frameworks operating across international borders must navigate diverse legal landscapes, encompassing various data protection regimes and cross-border data transfer regulations. Compliance with these legal frameworks demands robust data governance, ensuring that FL models operate within the bounds of regulatory guidelines, preserving data rights, and maintaining compliance with jurisdictional mandates. Furthermore, legal implications extend to liability concerns, requiring the delineating responsibilities among stakeholders involved in FL, establishing accountability for model decisions and outcomes, and mitigating potential legal risks associated with erroneous or biased predictions. By addressing these ethical considerations and legal implications, accountable FL endeavors to forge a path that not only aligns with ethical norms but also navigates the legal terrain, fostering responsible and legally compliant AI deployments within distributed learning environments.

13.5.2 Regulatory compliance

Regulatory frameworks and standards applicable to FL at the edge constitute a multifaceted landscape encompassing diverse legal and regulatory dimensions aimed at governing AI applications, data privacy, and security in decentralized learning environments. These frameworks include regional and global regulations, such as the General Data Protection Regulation (GDPR) in the European Union, which delineates stringent guidelines for data processing, storage, and transfer, imposing obligations on organizations handling personal data, including those engaged in FL activities. Additionally, sector-specific regulations like the Health Insurance Portability and Accountability Act (HIPAA) in the healthcare domain or the Gramm-Leach-Bliley Act (GLBA) in the financial sector mandate stringent data protection measures and privacy safeguards, influencing the conduct of FL activities in these sectors. Moreover, international standards such as ISO/IEC 27001 for information security management and ISO/IEC 27701 for privacy information management serve as benchmarks for ensuring robust data security and privacy practices within FL frameworks operating across distributed edge devices. Compliance with these regulatory frameworks and standards becomes imperative for FL initiatives, necessitating robust data governance, stringent security measures, and adherence to privacy-preserving techniques to ensure lawful and ethically sound AI deployments at the edge.

Furthermore, emerging regulations and evolving standards are shaping the regulatory landscape for FL at the edge, reflecting a dynamic environment that demands adaptability and compliance from FL stakeholders. The emergence of specific guidelines or industry consortia dedicated to AI ethics, data privacy, and responsible AI deployment further contributes to shaping the regulatory ecosystem for FL. For instance, initiatives like the IEEE Global Initiative on Ethics of Autonomous and Intelligent Systems and the Partnership on AI (PAI) advocate for ethical guidelines and standards for AI applications, including FL, emphasizing transparency, accountability, and fairness. Additionally, governments and regulatory bodies worldwide are actively considering AI-related legislation and guidelines, contemplating principles governing data usage, algorithmic accountability, and AI governance, all of which will significantly impact FL practices at the edge. Navigating this evolving regulatory landscape requires FL practitioners to remain cognizant of emerging regulations, adapt their practices to comply with evolving standards, and implement robust governance structures to ensure legal and ethical compliance in FL deployments across diverse edge environments.

Compliance requirements for data privacy and security in decentralized learning environments, such as those encountered in FL, are underpinned by a multitude of regulations and standards aimed at safeguarding sensitive information while facilitating collaborative learning across distributed edge devices. These compliance measures primarily hinge upon stringent data privacy regulations, including the GDPR in the European Union, which mandates strict requirements for the processing, storage, and transfer of personal data. Adherence to GDPR necessitates robust mechanisms within FL frameworks to protect personal information, including data minimization, purpose limitation, and ensuring lawful grounds for data processing across the distributed network. Furthermore, compliance with sector-specific regulations such as the HIPAA in healthcare or the Payment Card Industry Data Security Standard (PCI DSS) in financial sectors demands heightened data security measures and privacy safeguards within FL deployments operating in these domains. Compliance entails implementing encryption, access controls, and stringent data handling procedures to ensure data confidentiality and integrity while engaged in collaborative learning across decentralized devices.

Additionally, compliance requirements encompass international standards such as ISO/IEC 27001 for information security management and ISO/IEC 27701 for privacy information management, serving as benchmarks for ensuring robust data security and privacy practices within decentralized learning environments. These standards mandate the implementation of comprehensive security measures, risk assessments, and privacy-enhancing techniques throughout the FL workflow. Moreover, compliance frameworks emphasize the need for transparent data governance practices, consent management mechanisms, and the establishment of clear accountability structures among FL stakeholders. Adhering to these compliance requirements mandates a holistic approach, encompassing technical, organizational, and procedural measures within FL setups to ensure data privacy, security, and ethical use of information while fostering collaborative learning across diverse edge devices.

13.6 Trustworthy FL

This section delves into the pivotal domain of trustworthy FL, focusing on the multifaceted dimensions that underpin trust, reliability, and integrity within decentralized learning ecosystems. Trust is a cornerstone in the deployment of FL, ensuring that collaborative model training across distributed edge devices operates with reliability, transparency, and ethical principles.

13.6.1 Trust challenges in FL

Trust-related issues, encompassing data integrity, model security, and participant reliability, constitute crucial dimensions within FL ecosystems, shaping the reliability and dependability of collaborative learning across decentralized edge devices. Data integrity concerns revolve around ensuring the accuracy, consistency, and reliability of data used for collaborative model training across diverse devices. In FL setups, maintaining data integrity across distributed edge devices becomes paramount to mitigate the risks of unreliable or manipulated data, which could adversely impact the accuracy and fairness of the trained models. Techniques ensuring data integrity include cryptographic measures like secure aggregation protocols or verifiable computing methods that enable the verification of computations performed on decentralized data without exposing raw information, thereby fostering trust in the integrity of the contributed data. Simultaneously, model security remains a pivotal concern, requiring robust measures to safeguard FL models against unauthorized access, adversarial attacks, or data poisoning attempts. Techniques like FL with differential privacy or secure multiparty computation contribute to enhancing model security by preventing information leakage or manipulation during collaborative model training, thereby fortifying trust in the security of the learned models.

Moreover, ensuring participant reliability plays a crucial role in fostering trust within FL environments, necessitating mechanisms to assess and validate the credibility and reliability of participating devices. Reliable participants within FL frameworks contribute trustworthy and accurate updates, thereby enhancing the overall quality of the federated model. Techniques such as reputation-based mechanisms or FL with anomaly detection enable the assessment of participant behavior, identifying and mitigating unreliable or malicious contributions across the distributed network. By fostering trust in the reliability of participating devices, FL ecosystems strive to promote dependable collaborative learning, ensuring that contributions from diverse edge devices contribute positively to the overall model's integrity and accuracy, thereby enhancing the trustworthiness and reliability of FL models.

Establishing trust among distributed parties within FL environments presents multifaceted challenges stemming from the decentralized nature of collaborative learning across diverse edge devices. One significant challenge pertains to the disparate data distributions and characteristics across participating devices, leading to variations in data quality, quantity, and representativeness. The lack of uniformity in local datasets poses challenges in aligning and harmonizing data contributions, impacting the overall accuracy and fairness of the federated model. Additionally, the diversity in computational capabilities and reliability among edge devices introduces complexities in ensuring consistent and reliable contributions. Discrepancies in device capabilities, network connectivity, or computational resources among participants hinder the equitable participation of all devices, potentially affecting the reliability and quality of contributed updates. These disparities in data quality and device capabilities pose significant hurdles in establishing trust among distributed parties, necessitating mechanisms to address variations in data quality and reliability across the FL network.

Moreover, privacy and security concerns pose substantial challenges in fostering trust among distributed parties in FL setups. The imperative to preserve data privacy while engaging in collaborative model training restricts direct access to raw data, hindering the ability to validate the integrity and quality of individual contributions. Privacy-enhancing techniques, such as FL with differential privacy or homomorphic encryption, introduce complexities in assessing the reliability and trustworthiness of contributed updates without compromising individual data privacy. Furthermore, ensuring security against adversarial attacks or data breaches within decentralized learning environments adds layers of complexity in establishing trust. The need to protect FL models and data against potential vulnerabilities or malicious attacks demands robust security measures, necessitating continual monitoring, auditing, and safeguards without impeding collaborative learning. Addressing these privacy and security concerns while ensuring equitable participation and reliable contributions among distributed parties poses intricate challenges in fostering trust within FL ecosystems, underscoring the need for innovative approaches and robust mechanisms to overcome these hurdles and bolster trust among decentralized participants in FL.

13.6.2 Security and privacy measures

Security protocols and privacy-preserving techniques are pivotal in safeguarding sensitive information and ensuring secure collaborative learning within FL frameworks operating across distributed edge devices. Secure aggregation protocols represent one key security measure employed in FL, facilitating the aggregation of model updates from multiple devices while preserving the confidentiality of individual contributions. These protocols employ cryptographic techniques, such as secure multi-party computation (MPC) or homomorphic encryption, allowing the aggregation server to compute the global model parameters without directly accessing raw data. Homomorphic encryption enables computations on encrypted data without decrypting it, ensuring that individual contributions remain private during the aggregation process. Similarly, FL with differential privacy injects controlled noise or perturbations into local updates, preserving individual data privacy while ensuring that aggregated model parameters do not reveal sensitive information. These security protocols ensure data confidentiality, integrity, and authenticity across the FL network, mitigating the risks of information leakage or unauthorized access to individual contributions.

In addition to security protocols, privacy-preserving techniques form a cornerstone in FL to protect sensitive information while engaging in collaborative model training. Differential privacy, a fundamental privacy-preserving technique, ensures that statistical queries performed on a dataset do not compromise individual data privacy. FL with differential privacy introduces noise or perturbations to locally computed updates, enhancing privacy protection without sacrificing model accuracy. Secure FL frameworks employ techniques like FL with secure aggregation or federated encrypted computation, enabling model training while preserving the privacy of local data. Moreover, using FL with trusted execution environments (TEE) or FL with encrypted gradients ensures that sensitive information remains encrypted throughout the learning process, preventing unauthorized access to individual data while contributing to collaborative model training. These privacy-preserving techniques underpin the foundation of FL, allowing collaborative learning across decentralized devices while ensuring stringent privacy protections and fostering trust among participants in FL ecosystems.

Encryption methods, differential privacy, and secure aggregation are fundamental pillars in ensuring privacy, security, and confidentiality within FL frameworks operating across distributed edge devices. Encryption techniques, such as homomorphic encryption and FL with encrypted computation, enable computations to be performed on encrypted data without compromising individual privacy. Homomorphic encryption allows computations to be carried out on encrypted data, ensuring that sensitive information remains encrypted throughout the FL process. Similarly, FL with encrypted computation enables model training while preserving the privacy of local data by encrypting model parameters or gradients before transmission, preventing unauthorized access to individual contributions.

Differential privacy, a cornerstone of privacy-preserving techniques in FL, introduces controlled noise or perturbations into local updates to protect individual data privacy. This technique ensures that statistical queries or computations performed on aggregated data do not reveal sensitive information about any individual's contribution. By adding carefully calibrated noise to individual updates, differential privacy provides a mathematical guarantee of privacy protection, allowing collaborative learning across decentralized devices while safeguarding against the risk of information leakage.

Secure aggregation protocols further fortify privacy and security in federated settings by enabling the aggregation of model updates while preserving the confidentiality of individual contributions. Cryptographic techniques like secure multiparty computation (MPC) or FL with trusted execution environments (TEE) facilitate secure aggregation without requiring direct access to raw data. These protocols allow the aggregation server to compute global model parameters without exposing individual contributions, ensuring data confidentiality and integrity throughout the collaborative learning process. Implementing encryption methods, differential privacy, and secure aggregation mechanisms in FL frameworks forms the bedrock of privacy-preserving practices, enabling secure and confidential collaborative model training across diverse edge devices while maintaining stringent privacy protections.

Building trust in FL models tailored for edge IoT systems encompasses a series of strategies addressing the unique challenges of decentralized, resource-constrained environments while ensuring reliability, security, and transparency. One pivotal strategy involves robust privacy-preserving mechanisms tailored for edge IoT settings. Implementing differential privacy techniques optimized for edge devices mitigates privacy risks by injecting calibrated noise into local updates, ensuring individual data privacy while facilitating collaborative model training. Moreover, FL with federated encryption or FL with trusted execution environments (TEE) encrypts sensitive information throughout the FL process, safeguarding against unauthorized access to local data, thereby bolstering privacy and engendering trust among edge IoT participants. These privacy-centric strategies are pivotal in edge IoT environments, where data sensitivity and limited computational resources necessitate tailored approaches to ensure secure and privacy-preserving FL operations.

Furthermore, enhancing model security and reliability at the edge is critical in cultivating trust in FL models within IoT ecosystems. Employing secure aggregation protocols optimized for resource-constrained edge devices fortifies model security by enabling the aggregation of model updates while preserving the confidentiality of individual contributions. Techniques like FL with encrypted computation or homomorphic encryption allow model updates to be processed securely without revealing raw data, enhancing model security and engendering trust among edge IoT devices. Additionally, leveraging anomaly detection mechanisms or reputation-based systems tailored for edge devices enables the assessment and validation of participant reliability, ensuring that contributions from diverse edge IoT devices are trustworthy and reliable. These strategies bolster the trustworthiness of FL models in edge IoT systems, fostering reliability, security, and transparency within resource-constrained and decentralized environments.

Explainability and accountability are crucial in establishing trust within edge computing environments, playing pivotal roles in elucidating model decisions and ensuring responsible AI deployments across resource-constrained and decentralized settings. Explainability mechanisms tailored for the edge significantly enhance transparency and comprehension of model behavior, enabling stakeholders to understand the reasoning behind model predictions despite the limited computational resources. By employing lightweight and interpretable models or post-hoc explanation techniques optimized for edge devices, stakeholders gain insights into the decision-making processes of AI models operating at the edge. This transparency fosters trust by providing users, operators, and stakeholders with comprehensible explanations for model decisions, thereby ensuring confidence in the reliability and fairness of AI applications within edge computing environments.

Moreover, accountability mechanisms embedded within edge AI systems contribute to instilling trust by establishing responsible and transparent practices. Accountability frameworks designed for edge devices outline clear delineations of responsibilities among stakeholders, ensuring adherence to ethical guidelines, regulatory compliance, and the ethical use of AI models despite the challenges posed by limited computational capabilities. These frameworks enforce transparency in data handling, model training, and decision-making processes, enabling stakeholders to trace model behaviors and outcomes, thereby fostering trust in the reliability and ethical governance of AI systems at the edge. The synergy between explainability, which elucidates model behavior, and accountability, which ensures responsible AI practices, significantly contributes to instilling trust within edge computing ecosystems, thereby facilitating the adoption of AI technologies in resource-constrained and decentralized environments.

References

[1] Amin, S. U. and Hossain, M. S. (2020). Edge intelligence and Internet of Things in healthcare: a survey. *IEEE Access*, 9, 45–59.

[2] Jagatheesaperumal, S. K., Pham, Q. V., Ruby, R., Yang, Z., Xu, C., and Zhang, Z. (2022). Explainable AI over the IoT: overview, state-of-the-art and future directions. *IEEE Open Journal of the Communications Society*, 3, 2106–2136, doi: 10.1109/OJCOMS.2022.3215676.

[3] Nguyen, D. C., Ding, M., Pathirana, P. N., *et al.* (2021). 6G Internet of Things: a comprehensive survey. *IEEE Internet of Things Journal*, 9(1), 359–383.

[4] Alshehri, F. and Muhammad, G. (2020). A comprehensive survey of the IoT and AI-based smart healthcare. *IEEE Access*, 9, 3660–3678.

[5] Sfar, A. R., Natalizio, E., Challal, Y., and Chtourou, Z. (2018). A roadmap for security challenges in the Internet of Things. *Digital Communications and Networks*, 4(2), 118–137.

[6] Barbuto, V., Savaglio, C., Chen, M., and Fortino, G. (2023). Disclosing edge intelligence: a systematic meta-survey. *Big Data and Cognitive Computing*, 7(1), 44.

[7] Moustafa, N., Koroniotis, N., Keshk, M., Zomaya, A. Y., and Tari, Z. (2023). Explainable intrusion detection for cyber defences in the Internet of Things: opportunities and solutions. *IEEE Communications Surveys & Tutorials*, 25(3), 1775–1807, thirdquarter, doi: 10.1109/COMST.2023.3280465.

[8] Issa, W., Moustafa, N., Turnbull, B., Sohrabi, N., and Tari, Z. (2023). Blockchain-based FL for securing Internet of Things: a comprehensive survey. *ACM Computing Surveys*, 55(9), 1–43.

[9] Yu, W., Liang, F., He, X., *et al.* (2017). A survey on the edge computing for the Internet of Things. *IEEE Access*, 6, 6900–6919.

[10] Rahman, M. A. and Hossain, M. S. (2022). Design of a secure internet of artificially intelligent things in the 6G era. *IEEE Internet of Things Magazine*, 5(2), 49–54.

[11] Wu, H., Han, H., Wang, X., and Sun, S. (2020). Research on artificial intelligence enhancing Internet of Things security: a survey. *IEEE Access*, 8, 153826–153848.

[12] Arisdakessian, S., Wahab, O. A., Mourad, A., Otrok, H., and Guizani, M. (2022). A survey on IoT intrusion detection: FL, game theory, social psychology, and explainable AI as future directions. *IEEE Internet of Things Journal*, 10(5), 4059–4092.

[13] Alwarafy, A., Al-Thelaya, K. A., Abdallah, M., Schneider, J., and Hamdi, M. (2020). A survey on security and privacy issues in edge-computing-assisted Internet of Things. *IEEE Internet of Things Journal*, 8(6), 4004–4022.

[14] Nguyen, D. C., Ding, M., Pathirana, P. N., Seneviratne, A., Li, J., and Poor, H. V. (2021). FL for Internet of Things: a comprehensive survey. *IEEE Communications Surveys & Tutorials*, 23(3), 1622–1658.

[15] Molokomme, D. N., Onumanyi, A. J., and Abu-Mahfouz, A. M. (2022). Edge intelligence in smart grids: a survey on architectures, offloading models, cyber security measures, and challenges. *Journal of Sensor and Actuator Networks*, 11(3), 47.

[16] Zhang, J. and Tao, D. (2020). Empowering things with intelligence: a survey of the progress, challenges, and opportunities in artificial intelligence of things. *IEEE Internet of Things Journal*, 8(10), 7789–7817.

[17] Chang, Z., Liu, S., Xiong, X., Cai, Z., and Tu, G. (2021). A survey of recent advances in edge-computing-powered artificial intelligence of things. *IEEE Internet of Things Journal*, 8(18), 13849–13875.

[18] Dai, H. N., Zheng, Z., and Zhang, Y. (2019). Blockchain for Internet of Things: a survey. *IEEE Internet of Things Journal*, 6(5), 8076–8094.

[19] https://smijournal.org/index.php/smij/article/view/2

[20] Al-Garadi, M. A., Mohamed, A., Al-Ali, A. K., Du, X., Ali, I., and Guizani, M. (2020). A survey of machine and deep learning methods for IoT security. *IEEE Communications Surveys & Tutorials*, 22(3), 1646–1685.

[21] Alotaibi, B. (2023). A survey on industrial Internet of Things security: requirements, attacks, AI-based solutions, and edge computing opportunities. *Sensors*, 23(17), 7470.

Chapter 14
Explainable security intelligence for zero-trust IoT

14.1 Introduction

Security and privacy concerns in the IoT field have become paramount due to the proliferation of interconnected devices spanning various sectors like healthcare, smart homes, industrial automation, and more. The interconnected nature of IoT devices, while enhancing convenience and efficiency, introduces inherent vulnerabilities that malicious actors can exploit. IoT security issues encompass a wide array of threats, including unauthorized access, data breaches, device manipulation, and network infiltrations. Often, these devices have limited computational power and inadequate built-in security features, making them susceptible to attacks. Moreover, the vast amount of data generated by IoT devices, often personal and sensitive, raises significant privacy concerns. The collection, storage, and utilization of this data without robust privacy measures can lead to unauthorized data access, profiling, and misuse, jeopardizing individuals' privacy and even safety [1].

Addressing security and privacy challenges in IoT requires a multi-faceted approach. Robust encryption protocols, authentication mechanisms, and regular software updates are crucial to safeguard IoT ecosystems. Implementing a zero-trust security model, which assumes no implicit trust and constantly verifies devices and users, is increasingly essential. Additionally, ensuring privacy in IoT involves clear data governance frameworks, transparent data collection practices, and user consent mechanisms. Emerging technologies like blockchain offer potential solutions by providing decentralized and tamper-proof systems for data management. Regulatory frameworks that mandate security standards and privacy regulations further play a vital role in fostering a secure and privacy-respecting IoT landscape. The collaborative efforts of industry stakeholders, policymakers, and cybersecurity experts are imperative to navigate the evolving challenges and build a trustworthy and resilient IoT infrastructure.

Zero-trust IoT security represents a paradigm shift in cybersecurity strategies, advocating a fundamental change in how we perceive network security within the IoT ecosystem. Traditional security models typically relied on perimeter-based defenses, assuming that once inside the network, devices and users were inherently trusted. However, the dynamic and diverse nature of IoT devices has rendered this approach inadequate, leading to the evolution of zero-trust principles. At its core, zero-trust IoT security operates on the premise that no device, user, or entity should be automatically trusted, regardless of its location within the network. Instead, it emphasizes continuous verification and authentication, ensuring that every access request, communication, or interaction is rigorously authenticated and authorized. This approach significantly reduces the attack surface and mitigates potential threats, safeguarding IoT environments against unauthorized access, data breaches, and malicious activities [2].

Explainable artificial intelligence (XAI) plays a pivotal role in advancing the tenets of zero-trust within IoT security frameworks by enhancing transparency and interpretability in decision-making processes. In the context of IoT, where complex machine-learning algorithms and AI-driven systems make critical security decisions, the ability to understand and explain the rationale behind these decisions becomes crucial. XAI techniques enable the interpretation of AI models' outputs, providing understandable explanations for why specific decisions are made. In zero-trust IoT security, XAI contributes by shedding light on the reasoning behind access approvals, anomaly detections, or device trust evaluations. By offering transparent insights into AI-driven security mechanisms, XAI enhances user trust and acceptance and assists security professionals in identifying vulnerabilities or biases within the system. Ultimately, the integration of XAI ensures that security measures in zero-trust IoT environments are not only robust but also comprehensible and accountable, thereby reinforcing the overall security posture of IoT ecosystems.

This chapter explores the convergence between zero-trust IoT security and XAI, delving into foundational concepts, integration strategies, practical applications, and future implications. Beginning with an in-depth analysis of zero-trust principles tailored for IoT environments, the chapter navigates through the role of XAI within each layer of security, emphasizing device authentication, network segmentation, and anomaly detection. It further elucidates the challenges and opportunities in integrating XAI into zero-trust frameworks, addressing the intricacies of achieving transparency in decision-making processes. Through insightful case studies spanning healthcare IoT ecosystems and smart city infrastructure, the chapter illustrates real-world applications, demonstrating how the symbiotic relationship between zero-trust principles and XAI fortifies security measures while ensuring comprehensibility and accountability. Lastly, it explores future prospects, ethical considerations, and adoption challenges, projecting potential advancements and ethical frameworks necessary for fostering a trustworthy and adaptive IoT security landscape.

14.2 Background of history of zero-trust security

Securing enterprise systems poses formidable challenges owing to the intricate nature of IT infrastructures, the expansive scope and rapid pace of user access, and the inherent adversarial dynamics of information security. Compounded by the pervasive openness of most enterprise networks, the absence of enforcing the principle of least privilege at both network and application tiers renders organizations exceedingly susceptible to cyber threats. This vulnerability extends across internal networks and externally facing services like virtual private networks (VPNs), which, become targets for potential adversaries when exposed to the Internet. In today's threat landscape, the architectural design of systems resembling this open model would be deemed imprudent. Nonetheless, prevalent traditional security and networking systems persist in perpetuating this vulnerability despite the glaring risks they pose [3].

The focal point of this chapter revolves around zero-trust security, heralding a contemporary paradigm shift in security strategies, particularly vital within the context of IoT. This approach fundamentally alters the landscape by firmly implementing the principle of least privilege across both IoT networks and applications. It operates on the premise that unauthorized entities, whether users or systems, are categorically denied access to enterprise resources, while authorized users are meticulously granted only the bare minimum access essential for their functionalities within the IoT ecosystem. This stringent approach significantly fortifies the security posture of enterprises operating within the IoT domain, bolstering resilience against potential threats and vulnerabilities. Additionally, the adoption of zero-trust principles not only amplifies safety measures but also accentuates efficiency and effectiveness. This is achieved through the implementation of automated, dynamic, and identity-centric access policies, allowing for streamlined and secure interactions between IoT devices and networks while minimizing the risk of unauthorized access or breaches [4].

In the context of IoT security, the concept of "zero trust" embodies a significant departure from the conventional approach to trust within networks and applications. Despite its name, "zero trust" does not imply a complete absence of trust, but rather signifies the elimination of inherent or implicit trust assumptions. Instead, it emphasizes a meticulous process of establishing trust from the ground up, gradually allowing appropriate access levels at precise moments within the IoT ecosystem. While alternative terms like "earned trust" or "adaptive trust" might have been more fitting descriptors, the adoption of the term "zero trust" gained prominence due to its allure and resonance. It is crucial not to interpret the term "zero" literally; rather, it denotes the meticulous and discerning allocation of trust. This approach, while gaining traction as a prominent industry trend, transcends mere marketing hype, embodying substantial and meaningful principles. At its core, zero trust embodies a philosophy and a set of guiding principles adaptable to diverse enterprise settings, including those navigating IoT landscapes [5].

Zero trust encapsulates a philosophy that is crucial for establishing robust enterprise security, a notion particularly relevant within the IoT realm. Although historically associated predominantly with network security, zero trust extends its purview far beyond networks, encompassing applications, data, identities, operations, and policies. This broader perspective is vital in comprehensively safeguarding IoT ecosystems against multifaceted threats. In this chapter, we aim to offer practical insights into the integration of XAI and zero-trust journeys. It is essential to perceive zero trust as an ongoing and evolving initiative rather than a finite project, hence the deliberate use of the term "journey." Our objective is to serve as a guiding resource, sharing perspectives and advice on navigating the complexities of implementing and sustaining XAI-empowered zero-trust principles effectively within IoT environment. Embracing zero trust fundamentally enhances and redefines the approach toward achieving robust enterprise security, transcending network boundaries to fortify various facets integral to the IoT landscape.

The evolution of zero trust spans several decades, originating from a fundamental shift in cybersecurity paradigms and continuously evolving in response to the changing threat landscape. The concept emerged as a response to the inadequacies of traditional perimeter-based security models that presumed internal networks to be inherently secure, thereby fostering a false sense of trust. Although the term "zero trust" gained significant prominence in the 2000s, its core principles have historical antecedents [6].

In 2007, the Jericho Forum published a series of influential white papers, notably the "Jericho Forum Commandments," which articulated key principles that laid the foundation for the zero-trust concept. These principles aimed to address the inadequacies of perimeter-based security models and advocated for a more dynamic and adaptive approach to security. The commandments emphasized concepts such as "Assume the perimeter is breached," "No implicit trust in any network," and "Security must be pervasive, resilient, and adaptable."

The Jericho Forum's thought leadership and publications served as a catalyst for redefining security paradigms, prompting discussions around the need to move away from perimeter-centric models toward a more nuanced and context-aware approach. The ideas and principles articulated by the Jericho Forum were instrumental in shaping the philosophy behind zero trust, emphasizing the importance of continuous verification, stringent access controls, and a shift from implicit trust to a more cautious, risk-based trust model. While the term "zero trust" itself might not have been explicitly coined by the Jericho Forum, their contributions and vision laid a solid groundwork for the evolution of security paradigms, which eventually led to the emergence and refinement of the zero-trust security model we recognize today. The Jericho Forum's insights and principles continue to influence and resonate within the cybersecurity community, underscoring the need for adaptive, context-aware security measures in the face of evolving threats and digital transformations.

In 2010, Forrester Analyst John Kindervag played a pivotal role in introducing the term "zero trust" through the influential paper titled "No More Chewy Centers: Introducing the Zero Trust Model of Information Security." This significant whitepaper encapsulated concepts that had been circulating within the industry and were previously championed by groups like the Jericho Forum. It outlined a transformative departure from the traditional security approach reliant on a rigid perimeter, advocating instead for a model that scrutinized and comprehended elements within a network before granting them any level of trust or access. This marked a notable shift in security strategy, emphasizing the need for a continuous verification process for entities operating within the network environment. As time progressed, Forrester further refined and expanded this concept, culminating in what is now recognized as the zero-trust eXtended (ZTX) framework. This evolved framework incorporates essential elements such as Data, Workloads, and Identity, serving as the core components that underpin the zero-trust security approach [7].

Subsequently, in 2014, indeed, Google's BeyondCorp initiative stands as a pivotal milestone in the evolution and popularization of zero-trust principles. Introduced in 2013, BeyondCorp represented a groundbreaking departure from the traditional perimeter-based security model prevalent at that time. BeyondCorp was conceived in response to the recognition that the traditional approach, wherein trust was primarily based on network location (inside or outside the corporate network), was becoming increasingly ineffective and obsolete. Instead, BeyondCorp aimed to eliminate the concept of a trusted internal network entirely, adopting a zero-trust approach to security. At its core, the BeyondCorp initiative focused on shifting the security model from a network-centric to a user- and device-centric perspective. It emphasized verifying trust through meticulous device and user authentication processes and granting access based on various contextual factors, such as device health, user identity, location, and other attributes. This contextual access control approach allowed for a more granular and adaptive security posture, enabling secure access to resources regardless of the network from which the user or device originated. Google's successful implementation of the BeyondCorp model demonstrated that zero-trust principles could be effectively applied in real-world scenarios within a large-scale organization. By eschewing the reliance on traditional VPNs and perimeter-based defenses, BeyondCorp showcased the feasibility and advantages of a zero-trust architecture (ZTA) in enhancing security, particularly in the context of a modern, dynamic, and increasingly mobile workforce [8].

The Continuous Adaptive Risk and Trust Assessment (CARTA) framework represents a significant milestone in the evolution of zero-trust principles. Introduced by Gartner in 2017, CARTA redefined the approach to cybersecurity by advocating continuous monitoring, adaptive security measures, and risk assessment within ZTAs. CARTA emphasizes the importance of continuously assessing risks and trust levels across an organization's digital ecosystem. It acknowledges that traditional security approaches based on static, perimeter-centric controls are insufficient in today's dynamic threat landscape. Instead, CARTA promotes a strategy that involves continuously evaluating risks, monitoring user behavior, and adapting security measures accordingly. Within the zero-trust framework, CARTA's approach aligns with the core principles of assuming zero trust by default and continuously verifying trust in real-time. It emphasizes the need for adaptive security strategies that respond to evolving threats and changing user behaviors. This involves leveraging technologies such as behavioral analytics, machine learning, and automation to detect anomalies, predict potential threats, and dynamically adjust access controls based on risk assessments [9].

The National Institute of Standards and Technology (NIST) played a pivotal role in the evolution and standardization of zero-trust principles within cybersecurity. In September 2020, NIST released Special Publication 800-207 titled "Zero Trust Architecture," providing a comprehensive framework and guidelines for implementing zero-trust principles. This publication from NIST was a significant milestone in the history of zero trust, as it provided a detailed blueprint for organizations to understand, plan, and deploy ZTAs effectively. It emphasized the importance of assuming zero trust by default, continuously authenticating and authorizing users and devices, and applying strict access controls regardless of whether they are inside or outside the traditional network perimeter.

The evolution of zero trust continued with its integration into various cybersecurity frameworks, industry standards, and organizational strategies. It garnered further momentum with technology advancements, including the proliferation of IoT devices, which necessitated a more adaptive and resilient security approach. This evolution propelled zero trust from being primarily network-centric to encompassing a broader spectrum of assets, including applications, data, identities, and policies. Over the years, zero trust has evolved into a foundational cybersecurity philosophy rather than merely a specific technology or solution. Its evolution continues to adapt to the evolving threat landscape, emphasizing the need for continuous verification, dynamic access controls, and holistic security measures. Today, zero trust remains a critical and evolving approach, continually refined to meet the complex security challenges faced by modern enterprises, including those operating within IoT ecosystems [10].

14.3 Principles of zero-trust IoT security

Zero trust represents a comprehensive approach to securing not just networks but also application and data resources within the IoT domain. At its essence, zero trust pivots around an identity-centric policy model aimed at regulating access. While enterprises commonly possess a suite of IT and security tools, zero trust necessitates a holistic perspective and operation, centering on identity as the linchpin. It calls for the unified enforcement of policies that are sensitive to attributes and contexts across the entire IoT environment. This holistic view becomes imperative in the context of IoT, where diverse devices and systems interact, emphasizing the significance of identity-based policies that dynamically adapt to various IoT device attributes and contextual parameters. As we delve into the foundational principles of zero trust, categorized into primary and secondary principles, the significance of identity-centric policies in the IoT landscape will become evident. These principles are pivotal in fortifying the security posture of IoT ecosystems, ensuring that access controls adapt to the nuanced characteristics and contexts of interconnected IoT devices, applications, and data resources.

14.3.1 Primary principles

In the IoT domain, the bedrock of zero-trust principles rests upon three fundamental pillars widely acknowledged across the industry as indispensable. Initially outlined in the seminal "No More Chewy Centers" publication by Forrester, these principles form the cornerstone of any robust zero-trust implementation. They emphasize the necessity of reevaluating trust assumptions, advocating strict access controls, continuous verification of devices and users, and a shift away from perimeter-based security models. Additionally, the NIST ZTA document incorporates essential tenets that complement these core principles, further enriching the framework. Our interpretation, aligned with current industry perspectives, amalgamates these core principles and NIST's recommendations within the IoT landscape. This amalgamation underscores the criticality of continual scrutiny of trust relationships, real-time risk assessments, and the imperative need for adaptive security measures throughout IoT ecosystems. It reinforces the notion that trust should never be assumed but continuously verified, reflecting the dynamic nature of interactions within IoT environments.

When organizations embark on the transformative journey of embracing zero trust as a foundational security framework within the IoT domain, a core set of principles serves as a guiding beacon, offering a concise and universally understood framework to steer and unify the entire organization. These core principles, enumerated within specific guidelines, serve as fundamental directives for adopting and implementing ZTAs. They pivot around the integration of people, processes, and technology, serving as essential guidelines not only for newly introduced security initiatives but also for retroactively applying robust security practices to pre-existing systems and activities. These primary principles encompass several thematic categories that encompass various facets of zero-trust implementation [11].

Guardrails and governance principles play a crucial role in guiding compliance, risk management, and information security stakeholders toward adopting zero-trust practices. They ensure the sustainability of security assurances within IoT environments by addressing the dynamic nature of compliance and regulatory requirements, necessitating proactive integration of industry standards and organizational controls specific to IoT contexts.

- People guidance and inspiration: In the IoT domain, effective organizational governance frameworks play a pivotal role in guiding decisions related to people, processes, and technology within the ecosystem. Clear ownership and delineation of responsibilities regarding decisions, policies, and visionary aspirations are fundamental. Within IoT, where the interconnectedness of devices and systems necessitates harmonious collaboration between stakeholders, a well-defined governance framework inspires stakeholders to align their efforts toward common objectives. It is imperative to have governance structures that not only guide decision-making but also inspire individuals involved in IoT initiatives, fostering a shared vision and accountability across the organization.
- Risk and complexity reduction: Governance within IoT environments must strive to minimize complexity while concurrently reducing the threat surface area. Simplification of processes and systems enhances clarity and ease of management within the intricate IoT landscape, thereby reducing potential vulnerabilities. Reducing the threat surface area, especially in IoT, where interconnected devices amplify potential attack vectors, is crucial for bolstering security. Governance should aim to streamline IoT architectures, minimizing

complexities and vulnerabilities that adversaries might exploit, thereby fortifying the overall security posture.

- Alignment and automation: Policies and security metrics within IoT ecosystems should align directly with the organizational mission and risk requirements. It is essential that these policies and metrics seamlessly integrate with the broader objectives of the organization while favoring automated execution and reporting mechanisms within IoT environments. Automation aids in enforcing consistent policies, enhancing efficiency, and providing real-time insights into security metrics, aligning security practices with organizational goals effectively.
- Security for the full lifecycle: Governance within IoT necessitates sustaining risk analysis and ensuring confidentiality, integrity, and availability assurances throughout the entire lifecycle of data, transactions, or relationships. This approach includes continuous risk analysis and assurance measures, not only during data transmission or storage but also throughout its entire lifecycle. Within IoT, where data is dynamic and constantly in use, at rest, or in transit, it's crucial to ensure sensitive or regulated data is handled appropriately. Governance frameworks should prioritize reducing asset sensitivity where feasible, while providing robust assurances and protections for data under various states, reinforcing security at every phase of the IoT ecosystem.

Organizational value and risk alignment principles are pivotal in aligning the strategic goals of the business, IT, and security stakeholders, ensuring that the adoption of zero trust addresses overarching strategic drivers. Within the IoT landscape, this alignment is critical to integrating security seamlessly into the IoT ecosystem, harmonizing with business objectives while mitigating associated risks [12].

- **Modern work enablement—users:** In the context of IoT ecosystems, facilitating modern work enablement principles is essential, ensuring that users within organizational frameworks possess the flexibility to work across diverse networks and locations while maintaining consistent and robust security measures. Particularly within the intricate IoT landscape, where users interact with an array of interconnected devices and systems, the ability to seamlessly operate from any location becomes crucial. Enabling users to access and utilize IoT functionalities securely, irrespective of their network or location, not only fosters a more productive work environment but also ensures that the integrity and confidentiality of the interconnected IoT ecosystem remain intact, enhancing overall efficiency.
- **Goal alignment:** Security measures within IoT landscapes must seamlessly align with and contribute to attaining organizational objectives while operating within predefined risk thresholds. In the context of IoT, where connectivity and functionality are paramount, aligning security practices with organizational goals ensures that security efforts complement and bolster the achievement of overarching objectives. Ensuring that security measures do not impede but rather facilitate the fulfillment of organizational objectives within predefined

risk parameters is crucial. This alignment enables security efforts in IoT environments to be an enabler rather than a hindrance to organizational goals.

- **Risk alignment:** Within IoT ecosystems, effectively managing and consistently measuring security risks in alignment with the organization's risk framework is imperative. Security measures must align with the organization's risk tolerance and thresholds, providing a comprehensive approach to identifying, assessing, and mitigating risks inherent to the IoT landscape. Given the intricate and interconnected nature of IoT devices and systems, adherence to the organizational risk framework ensures that security protocols address inherent risks while aligning with the organization's risk management strategy. This approach ensures that the IoT environment operates within acceptable risk levels, and security efforts are commensurate with potential risks within the ecosystem.

Security control principles are geared toward security and IT architects, laying the groundwork for robust confidentiality, integrity, and availability assurances within IoT systems. These principles underscore the importance of establishing and maintaining strong security foundations across IoT devices and networks to uphold the integrity and resilience of the entire ecosystem.

- **Simple and pervasive:** Within the realm of IoT, implementing security mechanisms that are both simple and pervasive is paramount. Security measures should be straightforward, scalable, and easily manageable across the entire organizational ecosystem, encompassing both internal and external facets. Specifically, in the IoT domain, where diverse devices and systems interact within intricate networks, simplicity and scalability in security mechanisms are imperative. Simple yet effective security measures streamline implementation and management, ensuring seamless integration across the IoT landscape without compromising on robustness. Scalable security mechanisms facilitate the management of a vast array of IoT devices, contributing to the efficiency and reliability of the entire ecosystem.
- **Explicit trust validation:** Explicit trust validation principles within IoT landscapes emphasize the need to explicitly validate assumptions of integrity and trust levels against the organization's risk thresholds and tolerance. Before allowing assets or data systems to interact with other entities or devices within IoT ecosystems, a meticulous validation process is necessary. This process involves scrutinizing the integrity and trustworthiness of assets or systems against predetermined risk thresholds, ensuring that only validated and trusted entities can access or interact within the IoT environment. Particularly in the interconnected IoT landscape, where numerous devices and data sources communicate, explicit trust validation becomes crucial in safeguarding against potential threats or compromises, ensuring that interactions occur within specified levels of trust and integrity.

Technology principles hold significance for IT organizations, information security, and risk and compliance stakeholders, shaping technology decisions fundamental to constructing a ZTA in the IoT sphere. These principles focus on critical elements such as identity management, access control, and minimizing the threat surface area within IoT frameworks [13].

- **Asset-centric security:** In the context of IoT environments, adopting asset-centric security principles is crucial, advocating for security measures that closely align with the assets themselves rather than solely relying on network-centric strategies. Embracing data-centric and application-centric approaches within the IoT landscape ensures a tailored and precise security framework. This approach minimizes disruption to productivity by focusing on securing the core assets, such as data and applications, at the point of their existence and utilization. In IoT ecosystems where a multitude of interconnected devices interact and generate vast amounts of data, this principle emphasizes securing these assets at their core, ensuring that security measures are intricately woven into the data and applications themselves, thereby reducing potential vulnerabilities without impeding productivity.
- **Least privilege:** The principle of least privilege is critical within IoT frameworks, advocating for granting access to systems and data based solely on the specific requirements of individuals or entities. In the context of IoT, where numerous devices and systems are interconnected, restricting access to the bare essentials necessary for functionalities becomes paramount. This approach ensures that access rights are strictly aligned with operational needs, minimizing the risk of unauthorized access or misuse of sensitive data within IoT environments. Moreover, timely removal of access privileges once they are no longer necessary further mitigates potential risks, enhancing the overall security posture without compromising the essential functionalities of IoT systems.

14.3.2 Secondary principles

Apart from the fundamental zero-trust principles previously examined, we contend that three supplementary principles hold equal significance and indispensability within any enterprise-grade zero-trust environment [14].

14.3.2.1 Ensure all components support APIs for event and data exchange

In fostering a robust zero-trust environment within the realm of IoT, it is imperative that all interconnected components seamlessly support Application Programming Interfaces (APIs) for efficient event and data exchange. This principle emphasizes the necessity of establishing a cohesive communication framework across various devices, systems, and applications within the IoT ecosystem. APIs serve as the linchpin for interoperability, facilitating smooth and standardized communication pathways between disparate components. By ensuring that every device and system within the IoT network adheres to API support for event and data exchange, organizations lay the groundwork for streamlined data flow, enabling efficient sharing of information and triggering automated actions based on real-time insights. This approach fosters a cohesive IoT environment where devices and systems communicate fluidly, enhancing the overall agility and responsiveness of the ecosystem to dynamic events and contextual changes.

Furthermore, API-driven data exchange offers the advantage of enabling seamless integration of diverse devices and applications, propelling innovation and scalability within IoT ecosystems. This interoperability not only simplifies the integration process but also empowers organizations to leverage a broader array of devices and technologies, thereby enhancing the ecosystem's flexibility and adaptability. Through standardized APIs for event and data exchange, organizations not only establish a foundation for real-time communication but also pave the way for the integration of emerging technologies, ensuring that the zero-trust environment remains agile, future-proof, and capable of evolving with the dynamic demands of the IoT landscape.

14.3.2.2 Automate actions across environments and systems, driven by context and events

Within the context of zero trust in IoT, automating actions across diverse environments and systems based on contextual insights and events holds pivotal importance. This principle underscores the need for automated responses triggered by real-time contextual data and events occurring within the IoT ecosystem. Automation driven by contextual insights ensures that responses to security incidents or changes in the environment occur swiftly and accurately. By harnessing contextual information derived from diverse IoT devices and systems, automated actions can dynamically adjust access controls, enforce security policies, or initiate responses to potential threats in real time. This proactive approach significantly enhances the responsiveness and agility of the zero-trust environment, enabling it to adapt swiftly to evolving situations and mitigate risks promptly. Moreover, automation based on contextual awareness within IoT environments not only fortifies security measures but also amplifies operational efficiency. By autonomously responding to predefined events or changes in context, organizations reduce the burden on manual intervention, allowing resources to focus on strategic initiatives. Additionally, automated actions driven by contextual data facilitate a more adaptive security posture within the IoT landscape. As the IoT ecosystem continues to evolve and grow, automation fueled by contextual awareness becomes indispensable in ensuring that security measures remain aligned with the dynamic nature of IoT environments, effectively safeguarding against emerging threats and vulnerabilities.

14.3.2.3 Deliver tactical and strategic value

In the sphere of zero trust within IoT environments, the principle of delivering both tactical and strategic value is foundational. This principle underscores the importance of not only addressing immediate security concerns tactically but also aligning security initiatives strategically with broader organizational objectives. Tactical value refers to the immediate benefits derived from implementing security measures, such as enhanced threat mitigation and improved incident response within the IoT ecosystem. These tactical advantages ensure the immediate safeguarding of IoT assets and data, mitigating risks effectively in real-time scenarios.

However, beyond immediate tactical gains, aligning security efforts with strategic organizational goals is equally imperative. Strategic value emphasizes how security initiatives within the IoT landscape contribute to overarching business objectives. This alignment ensures that security measures not only protect IoT assets but also augment organizational efficiency, innovation, and competitiveness. By aligning security efforts with broader strategic goals, such as improving operational efficiency, fostering innovation, and ensuring regulatory compliance within IoT frameworks, organizations derive long-term strategic value from their security investments. This strategic alignment ensures that security initiatives contribute meaningfully to the organization's success, promoting growth and sustainability within the rapidly evolving IoT landscape. Ultimately, by delivering both tactical and strategic value, zero trust in IoT environments becomes an enabler of not just security but also business resilience and success.

14.4 Zero-trust architecture

As ZTA has evolved, fundamental concepts have emerged, reshaping security paradigms. One of these key concepts revolves around the notion of a "Subject," which encompasses users, applications, or devices operating within a computer system and seeking access to an enterprise resource. These enterprise resources could span applications, data repositories, documents, or workloads within the enterprise's control and protected by the zero-trust system (refer to Figure 14.1). Throughout this discussion, these resources are often denoted simply as "Resources."

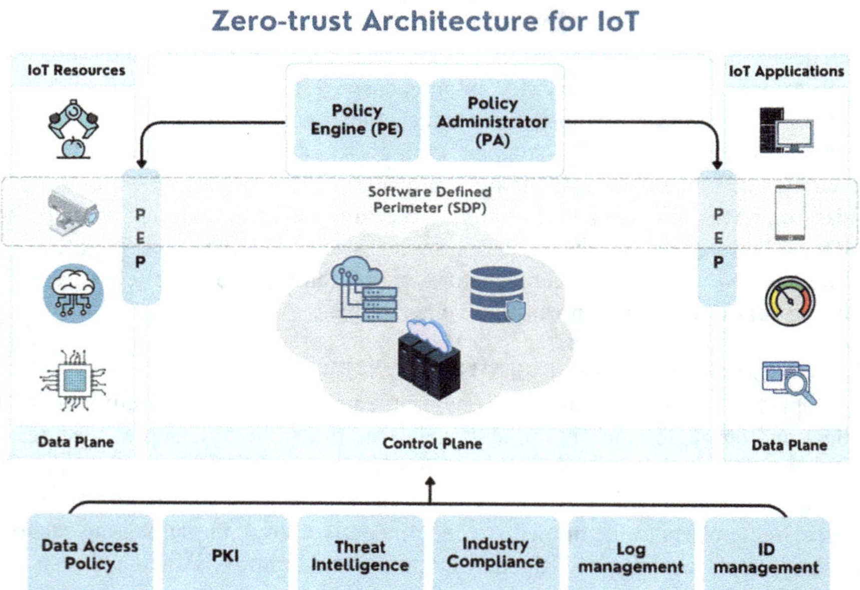

Figure 14.1 Visualization of ZTA

Operating within an untrusted environment and traversing untrusted networks, the subject's access to these resources is strictly governed. Access is exclusively granted through a designated policy enforcement point (PEP). This PEP assumes a pivotal role in regulating and controlling subject access to resources within what NIST terms as the "implicit trust zone." This concept delineates an area where access permissions are enforced and verified. However, it is important to note that the PEP itself does not house or formulate policy determinations. Instead, this responsibility lies with the policy decision point (PDP), which is entrusted with the task of formulating and enforcing access policies based on established parameters and security protocols. This separation of duties ensures a layered approach to access control, with the PEP serving as the gateway and the PDP orchestrating the policies governing access to resources within the zero-trust framework [15].

Within the ZTA framework, the interaction between the subject and the enterprise resource takes place within what is termed the data plane—a distinct entity separate from the control plane. According to NIST, the PDP and PEP communicate across a network that is logically detached and inaccessible directly by enterprise assets and resources. This separation delineates the control plane for policy decision-making and the data plane for the transmission of application data traffic.

Moreover, it is critical to reframe the depiction of certain additional elements, previously perceived as external to the system. Elements such as the continuous diagnostics and mitigation (CDM) framework and public key infrastructure (PKI) play crucial roles within the zero-trust system. These elements should ideally be viewed as integrated components within the framework or, at the least, illustrated with bidirectional arrows indicating varying degrees of integration. These elements, despite appearing external, are fundamentally integrated into the zero-trust system, providing essential contextual information. Their inclusion significantly influences policy decisions within the zero-trust framework, contributing valuable input and contextual understanding that directly impacts the system's policy formulations and operational decisions. Therefore, recognizing their inherent integration within the zero-trust ecosystem is essential for a comprehensive understanding of the system's functionality and policy decision-making processes.

In the framework of zero trust, a policy is articulated as a declarative statement that precisely specifies the allowance for a subject to execute a particular action upon a designated target, contingent upon the fulfillment of specific conditions. This policy component embodies critical elements, including subjects, criteria, action, target, and conditions, forming the bedrock of access control within the ZTA.

Subjects within policies represent the entities or individuals initiating actions. These subjects must undergo authentication, ensuring their legitimacy within the system. Subject criteria are integral components of policies, delineating the specific subjects to which the policy applies, thereby enforcing precise access controls based on authenticated identities.

Actions encapsulate the activities carried out by the subjects. An action must distinctly involve either a network or application component, and in some instances, it may incorporate both, defining the nature of the activity being undertaken within the zero-trust environment.

Targets, constituting the objects or resources upon which the action is directed, may exhibit static or dynamic characteristics within the policy framework. The definition of the target may span a broad or narrow scope; however, a more tailored and precise definition is preferred within zero-trust policies to enhance specificity and minimize potential security risks [16].

Conditions serve as the delineating factors dictating when a subject is authorized to execute a specific action upon the designated target. In the zero-trust system, the definition of conditions extends across various attributes, encompassing subject attributes, environmental factors, and attributes associated with the designated target. These conditions are pivotal in establishing the circumstances under which access is permitted, allowing for a granular and contextually aware access control mechanism within the zero-trust framework.

PEPs serve as critical components within security architectures, implementing and enforcing access control policies. In the realm of zero trust and broader cybersecurity contexts, different types of PEPs cater to diverse environments and systems, each playing a unique role in ensuring secure access. Here, we discuss various types of PEPs:

- **Network PEPs (N-PEPs):** N-PEPs focus on controlling access to network resources and infrastructure. These enforcement points typically reside within network devices like routers, switches, firewalls, or proxies. They regulate traffic flow, inspect packets, and enforce access control policies based on predefined rules. N-PEPs are instrumental in implementing network segmentation, inspecting data traffic, and restricting or allowing access to resources based on network-centric policies.
- **Host-based PEPs (H-PEPs):** H-PEPs operate at the endpoint or host level, controlling access to resources based on local policies. They are embedded within individual devices or systems, such as endpoint protection platforms, operating system security controls, or agent-based security software. H-PEPs manage access based on the specific attributes and configurations of the device, applying policies that dictate actions or limitations on resource access at the endpoint level.
- **Application PEPs (A-PEPs):** A-PEPs are deployed within application environments to regulate access control based on application-specific policies. These enforcement points are integrated into the application architecture or application-level gateways. A-PEPs manage and enforce fine-grained access control policies concerning application resources, functions, or data, ensuring that only authorized entities can interact with specific application components.
- **Cloud PEPs (C-PEP):** C-PEPs govern access control within cloud environments, ensuring adherence to security policies in cloud-based services, platforms, or infrastructures. They operate within cloud security frameworks, implementing policies that govern access to cloud resources, data, or services. C-PEPs facilitate secure access and mitigate risks within the dynamic and distributed nature of cloud-based environments.

- **Data PEPs (D-PEP)**: D-PEPs focus on governing access to data repositories and databases. These enforcement points regulate access to sensitive data, ensuring compliance with data protection policies, and enforcing access controls based on data attributes, user permissions, or contextual factors. D-PEPs often integrate with data management solutions or database security tools to enforce granular data access policies.

14.5 Synergies between zero trust and XAI

14.5.1 Explainability in zero-trust systems

Understanding the importance of explainability within the domain of zero-trust IoT is fundamental to establishing robust security paradigms. In ZTAs applied to IoT ecosystems, the necessity for transparency and comprehensibility in decision-making processes holds paramount significance. Given the interconnected and complex nature of IoT environments, where diverse devices interact and exchange data within stringent security frameworks, explainability becomes a cornerstone. XAI techniques play a pivotal role in elucidating the rationale behind access decisions, providing insights into why certain actions are permitted or denied within the zero-trust IoT framework. This transparency fosters trust and confidence among stakeholders, including users, administrators, and system auditors.

By enhancing the interpretability of access control decisions, explainability contributes significantly to the credibility and reliability of zero-trust IoT systems, empowering stakeholders to understand, validate, and ensure the integrity of security measures implemented within intricate IoT networks. Moreover, within zero-trust IoT architectures, explainability serves as a catalyst for effective risk management and incident response. Through transparent decision-making processes elucidated by XAI, security administrators can swiftly comprehend the underlying reasons behind access decisions in IoT environments. This comprehension aids in identifying potential vulnerabilities or anomalous behavior more efficiently, enabling proactive responses to security incidents. By providing insights into access control policies and their operational implications in IoT contexts, explainability empowers security teams to fine-tune policies, refine security protocols, and effectively adapt strategies to mitigate risks. Therefore, within zero-trust IoT frameworks, the integration of explainability not only enhances overall system trustworthiness but also fortifies the ability to proactively manage and mitigate security threats in dynamically evolving IoT landscapes [17,18].

14.5.2 Integration of XAI in zero trust

Within the realm of IoT, the role of transparent decision-making and auditability in zero-trust policies stands as an integral facet shaping the security posture of interconnected systems. Transparent decision-making mechanisms within zero-trust policies for IoT environments are pivotal in enabling explicit and accountable access control. These mechanisms elucidate the rationale behind access decisions, granting stakeholders insights into the authorization process and the factors influencing access approvals or denials across diverse IoT devices and data streams. This transparency fosters a deeper understanding of the principles guiding access control, ensuring that decisions align with established policies and predefined parameters. Additionally, auditability, when integrated into zero-trust IoT policies, facilitates the traceability and scrutiny of access decisions. It enables the systematic recording and analysis of access control activities, allowing for comprehensive assessments of system behavior, access patterns, and potential anomalies. By embracing auditability, zero-trust policies in IoT environments ensure that access decisions are not only transparent but also subject to systematic scrutiny, thereby enhancing the overall accountability and trustworthiness of the security measures governing IoT ecosystems. Furthermore, within zero-trust policies applied to IoT contexts, the role of transparent decision-making and auditability extends beyond mere compliance with established protocols. These elements serve as foundational pillars in facilitating adaptive and responsive security measures. Transparent decision-making empowers stakeholders to comprehend the intricacies of access controls in IoT environments, allowing them to validate the alignment of decisions with overarching security goals and policies. Concurrently, auditability serves as a critical tool for continuous monitoring and refinement of access policies within evolving IoT landscapes.

Integrating XAI techniques within the framework of ZTA in IoT environments presents a promising avenue to enhance transparency and comprehension in access control mechanisms. The application of XAI methodologies within ZTA for IoT involves leveraging interpretable models and algorithms to elucidate the decision-making processes underlying access control policies. By employing XAI techniques, such as rule-based systems or interpretable machine learning models, ZTA in IoT can offer detailed explanations for access decisions made across interconnected devices and systems. These techniques aid in revealing the rationale behind authorization outcomes, shedding light on the attributes, features, or contextual factors that influenced access approvals or denials within the IoT ecosystem. This integration fosters a deeper understanding of the intricate relationships between access control policies and IoT environments, empowering stakeholders to comprehend the logic driving access decisions and fostering trust in the security measures implemented within the zero-trust framework. Moreover, the integration of XAI techniques into ZTA for IoT environments not only enhances transparency but also facilitates the proactive identification and mitigation of potential security risks. XAI enables the identification of patterns or factors contributing to security incidents, allowing for the refinement of access control policies in IoT settings. These techniques support the continuous improvement of ZTA in IoT landscapes, enabling adaptive responses to emerging threats and evolving security challenges. The integration of XAI methodologies into ZTA for IoT holds immense promise in not only enhancing transparency but also bolstering the resilience and agility of security measures within interconnected IoT ecosystems [19].

The application of XAI methodologies stands as a cornerstone in enhancing transparency and fostering accountability within zero-trust policy enforcement mechanisms in IoT landscapes. With the use of XAI techniques, zero-trust policies governing access control in IoT environments can offer comprehensible explanations for the decisions made regarding resource access and usage. Through interpretable machine-learning models or rule-based systems, the underlying logic behind access control decisions across diverse IoT devices and networks becomes elucidated. This transparency ensures that stakeholders, including administrators and users, gain insights into the factors influencing access approvals or denials within the zero-trust framework. Additionally, these explainability methods enable comprehensive audits of access decisions, fostering accountability by allowing scrutiny and validation of the adherence of access controls to established policies and regulatory compliance frameworks in IoT settings. Such clarity and auditability brought forth by XAI methodologies contribute significantly to bolstering trust and confidence in zero-trust policy enforcement within complex and interconnected IoT ecosystems. Furthermore, the integration of XAI methodologies in zero-trust policy enforcement for IoT environments not only enhances transparency but also serves as a catalyst for fostering proactive risk management strategies. By offering insights into the decision-making processes underlying access control, XAI enables the identification of potential vulnerabilities, anomalies, or deviations from established security norms within IoT networks. This proactive approach empowers security teams to pre-emptively detect and address emerging threats, facilitating timely adjustments and refinements to access control policies. Moreover, the transparency provided by XAI methods facilitates the ongoing optimization of zero-trust policy enforcement measures in IoT, ensuring their alignment with evolving security requirements and mitigating potential risks effectively. As a result, the application of XAI methodologies within zero-trust policy enforcement mechanisms significantly augments transparency, accountability, and the resilience of security measures governing IoT ecosystems.

14.5.3 Enhancing trust through XAI

The integration of XAI methodologies within zero-trust IoT environments plays a pivotal role in fostering trust and credibility by providing insights into access control decisions across interconnected systems. XAI serves as a linchpin in enhancing transparency, enabling stakeholders to understand and interpret the logic behind authorization decisions within the zero-trust framework applied to IoT. By employing interpretable machine learning models or rule-based systems, XAI offers transparent explanations for access control outcomes, elucidating the factors, attributes, or contextual elements influencing decisions across diverse IoT devices and networks. This transparency instills confidence among users, administrators, and auditors, allowing them to comprehend and verify the integrity and fairness of access decisions, thereby fostering trust in the zero-trust IoT environment. Moreover, XAI techniques contribute significantly to bolstering credibility within zero-trust IoT environments by facilitating comprehensive audits and validations of access control mechanisms. The transparency and interpretability provided by XAI enable stakeholders to conduct systematic reviews and evaluations of access decisions, ensuring compliance with established policies, regulatory standards, and ethical considerations. This fosters credibility as stakeholders gain confidence in the alignment of access control decisions with organizational security objectives and legal frameworks governing IoT systems. Furthermore, the ability to trace and comprehend access control rationales through XAI methodologies enhances the credibility of zero-trust IoT systems by ensuring that access decisions are consistent, justified, and accountable. Consequently, the integration of XAI within zero-trust environments for IoT not only augments transparency but also substantiates the credibility of access control mechanisms, reinforcing trust among stakeholders in the security measures governing interconnected IoT ecosystems [20].

Transparent decision-making within IoT systems operating under zero-trust paradigms exerts profound implications on user trust and system reliability. When users have visibility into the reasoning behind access control decisions across interconnected IoT devices, it engenders a sense of confidence and trust in the security measures implemented within the system. By employing transparent decision-making mechanisms facilitated by XAI techniques or interpretable models, users gain insights into why certain actions are permitted or denied within the zero-trust framework applied to IoT environments. This transparency fosters a stronger sense of trust, as users understand the logic governing access to sensitive resources, enhancing their confidence in the security and reliability of the system. Moreover, transparent decision-making engenders a proactive security culture among users within IoT ecosystems, empowering them to adhere to security protocols, fostering responsible usage practices, and contributing to the overall reliability of the interconnected IoT infrastructure. Furthermore, the implications of transparent decision-making on system reliability within IoT environments under a zero-trust model extend beyond user trust to encompass the robustness and dependability of the entire system. Transparent access control decisions enable stakeholders to scrutinize and validate the integrity and fairness of the system's security measures. This scrutiny contributes to enhancing system reliability by identifying and addressing potential vulnerabilities or irregularities in access control policies governing IoT devices and networks. As users and administrators gain insights into the factors influencing access decisions, the system becomes more adaptable and resilient. Transparent decision-making fosters a culture of accountability and continuous improvement, leading to refinements and optimizations in access control policies within IoT systems. Ultimately, this contributes to the reliability of the zero-trust IoT environment, ensuring consistent and dependable access control mechanisms that align with organizational security goals, regulatory requirements, and user expectations.

14.5.4 Challenges and opportunities

Integrating XAI into zero-trust frameworks within IoT environments presents several challenges that need careful consideration and resolution to ensure effective implementation. One of the primary challenges lies in the complexity and heterogeneity of IoT systems, comprising diverse devices, protocols, and data formats. Given this diversity, creating XAI models compatible with varying IoT architectures and technologies proves challenging. Achieving interoperability and scalability across different IoT devices while maintaining the transparency and interpretability of AI models becomes a significant hurdle. Additionally, the resource-constrained nature of many IoT devices, characterized by limited computational power and memory, poses challenges in implementing complex XAI algorithms on these devices. Balancing the need for robust explanation capabilities with the resource limitations of IoT devices remains a crucial challenge in achieving XAI within zero-trust frameworks for IoT.

Moreover, ensuring the privacy and security of sensitive data within the context of XAI implementation in zero-trust IoT environments emerges as a critical concern. The integration of XAI may involve the collection and processing of vast amounts of data from IoT devices to generate transparent decision-making insights. However, this data handling requires stringent privacy protections to safeguard sensitive information, considering the diversity of data types and the potential risks associated with unauthorized access. Balancing the trade-off between transparency and data privacy becomes pivotal, as the extraction of explainability may inadvertently reveal sensitive or personally identifiable information. Therefore, devising robust mechanisms for data anonymization, encryption, and secure data sharing while maintaining the explainability of AI models poses a significant challenge in the integration of XAI into zero-trust IoT frameworks. Addressing these challenges necessitates a multidisciplinary approach involving expertise in AI, cybersecurity, IoT technologies, and privacy regulations to ensure successful and ethically sound integration of XAI within ZTAs for IoT. Table 14.1 delineates the multifaceted challenges inherent in integrating XAI into zero-trust frameworks within IoT environments, highlighting various aspects across layers, functionalities, planes, deployment models, and proposed mitigation strategies.

Leveraging XAI within zero-trust security measures presents a multitude of opportunities to enhance the efficacy and resilience of IoT environments. One notable opportunity lies in bolstering anomaly detection and threat identification capabilities. XAI models can elucidate the underlying factors contributing to anomalous behavior within IoT networks, facilitating the timely detection and mitigation of potential security threats. By providing interpretable insights into access decisions and system behaviors, XAI enables security teams to discern patterns or deviations indicative of malicious activities. This proactive identification of anomalies, coupled with transparent decision-making, empowers security analysts to take prompt and informed actions, thereby fortifying the security posture of zero-trust IoT environments. Furthermore, XAI offers substantial potential in facilitating adaptive and context-aware access control within IoT systems under the zero-trust model [21]. By employing interpretable machine learning algorithms, the integration of XAI enables the development of adaptive access control policies that dynamically adjust based on contextual factors. These contextual decisions consider attributes such as device behavior, user context, environmental conditions, and network patterns, enhancing the granularity and accuracy of access control decisions within IoT ecosystems. Through the transparency provided by XAI, the rationale behind adaptive access decisions becomes understandable and auditable, enabling stakeholders to validate the appropriateness of access privileges granted or denied. Consequently, leveraging XAI to foster adaptability and context sensitivity

Table 14.1 Challenges and aspects of integrating XAI into zero-trust frameworks in IoT environments

Challenges	Layer	Functionality	Plane	Deployment model	Mitigation strategies	Complexity
Diversity of IoT ecosystems	Data	Access control	Control	Edge computing	Standardized interfaces and modular models	High
Resource constraints	Device	Authentication	Data	Fog computing	Lightweight algorithms and model compression	Medium
Privacy and data security	Network	Authorization	Control	Cloud computing	Privacy-preserving techniques and encryption	High
Interpretability and complexity	Application	Anomaly detection	Control	Hybrid deployment	Simplified models and visualization tools	High
Regulatory compliance	Gateway	Compliance monitoring	Data	Distributed	Compliance framework integration	Medium
Human–computer interaction	Infrastructure	User interface	Control	Centralized	User-centric designs and explainable UIs	Low
Training and education	Cloud	Model training	Data	Decentralized	Stakeholder workshops and training programs	Low
Ethical considerations	Edge	Bias mitigation	Control	Multi-cloud	Ethical AI guidelines and fairness	

in access control mechanisms strengthens the zero-trust security model, ensuring robust protection for interconnected IoT devices and networks.

References

[1] Teerakanok, S., Uehara, T., and Inomata, A. (2021). Migrating to zero trust architecture: reviews and challenges. *Security and Communication Networks*, 2021, 1–10.

[2] Ramezanpour, K. and Jagannath, J. (2022). Intelligent zero trust architecture for 5G/6G networks: principles, challenges, and the role of machine learning in the context of O-RAN. *Computer Networks*, 217, 109358.

[3] Guembe, B., Azeta, A., and Osamor, V. (2022). Explainable Artificial Intelligence, the Fourth Pillar of Zero Trust Security. Available at SSRN 4331547.

[4] Buck, C., Olenberger, C., Schweizer, A., Völter, F., and Eymann, T. (2021). Never trust, always verify: a multivocal literature review on current knowledge and research gaps of zero-trust. *Computers & Security*, 110, 102436.

[5] Colombo, P., Ferrari, E., and Tümer, E. D. (2021). Access control enforcement in IoT: state of the art and open challenges in the zero trust era. In *2021 Third IEEE International Conference on Trust, Privacy and Security in Intelligent Systems and Applications (TPS-ISA)* (pp. 159–166). Piscataway, NJ: IEEE.

[6] Chen, B., Qiao, S., Zhao, J., *et al.* (2020). A security awareness and protection system for 5G smart healthcare based on zero-trust architecture. *IEEE Internet of Things Journal*, 8(13), 10248–10263.

[7] Syed, N. F., Shah, S. W., Shaghaghi, A., Anwar, A., Baig, Z., and Doss, R. (2022). Zero trust architecture (ZTA): a comprehensive survey. *IEEE Access*, 10, 57143–57179.

[8] Ray, P. P. (2023). Web3: a comprehensive review on background, technologies, applications, zero-trust architectures, challenges and future directions. *Internet of Things and Cyber-Physical Systems*, 3, 213–248.

[9] Laplante, P. and Voas, J. (2022). Zero-trust artificial intelligence? *Computer*, 55(2), 10–12.

[10] Liu, Y., Hao, X., Ren, W., *et al.* (2022). A blockchain-based decentralized, fair and authenticated information sharing scheme in zero trust Internet-of-Things. *IEEE Transactions on Computers*, 72(2), 501–512.

[11] Kondaveety, V. B., Lamkuche, H., and Prasad, S. (2022). A zero trust architecture for next generation automobiles. *AIP Conference Proceedings* 2519(1). AIP Publishing.

[12] Kobayashi, N. (2023). Zero trust security framework for IoT actuators. In *2023 IEEE 47th Annual Computers, Software, and Applications Conference (COMPSAC)* (pp. 1285–1292). Piscataway, NJ: IEEE.

[13] do Amaral, T. M. S. and Gondim, J. J. C. (2021). Integrating zero trust in the cyber supply chain security. In *2021 Workshop on Communication Networks and Power Systems (WCNPS)* (pp. 1–6). Piscataway, NJ: IEEE.

[14] Adahman, Z., Malik, A. W., and Anwar, Z. (2022). An analysis of zero-trust architecture and its cost-effectiveness for organizational security. *Computers & Security*, 122, 102911.

[15] Hale, B., Van Bossuyt, D. L., Papakonstantinou, N., and O'Halloran, B. (2021). A zero-trust methodology for security of complex systems with machine learning components. In *International Design Engineering Technical Conferences and Computers and Information in Engineering Conference* (Vol. 85376, p. V002T02A067). New York: American Society of Mechanical Engineers.

[16] Wylde, A. (2023). Intelligent virtual assistants (IVAs): trust and zero trust. In *Science and Information Conference* (pp. 101–108). Cham: Springer Nature Switzerland.

[17] Talan, A. (2022). *Zero Trust Network Access with Cybersecurity Challenges and Potential Solutions* (Doctoral dissertation, Dublin, National College of Ireland).

[18] Zayed, M., Anwar, A., Rahman, Z., Arefin, S. S., and Islam, R. (2022). Owner identity verification in the internet of connected vehicles: zero trust based solution. *Cryptology ePrint Archive*.

[19] Singh, R., Srivastav, G., Kashyap, R., and Vats, S. (2023). Study on zero-trust architecture, application areas & challenges of 6G technology in future. In *2023 International Conference on Disruptive Technologies (ICDT)* (pp. 375–380). Piscataway, NJ: IEEE.

[20] Deshpande, A. (2021). A study on rapid adoption of zero trust network architectures by global organizations due to COVID-19 pandemic. *New Visions in Science and Technology*, 1, 26–33.

[21] Muthuswamy, M. and Ali, A. M. (2023). Sustainable supply chain management in the age of machine intelligence: addressing challenges, capitalizing on opportunities, and shaping the future landscape. *Sustainable Machine Intelligence Journal*, 3, 1–14. doi: 10.61185/SMIJ.2023.33103.

Chapter 15
Explainable security intelligence in IoT applications

15.1 Introduction

The advancement of technology has created opportunities for significant breakthroughs in the field of XAI and other industries, resulting in a multitude of transformational possibilities. This chapter delves into a thorough examination of the crucial function of XAI in different fields, where its incorporation transforms traditional methods, enhances decision-making, and clarifies the logic behind intricate AI systems. The introduction of XAI represents a significant change in various growing industries, including smart healthcare, transportation, manufacturing, agriculture, smart grids, retail, water industry, smart buildings, nanotechnology, and more.

Within the sphere of smart healthcare, XAI emerges as a game-changer, aiding in disease diagnosis, personalized treatment plans, and securing patient data. The transformative impact extends to smart transportation, optimizing traffic flow, ensuring autonomous vehicle safety, and fortifying connected vehicle systems against cyber threats. In smart manufacturing, XAI revolutionizes production processes, enhancing quality control and predictive maintenance. Similarly, in smart agriculture, XAI aids in crop monitoring and precision irrigation and ensures data security in IoT devices. Moreover, XAI optimizes energy demand within smart grids bolsters grid stability, and counteracts cyber threats. It plays a pivotal role in smart retail, aiding in customer behavior analysis, inventory management, and cybersecurity. In the water industry, XAI enhances water quality monitoring, leak detection, and water infrastructure security. Smart buildings witness XAI's influence on energy efficiency, occupancy monitoring, and security control systems. Furthermore, XAI's integration within nanotechnology transforms material design, drug delivery systems, and environmental applications, fosters sustainable solutions, and enhances data security and privacy.

This chapter delves deep into each sector's nuances, exploring five prominent use cases within every domain alongside an additional security application, unraveling XAI's significance. Within these subsections, XAI's relevance and promise are expounded, highlighting how it augments decision-making, enhances predictability, and assures transparency within these sectors. From leveraging, XAI to optimize traffic management in smart transportation to employing it for precision irrigation in smart agriculture, each section unveils the potential and challenges associated with integrating XAI into diverse domains. The chapter navigates through the applications, delves into the data intricacies, assesses the opacity levels, delineates the necessity for XAI integration, and scrutinizes the challenges, offering a comprehensive understanding of the transformative role of XAI in shaping the future of various industries.

15.2 Smart healthcare: leveraging XAI for enhanced services

In the domain of disease diagnosis and prediction, XAI has emerged as a transformative tool, offering novel avenues for enhanced accuracy and interpretability in healthcare diagnostics. XAI algorithms, designed to unravel the decision-making processes of complex machine-learning models, provide invaluable insights into the rationale behind diagnostic outcomes. By employing transparent and interpretable models such as decision trees, rule-based systems, or explainable neural networks, XAI aids healthcare professionals in understanding the intricate relationships between various medical parameters and diagnostic conclusions. For instance, in medical imaging, XAI algorithms elucidate the key features or regions in diagnostic images that contribute to the identification of pathologies, thus empowering radiologists to make more informed and confident decisions. This transparent nature of XAI not only enhances diagnostic accuracy but also fosters trust among clinicians by offering clear explanations, consequently facilitating collaborative patient-doctor discussions and more precise treatment strategies.

Furthermore, the application of XAI in disease diagnosis and prediction extends to predictive analytics, where these models leverage extensive patient data to anticipate potential health risks and diseases. By scrutinizing diverse datasets encompassing patient medical history, genetic information, lifestyle factors, and environmental influences, XAI-driven predictive models identify subtle patterns and correlations that may elude human observation. These models offer insights into the likelihood of developing specific diseases or conditions, allowing for early risk stratification and personalized preventive interventions. The transparency and interpretability of XAI-driven predictions empower healthcare professionals to comprehend the underlying reasons for risk estimations, thereby facilitating informed decision-making and enabling proactive measures to mitigate disease progression or prevent the onset of certain health issues, ultimately leading to improved patient outcomes and a more proactive healthcare paradigm [1,2].

Personalized medicine and treatment plans benefit immensely from integrating of XAI, ushering in a new era of healthcare tailored to individual patient needs. XAI algorithms, equipped with interpretability features, analyze vast and diverse datasets encompassing patient health records, genetic information, treatment outcomes, and lifestyle factors. By discerning intricate patterns and correlations within this wealth of data, XAI facilitates the identification of precise biomarkers, genetic signatures, or disease subtypes that influence an individual's response to medications or therapies. For instance, in oncology, XAI-powered models decipher genetic variations or molecular markers to predict the efficacy of targeted therapies, enabling oncologists to design personalized treatment plans. These transparent and interpretable insights provided by XAI not only refine treatment decisions but also foster a deeper understanding of how specific patient attributes influence treatment responses, thereby enabling clinicians to tailor interventions for optimized outcomes while minimizing adverse effects.

Furthermore, XAI in personalized medicine extends its reach to therapeutic dosage optimization and adverse reaction prediction, empowering healthcare providers to customize medication dosages based on individual patient characteristics. XAI-driven models, elucidating the complex relationships between patient demographics, genetics, and drug responses, aid in determining optimal dosages to achieve therapeutic efficacy while mitigating potential adverse reactions. The transparency XAI offers in these dosage recommendations provides healthcare professionals with valuable explanations regarding the reasoning behind dosage decisions, instilling confidence in treatment plans and supporting informed discussions between patients and clinicians. Ultimately, the integration of XAI in personalized medicine heralds a transformative shift toward tailored treatments, optimizing patient outcomes by providing precise and targeted therapeutic interventions aligned with individual patient profiles.

Remote patient monitoring and telemedicine witnessed a significant transformation with the integration of XAI, revolutionizing the delivery of healthcare services and enabling more proactive and personalized care. XAI-driven algorithms, designed to interpret vast streams of patient-generated data from wearable devices, sensors, and remote monitoring tools, facilitate real-time analysis of physiological parameters and health trends. These interpretable models detect subtle changes or anomalies in patient data, allowing healthcare providers to identify potential health risks or deteriorations promptly. For instance, in chronic disease management, XAI-enabled remote monitoring systems can recognize deviations from baseline health metrics, alerting healthcare professionals to intervene before critical health issues arise. The transparency offered by XAI in these monitoring systems enables clinicians to understand the rationale behind alerts or predictive insights, enhancing trust and encouraging timely interventions, thereby improving patient outcomes and reducing hospitalizations through proactive care strategies.

Moreover, XAI in remote patient monitoring plays a crucial role in telemedicine consultations by aiding in the interpretation of patient data remotely and guiding clinical decision-making. Interpretable AI models analyze patient-reported symptoms, vital signs, and historical medical data, assisting healthcare professionals in diagnosing and treating patients virtually. By providing transparent explanations for diagnostic recommendations or treatment decisions, XAI fosters collaboration between remote healthcare providers and patients, ensuring a shared understanding of the reasoning behind suggested interventions. This interpretability strengthens the credibility of telemedicine services and empowers patients to actively engage in their healthcare, facilitating informed discussions and treatment adherence. Ultimately, integrating XAI in remote patient monitoring and telemedicine establishes a more accessible and efficient healthcare framework, enabling remote care delivery with heightened accuracy and transparency.

XAI is crucial in transforming the field of drug discovery and development by providing an exceptional understanding of the deep complexity of molecular interactions and therapeutic targets. Within the field of pharmaceutical research, XAI-powered algorithms examine vast biological and chemical datasets to uncover the connections among molecular structures, biological targets, and medication reactions. These interpretable AI models unravel the mechanisms of action behind potential drug candidates, providing transparent explanations for predictions and highlighting the critical features contributing to their effectiveness. For instance, XAI algorithms help researchers understand the specific molecular interactions or pathways that influence a drug's efficacy or potential side effects, guiding informed decisions in selecting promising candidates for further experimental validation. This transparency and interpretability offered by XAI not only expedite the drug discovery process but also foster a deeper understanding of pharmacological mechanisms, leading to the development of safer and more effective therapeutics [1].

XAI in drug discovery extends its scope to optimize drug design and repurpose existing compounds for novel therapeutic applications. Interpretable AI models analyze vast repositories of chemical data and biological assays, identifying patterns and correlations that reveal promising drug candidates or potential drug interactions. By explaining the rationale behind predictions, XAI assists researchers in predicting the likelihood of success in repurposing existing drugs for new indications or optimizing chemical structures for improved therapeutic outcomes. This transparency in decision-making supports researchers in making informed choices regarding the prioritization of drug candidates, thereby reducing the time and resources needed for preclinical and clinical trials. Ultimately, the integration of XAI in drug discovery and development revolutionizes the pharmaceutical industry by expediting the identification and optimization of novel therapeutics, offering a more efficient and informed approach to bringing innovative drugs to market.

In the domain of health records analysis and management, XAI serves as a cornerstone in unlocking the potential of vast volumes of healthcare data, enabling efficient extraction of valuable insights and facilitating data-driven decision-making. XAI-driven algorithms, equipped with interpretability features, navigate through extensive electronic health records (EHRs), clinical notes, lab results, and other unstructured healthcare data. These models decipher complex patterns and correlations within the data, elucidating clinically relevant information and enabling healthcare professionals to extract actionable insights. For instance, XAI aids in categorizing and organizing patient data, allowing for more efficient data mining for research purposes and clinical decision support. Moreover, by providing transparent explanations for data processing and decision-making, XAI ensures the integrity and trustworthiness of the information derived from health records, thereby enhancing data reliability and fostering more accurate diagnoses and treatment planning.

Furthermore, XAI in health records management extends its utility to predictive analytics and risk stratification by leveraging historical patient data to anticipate health outcomes and potential disease risks. Interpretable AI models analyze comprehensive patient records coupled with real-time information, predicting the likelihood of specific health events or conditions. This predictive capability enables healthcare providers to proactively identify high-risk patients who might benefit from preventive interventions or targeted monitoring. The transparency offered by XAI in these predictions enables clinicians to understand the rationale behind risk estimations, facilitating informed discussions with patients regarding potential health risks and interventions. Ultimately, the integration of XAI in health records analysis and management streamlines healthcare operations, improves diagnostic accuracy, and empowers clinicians to make more informed decisions, leading to enhanced patient care and outcomes [3].

In healthcare security, XAI emerges as a cornerstone in safeguarding patient data and ensuring privacy in an increasingly digitized healthcare landscape. XAI-powered security measures employ interpretable algorithms that detect and mitigate potential vulnerabilities or threats within healthcare systems. These models scrutinize network activities, access logs, and system behaviors, identifying anomalous patterns that may signify cybersecurity risks or breaches. Additionally, XAI contributes to enhancing patient data security and privacy through advanced encryption techniques and robust access control mechanisms. Interpretable AI models facilitate the identification of potential weak points in data security infrastructure, enabling proactive measures to encrypt sensitive data and strengthen access controls. By explaining the rationale behind security protocols, XAI aids in educating healthcare stakeholders about the importance of security measures and their role in safeguarding patient information. This transparency strengthens cybersecurity measures and instills confidence among patients and healthcare professionals in securely handling sensitive medical data. Ultimately, integrating XAI in healthcare security fortifies data protection measures, ensures patient confidentiality, and contributes to a resilient and trustworthy healthcare ecosystem (refer to Table 15.1).

15.3 Smart transportation: optimizing systems with XAI

In the rapidly evolving transportation landscape, the integration of XAI marks a pivotal juncture, presenting a transformative pathway toward efficiency, safety, and innovation. This section delves into the dynamic realm of smart transportation, where the fusion of cutting-edge AI technologies with transportation systems stands poised to revolutionize how we navigate and optimize transportation networks. The advent of XAI in this domain holds promise in refining operational efficiency and in fostering a safer and more responsive transportation ecosystem. From traffic management and autonomous vehicles to predictive maintenance and logistics optimization, XAI emerges as a potent tool, offering transparent insights into transportation systems, empowering decision-makers, and revolutionizing the

Table 15.1 Exploring XAI attributes across healthcare domains

Applications	Data	Level of opaqueness	Need for XAI	Challenges	XAI techniques	Ethical considerations
Clinical pathology	Medical imaging (X-rays, CT scans, etc.)	High	Explanation of diagnoses	Interpretability of AI models	Model interpretability	Patient data confidentiality
Cardiology	ECG data and patient records	Medium–high	Personalized treatment plans	Complex data integration	Decision trees, LIME	Patient consent
Oncology	Genomic data and imaging	High	Precision medicine	Interpretable genomic analysis	SHAP, attention maps	Genetic privacy
Neurology	MRI, EEG, and patient history	High	Early disease detection	Complex brain data interpretation	Rule-based systems	Sensitive data handling
Telemedicine	Remote patient monitoring data	Medium	Proactive health monitoring	Reliability in remote diagnostics	Predictive models	Telehealth ethics
Chronic disease management	Wearable device data and patient records	Medium–high	Personalized disease management	Long-term data analysis	Time series analysis	Longitudinal data usage
Radiology	Medical imaging (MRI, CT, and ultrasound)	High	Accurate image interpretation	Validation of AI interpretations	Neural networks	Image data security
Pharmacovigilance	Electronic health records, and adverse events	Medium	Drug safety monitoring	Real-time surveillance	Anomaly detection	Pharmacovigilance ethics
Population health management	Demographic and clinical data	Low–medium	Predictive analytics	Ethical data use	Machine learning	Data anonymization
Mental health diagnosis	Patient interviews and behavioral data	Medium–high	Precise diagnosis	Sensitive data handling	Natural language processing	Mental health ethics

Table 15.2 Exploring XAI applications in smart transportation: data sources, opaqueness levels, necessity, and challenges

Applications	Data	Level of opaqueness	Need for XAI	Challenges
Traffic flow optimization	Traffic sensors, GPS, and historical traffic	Low–medium	High	Limited interpretability of traffic patterns
Autonomous vehicles	LiDAR, cameras, radar, and sensor data	High	Critical	Safety verification and human–machine interaction
Predictive maintenance	Sensor data and maintenance logs	Low–medium	High	Integration of multiple data sources and false alarms
Route optimization	GPS and historical route data	Low–medium	High	Real-time adaptation and scalability of algorithms
Public transit management	Passenger data, schedules, and traffic patterns	Medium	High	Reliability and congestion management
Vehicle-to-infrastructure	Vehicle data and infrastructure sensors	Medium	High	Standardization, data privacy, and network security
Fleet management	Telematics, GPS, and vehicle performance data	Medium	High	Integration of legacy systems and cost of implementation
Intelligent traffic lights	Traffic flow, sensors, and historical data	Low–medium	High	Adaptation to dynamic traffic and energy efficiency
Parking management	Parking sensor data and occupancy information	Low–medium	High	Accuracy of occupancy detection and scalability
Demand-responsive transport	Passenger data and traffic patterns	Medium	High	Personalization and efficient routing

way we conceive, manage, and interact with modern transportation infrastructure. Join us in exploring the myriad applications and transformative potential of XAI in reshaping the future of smart transportation.

In Table 15.2, we present an expansive overview of various XAI applications within the realm of smart transportation. Each application is meticulously examined across five critical dimensions: the types of data utilized, the level of opaqueness characterizing the application, the imperative need for XAI integration, and the prevalent challenges encountered. This comprehensive breakdown provides a holistic perspective on how XAI is leveraged across diverse facets of smart transportation, elucidating the significance of interpretability and transparency in optimizing transportation systems and overcoming challenges in the ever-evolving mobility landscape.

Transportation evolution is greatly influenced by traffic flow optimization and management, with XAI playing a crucial part in changing these important areas.

XAI models analyze complex traffic dynamics by integrating multiple data sources, including traffic sensors, GPS, historical traffic patterns, and weather data. These models offer insights into congestion patterns, peak traffic hours, and recurring bottlenecks, providing transportation authorities with interpretable analyses to enhance traffic flow efficiency. By offering transparency in decision-making, XAI enables authorities to comprehend the factors contributing to traffic congestion, empowering them to strategize interventions that alleviate bottlenecks and optimize traffic flow. However, challenges persist due to the complexity of traffic patterns, requiring advanced XAI algorithms capable of interpreting multifaceted data and adapting to dynamic conditions to significantly improve in traffic management. Moreover, XAI in traffic flow optimization addresses the need for real-time adaptability in traffic systems. Interpretable AI models continuously process and analyze incoming data streams, facilitating predictive analytics to foresee traffic disruptions or congestion in advance. This foresight enables authorities to proactively implement dynamic traffic management strategies, such as adaptive signal control or route diversions, mitigating congestion before it escalates. Despite the potential benefits, challenges include the necessity for seamless integration of multiple data sources and the accurate interpretation of rapidly changing traffic patterns in real time. The interpretability of XAI models becomes pivotal in assisting transportation authorities in comprehending the reasoning behind traffic predictions, fostering trust in decision-making, and enhancing the overall efficiency and safety of transportation networks [4].

Autonomous vehicle navigation and safety represent a transformative frontier where XAI stands as a cornerstone, driving innovation and reliability in self-driving systems. By integrating various sensors like LiDAR, cameras, radar, and sophisticated XAI algorithms, autonomous vehicles decipher their surroundings, enabling them to make real-time decisions. XAI models elucidate the decision-making process within these vehicles, offering interpretable insights into how the AI systems perceive and respond to dynamic environments. This transparency is crucial not only for understanding the vehicle's navigation decisions but also for ensuring passenger safety and building trust in autonomous driving technology. However, the complexity of real-world scenarios and the need for XAI to comprehend and interpret diverse and rapidly changing environments pose significant challenges. Safety verification remains a key challenge, demanding robust XAI algorithms capable of justifying decision-making, especially in complex scenarios where human-machine interaction and unexpected events occur.

Furthermore, XAI in autonomous vehicle navigation focuses on the interpretability of AI-driven navigation decisions to ensure safety and enhance human-machine collaboration. These interpretable AI models provide insights into why a specific action was taken by an autonomous vehicle, thereby allowing for better human understanding and potential intervention when necessary. Challenges persist in achieving seamless human-AI interaction, including the development of user-friendly interfaces that convey AI-driven decisions transparently to passengers or remote operators. Moreover, ensuring regulatory compliance and addressing ethical considerations regarding AI decisions in critical situations remain paramount. The need for XAI in autonomous vehicles lies in ensuring safety and in instilling confidence among passengers, regulators, and the public about the reliability and ethical decision-making of autonomous systems in navigating complex and diverse environments.

Predictive maintenance for vehicles and infrastructure harnesses the power of XAI to revolutionize maintenance strategies, ensuring the optimal performance and longevity of vehicles and critical infrastructure. XAI-driven predictive models analyze extensive sensor data, maintenance logs, and historical performance records to detect patterns indicative of potential faults or equipment degradation. These interpretable AI models predict potential breakdowns or maintenance needs and provide transparent explanations for these predictions. By offering insights into the reasoning behind maintenance recommendations, XAI empowers maintenance teams to prioritize and plan interventions effectively, minimizing unplanned downtime and reducing operational costs. However, challenges persist in integrating data from various sources and accurately interpreting complex data patterns to ensure early fault detection and prevent false alarms, thereby optimizing predictive maintenance strategies for vehicles and infrastructure [5].

Moreover, XAI in predictive maintenance fosters a proactive approach, enabling preemptive actions to address maintenance needs before failures occur. These interpretable AI models facilitate understanding the underlying factors contributing to equipment degradation or failure, allowing maintenance teams to make informed decisions on resource allocation and intervention strategies. Challenges lie in scaling these predictive models across diverse equipment types and ensuring their adaptability to evolving operational conditions. Furthermore, the implementation of XAI-driven maintenance strategies requires collaboration between data scientists and maintenance professionals to ensure the seamless integration of AI recommendations into existing maintenance workflows.

XAI plays a transformative role in route optimization and logistics by offering interpretable insights that revolutionize transportation planning and logistics management. By analyzing GPS data, historical route information, traffic patterns, and real-time conditions, XAI models decipher complex logistics networks. These models provide transparent explanations for route recommendations, considering multiple factors like traffic congestion, delivery schedules, and fuel efficiency. This transparency empowers logistics managers to make informed decisions, optimizing delivery routes, and minimizing operational costs. However, challenges persist in managing dynamic real-time data and ensuring the adaptability of XAI algorithms to unforeseen disruptions or changing logistics scenarios, requiring robust models capable of dynamically optimizing routes while maintaining interpretability and efficiency [6].

Furthermore, XAI in route optimization addresses the need for adaptable and efficient transportation networks. These interpretable AI models not only optimize routes but also provide insights into the reasoning behind route selections. This transparency aids logistics managers in understanding how various factors, such as delivery deadlines or vehicle capacities, influence route choices, enabling them to fine-tune logistics strategies. Challenges include the need for real-time adaptation to dynamic conditions, accurate interpretation of diverse data sources, and the scalability of algorithms to manage large-scale logistics operations efficiently.

XAI offers transformative capabilities in public transit management and scheduling, optimizing operations and improving the efficiency of public transportation systems. Interpretable AI models provide insights into public transit operations by analyzing vast amounts of passenger data, traffic patterns, historical schedules, and real-time conditions. These models offer transparent explanations for scheduling decisions, considering factors such as passenger demand, peak travel times, and route efficiencies. This transparency enables transportation authorities to make informed decisions, enhancing service reliability, optimizing routes, and improving the overall passenger experience. However, challenges persist in managing diverse datasets, ensuring timely scheduling adaptations to dynamic conditions, and maintaining reliability while accommodating unexpected disruptions.

XAI in public transit management empowers transportation authorities to address complex challenges in transit systems. These interpretable AI models not only optimize schedules but also provide insights into the rationale behind scheduling decisions. This transparency aids in comprehending how various factors, such as passenger demand patterns or service gaps, influence scheduling strategies, enabling authorities to adapt schedules to better meet commuter needs. Challenges include the need for real-time adjustments to unexpected events, accurate interpretation of complex data sets, and integrating XAI recommendations into existing transit management systems.

XAI plays a crucial role in fortifying the security of connected vehicles and infrastructure, ensuring the resilience and integrity of transportation systems. By analyzing diverse data sources such as vehicle data, infrastructure sensors, and communication networks, interpretable AI models scrutinize potential vulnerabilities and threats within connected transportation ecosystems. The transparency offered by these models aids in the identification of anomalous patterns or cyber threats that could compromise vehicle safety or critical infrastructure. Nonetheless, challenges persist in addressing evolving cybersecurity threats, ensuring seamless integration of security measures across interconnected systems, maintaining robustness against sophisticated attacks, and demanding AI algorithms capable of adaptive defense mechanisms and interpretability in cybersecurity decision-making.

Furthermore, XAI in securing connected vehicles and infrastructure focuses on enhancing the resilience of communication networks and safeguarding against potential intrusions or data breaches. These interpretable AI models provide insights into the reasoning behind security protocols and decisions, facilitating the understanding of how various factors, such as network anomalies or potential vulnerabilities, are addressed in cybersecurity strategies. Challenges include the need for standardized security protocols across interconnected systems, the complexity of securing communication networks while maintaining efficiency, and the continuous monitoring of threats in a rapidly evolving cybersecurity landscape.

15.4 Smart manufacturing: implementing XAI for improved processes

Predictive maintenance for machinery undergoes a profound transformation with the integration of XAI, revolutionizing the approach toward equipment upkeep and operational efficiency. Employing a spectrum of data sources encompassing sensor data, operational logs, and historical performance records, XAI-driven predictive models unearth patterns indicative of potential machinery faults or degradation. These interpretable AI models delve into the decision-making process behind maintenance predictions, offering transparent insights into the reasoning behind impending maintenance recommendations. By illuminating the factors influencing predictive maintenance, XAI empowers maintenance teams to proactively plan and execute interventions effectively, minimizing downtime, optimizing machinery performance, and curbing operational costs. Yet, challenges persist in efficiently assimilating multifaceted data sources, accurately interpreting complex patterns, and ensuring early fault detection to fine-tune predictive maintenance strategies for machinery across diverse industrial settings.

XAI in predictive maintenance for machinery orchestrates a paradigm shift toward preemptive maintenance strategies. The interpretable nature of AI models aids in comprehending the underlying causes of machinery degradation or imminent failures, allowing maintenance teams to strategize resource allocation and plan interventions ahead of critical malfunctions. Challenges encompass scaling XAI models across various machinery types, adapting them to evolving operational environments and integrating them seamlessly into existing maintenance workflows.

XAI emerges as a pivotal tool in quality control and defect detection, redefining precision and reliability in the manufacturing domain. Leveraging diverse data sources such as sensor readings, production logs, and historical defect patterns, interpretable AI models scrutinize manufacturing processes to detect anomalies and identify potential defects in products. These models provide transparent insights into the decision-making process behind defect identifications, offering explanations for why certain anomalies are flagged, thereby assisting quality control teams in understanding and addressing manufacturing flaws. By illuminating the factors influencing defect detection, XAI empowers manufacturing units to take corrective actions promptly, improving product quality, reducing waste, and enhancing overall production efficiency. However, challenges persist in accurately interpreting complex patterns within manufacturing data and ensuring the adaptability of XAI algorithms to diverse production environments and evolving defect types, necessitating robust models capable of dynamic and interpretable defect detection across manufacturing sectors.

Furthermore, XAI in quality control and defect detection fosters a proactive approach by facilitating a deeper understanding of the root causes of defects or discrepancies in manufacturing processes. These interpretable AI models not only detect anomalies but also provide insights into the reasoning behind identified defects, enabling manufacturers to refine production processes and mitigate potential sources of errors. Challenges encompass the need for real-time adaptability to changing production conditions, the integration of XAI-driven insights into existing quality control systems, and ensuring interpretability while handling large volumes of manufacturing data.

By revolutionizing operational efficiency and resource usage in industrial ecosystems, XAI is playing a key role in production optimization and resource allocation. Interpretable AI models analyze intricate production processes by integrating data from a variety of sources, including equipment performance, production records, and supply chain dynamics. These models provide clear and comprehensive insights into decision-making, offering explanations for the most effective allocation of resources and methods for production. This level of openness facilitates the comprehension of resource allocation recommendations by production managers, consequently allowing for the optimization of production schedules, inventory management, and manpower allocation. Nevertheless, there are ongoing difficulties in accurately deciphering complex data patterns, guaranteeing flexibility in ever-changing production environments, and enabling the smooth incorporation of XAI recommendations into current production workflows. This necessitates the use of agile and interpretable models that can dynamically optimize production across various manufacturing settings.

Moreover, XAI in production optimization focuses on enhancing resource utilization and production efficiency by providing insights into the reasoning behind resource allocation decisions. These interpretable AI models not only optimize production schedules but also assist in comprehending how various factors, such as machine downtime or fluctuating demand, influence resource allocation strategies. Challenges encompass the need for real-time adaptation to changing production conditions, accurate interpretation of diverse data sources, and ensuring interpretability while handling large volumes of production data.

XAI serves as a linchpin in supply chain management and inventory forecasting, reshaping the dynamics of inventory management and optimizing the efficiency of supply chains. Leveraging an array of data sources including historical sales records, market trends, supplier data, and logistics information, interpretable AI models analyze the intricacies of supply chains. These models elucidate the reasoning behind inventory forecasts, demand predictions, and optimal inventory levels. This transparency empowers supply chain managers to comprehend the factors driving inventory fluctuations, enabling proactive adjustments in procurement, inventory storage, and distribution strategies. Nonetheless, challenges persist in accurately interpreting complex data patterns, ensuring adaptability to evolving market dynamics, and integrating XAI-driven recommendations into existing supply chain systems, necessitating flexible and interpretable models capable of dynamic supply chain optimization across diverse industries and markets.

Furthermore, XAI in supply chain management focuses on enhancing inventory forecasting accuracy and supply chain resilience by providing insights into the rationale behind inventory allocation decisions. These interpretable AI models not only forecast demand but also assist in understanding how factors such as seasonality, market shifts, or disruptions impact inventory requirements. Challenges encompass the need for real-time adaptation to changing market conditions, accurate interpretation of diverse data sources, and ensuring interpretability while handling large volumes of supply chain data [7].

XAI is a game-changer when it comes to worker safety monitoring and accident prevention. By using interpretable AI models, they improve safety protocols and reduce risks, completely changing the way workplace safety is seen. By amalgamating data from various sources like sensors, environmental readings, and historical safety records, these models analyze patterns to identify potential hazards or risky situations within the workplace. The interpretability of these AI models provides transparent insights into the decision-making process behind safety monitoring, offering explanations for the identification of potential risks and hazardous conditions. This transparency assists safety supervisors in understanding the rationale behind safety alerts, enabling proactive interventions to prevent accidents and ensure a safer work environment. Nonetheless, challenges persist in ensuring real-time adaptability of safety protocols, accurately interpreting complex data patterns, and integrating XAI-driven insights into existing safety management systems while maintaining compliance with safety regulations across diverse work environments.

Moreover, XAI in worker safety monitoring and accident prevention focuses on providing insights into the underlying causes of potential accidents or safety hazards. These interpretable AI models not only detect risks but also assist in understanding how various factors, such as environmental conditions or equipment malfunctions, influence safety risks within workplaces. Challenges encompass the need for interpretability in safety alerts and interventions, accurate analysis of diverse safety data sources, and the seamless integration of XAI-driven insights into safety management practices. Ultimately, the integration of XAI in worker safety monitoring and accident prevention minimizes workplace hazards, enhances safety protocols, and fosters a culture of proactive risk mitigation, ensuring the well-being and safety of workers across various occupational settings.

XAI represents a cornerstone in fortifying cybersecurity within smart manufacturing systems, empowering organizations to safeguard against cyber threats in an era of interconnected industrial systems. By scrutinizing an array of data sources such as network logs, system activities, and historical cyber incidents, interpretable AI models analyze potential vulnerabilities and anomalous patterns within manufacturing networks. These models provide transparent insights into the decision-making process behind cybersecurity alerts and threat detections, offering explanations for the identification of potential cyber risks. This transparency assists cybersecurity experts and system administrators in comprehending the reasoning behind security alerts, enabling proactive measures to fortify network defenses, detect intrusions, and mitigate potential cyberattacks. Yet, challenges persist in accurately interpreting intricate data patterns, ensuring adaptability to evolving cyber threats, and integrating XAI-driven cybersecurity solutions into existing manufacturing systems while ensuring minimal disruptions to operations.

Furthermore, XAI in cybersecurity for smart manufacturing systems focuses on enhancing the resilience of industrial networks and protecting against potential cyber intrusions or data breaches. These interpretable AI models not only detect cyber threats but also provide insights into the rationale behind identified vulnerabilities and security breaches. Challenges encompass the need for real-time adaptability to emerging cyber threats, the accuracy of diverse data interpretation, and ensuring interpretability while handling large volumes of cybersecurity data. Ultimately, the integration of XAI in cybersecurity strategies for smart manufacturing systems fortifies network security, enhances cyber threat detection capabilities, and ensures the integrity and confidentiality of critical industrial data, safeguarding against potential disruptions in manufacturing operations and preserving the reliability of smart manufacturing infrastructures.

15.5 Smart agriculture: XAI for precision farming

Amid the burgeoning challenges faced by modern agriculture, the fusion of XAI with precision farming techniques stands poised to revolutionize the agricultural landscape. This section delves into the realm of smart agriculture, where the integration of advanced AI technologies with agricultural practices promises to optimize productivity, sustainability, and resource utilization. XAI in precision farming serves as a beacon of innovation, offering transparent and interpretable insights into crop management, livestock monitoring, and sustainable agricultural practices. From leveraging data-driven insights to enhance crop yields and mitigate environmental impact to revolutionizing farming methods, XAI represents a transformative force in reshaping the future of agriculture. Join us in exploring the multifaceted applications and promising advancements of XAI within smart agriculture, where transparency, precision, and sustainability converge to address the evolving challenges faced by the agricultural sector.

The comprehensive insights presented in Table 15.3 underscore the diverse applications of XAI within smart agriculture. This table encapsulates a comprehensive overview of 10 key areas where XAI intersects with agricultural practices, shedding light on the specific types of data utilized, the varying levels of opaqueness inherent in each application, the critical necessity for XAI integration, and the primary challenges faced within the realm of smart agriculture.

XAI is revolutionizing traditional agriculture methods by providing improved insights into crop health and early disease detection. Interpretable AI models monitor agricultural conditions and discover anomalies suggestive of probable diseases or stress by integrating diverse data sources such as satellite images, sensor data, and historical records. These models provide transparent insights into decision-making processes, elucidating the reasoning behind disease detection and flagging specific areas of concern. This transparency assists farmers in understanding the factors influencing crop health, enabling timely interventions to mitigate disease outbreaks and optimize agricultural practices. However, challenges persist in accurately interpreting complex crop patterns, ensuring adaptability to diverse environmental conditions, and integrating XAI-driven insights into existing agricultural workflows, necessitating robust models capable of dynamic and interpretable crop monitoring across different crop types and environments.

Table 15.3 Overview of XAI applications in smart agriculture: data sources, opaqueness levels, necessity, and challenges

Applications	**Data**	**Level of opaqueness**	**Need for XAI**	**Challenges**
Crop yield prediction	Weather data and soil information	Low–medium	High	Data quality and weather variability
Pest and disease detection	Imagery and sensor data	Low–medium	High	Accuracy and diverse pest identification
Precision irrigation	Soil moisture and weather forecasts	Low–medium	High	Real-time adaptation and resource optimization
Livestock monitoring	Sensor data and health records	Low–medium	High	Data integration and animal behavior understanding
Automated harvesting	Robotics and image analysis	High	Critical	Adaptability to crop variability
Soil health assessment	Soil composition and sensor data	Low–medium	High	Interpretation of soil data and soil heterogeneity
Crop management and planning	Field data and historical records	Low–medium	High	Integration of diverse data sources
Supply chain optimization	Logistics data and market trends	Low–medium	High	Data interoperability and real-time adjustments
Autonomous farm equipment	Robotics and sensor data	High	Critical	Safety verification and scalability of AI-driven machinery
Climate adaptation strategies	Climate models and historical data	Low–medium	High	Adaptability to changing climate conditions

Moreover, XAI in crop monitoring and disease detection focuses on enhancing precision agriculture by providing insights into the underlying causes of potential crop diseases or stress factors. These interpretable AI models not only detect anomalies but also assist in understanding how various factors, such as weather fluctuations or pest infestations, impact crop conditions. Challenges encompass the need for real-time adaptability to changing environmental conditions, accurate interpretation of diverse data sources, and ensuring interpretability while handling large volumes of crop health data [8].

XAI stands as a critical enabler in precision irrigation and water management, revolutionizing agricultural water usage and optimizing resource allocation. By assimilating data from soil moisture sensors, weather forecasts, and crop water requirements, interpretable AI models analyze intricate patterns to optimize irrigation schedules. These models provide transparent insights into decision-making processes, offering explanations for the timing and volume of water application in specific agricultural areas. This transparency empowers farmers to understand the factors influencing water needs, enabling precise and efficient water allocation while minimizing wastage. However, challenges persist in accurately interpreting diverse data sources, ensuring real-time adaptability to changing weather conditions, and integrating XAI-driven insights into irrigation systems, demanding adaptable and interpretable models capable of dynamic precision irrigation across varying soil types and crop varieties.

Moreover, XAI in precision irrigation and water management focuses on enhancing sustainability by providing insights into the rationale behind water allocation decisions. These interpretable AI models not only optimize irrigation but also assist in understanding how factors like soil types or crop growth stages influence water requirements. Table 15.4 serves as a comprehensive compendium showcasing the diverse applications where XAI intersects with the water industry, elucidating the types of data employed, the varying degrees of opaqueness associated with each application, the critical necessity for XAI integration, and the primary challenges inherent in implementing XAI-driven solutions.

Table 15.4 Overview of XAI applications in the water management

Applications	**Data**	**Level of opaqueness**	**Need for XAI**	**Challenges**
Water quality assessment	Sensor data and historical water records	Low–medium	Critical	Data accuracy and real-time monitoring
Leak detection	Flow data and pressure readings	Low–medium	High	False alarms and accuracy in identifying leaks
Predictive maintenance	Equipment sensor data and maintenance logs	Low–medium	High	Data integration and predicting failure patterns
Water distribution optimization	Flow data and network topology	Medium	Critical	Complex networks and real-time adaptability
Flood prediction and management	Weather data and river levels	Medium	High	Accuracy in prediction and rapid response
Water usage monitoring	Meter data and consumption patterns	Low–medium	High	Data privacy and interpreting consumption patterns
Reservoir management	Reservoir levels and weather forecasts	Low–medium	Critical	Forecasting accuracy and optimizing water release
Water conservation	Usage patterns and conservation initiatives	Low–medium	High	Behavior change and optimizing conservation strategies
Energy optimization in water treatment	Treatment plant data and energy consumption	Low–medium	Critical	Efficiency improvement and optimizing energy usage
Regulatory compliance monitoring	Compliance data and regulatory standards	Low–medium	High	Data accuracy and compliance with evolving regulations

XAI plays a transformative role in weather forecasting and crop yield prediction, offering unparalleled insights into agricultural production by leveraging diverse data sources and intricate analyses. By assimilating historical weather patterns, soil conditions, satellite imagery, and crop-specific data, interpretable AI models scrutinize intricate relationships between environmental factors and crop yields. These models elucidate the reasoning behind crop yield predictions based on weather forecasts and soil conditions. This transparency assists farmers and agricultural stakeholders in understanding the variables influencing crop productivity, aiding in informed decision-making regarding planting strategies, crop selection, and resource allocation for maximizing yields. However, challenges persist in accurately interpreting complex data interactions, ensuring adaptability to diverse climatic conditions, and integrating XAI-driven insights into existing agricultural systems, necessitating robust and interpretable models capable of dynamic crop yield prediction across varying geographies and climatic zones.

Furthermore, XAI in weather forecasting and crop yield prediction focuses on enhancing precision agriculture by providing insights into the underlying factors affecting crop performance. These interpretable AI models not only predict crop yields but also assist in understanding how various environmental factors, such as rainfall patterns or temperature fluctuations, impact crop growth. Challenges encompass the need for real-time adaptability to changing weather conditions, accurate interpretation of diverse data sources, and ensuring interpretability while handling vast volumes of weather and agricultural data.

XAI serves as a pivotal tool in livestock monitoring and management, revolutionizing traditional practices by offering comprehensive insights into animal health, behavior, and welfare. Through the integration of diverse data sources such as sensor readings, health records, and environmental conditions, interpretable AI models analyze patterns to monitor livestock health and behavior. These models provide transparent insights into decision-making processes, elucidating the reasoning behind livestock management strategies and health assessments. This transparency aids farmers in understanding the factors influencing livestock well-being, enabling proactive interventions for health issues, optimizing feeding schedules, and ensuring optimal management practices. However, challenges persist in accurately interpreting complex animal behavior patterns, ensuring adaptability to diverse livestock species, and integrating XAI-driven insights into existing livestock management systems, demanding robust models capable of dynamic and interpretable livestock monitoring across various animal husbandry environments.

Moreover, XAI in livestock monitoring and management focuses on enhancing precision husbandry by providing insights into the underlying factors affecting animal health and productivity. These interpretable AI models not only detect health anomalies but also assist in understanding how various factors, such as environmental conditions or feeding regimes, impact animal welfare and production. Challenges encompass the need for real-time adaptability to changing animal behavior, accurate interpretation of diverse data sources, and ensuring interpretability while handling vast volumes of livestock health and management data [9].

XAI holds significant promise in soil health monitoring and fertilizer optimization, offering profound insights into soil conditions and facilitating precise resource management in agriculture. Through the analysis of diverse data sources such as soil composition data, moisture levels, nutrient content, and historical crop yields, interpretable AI models scrutinize soil health parameters and identify optimal fertilizer requirements. These models provide transparent insights into decision-making processes, elucidating the reasoning behind soil health assessments and fertilizer recommendations. This transparency aids farmers in comprehending the factors influencing soil quality, enabling targeted interventions to improve soil health, optimize fertilizer application, and enhance crop productivity sustainably. However, challenges persist in accurately interpreting complex soil data, ensuring adaptability to diverse soil types and conditions, and integrating XAI-driven insights into existing agricultural practices, necessitating adaptable and interpretable models capable of dynamic soil health monitoring and fertilizer optimization across varying terrains and crop varieties.

Furthermore, XAI in soil health monitoring and fertilizer optimization focuses on enhancing precision agriculture by providing insights into the underlying factors affecting soil nutrient levels and crop growth. These interpretable AI models not only optimize fertilizer application but also assist in understanding how various factors, such as soil composition or climatic conditions, impact soil fertility and plant nutrition. Challenges encompass the need for real-time adaptability to changing soil conditions, accurate interpretation of diverse soil data sources, and ensuring interpretability while handling large volumes of soil health and agricultural data.

XAI stands as a pivotal guardian in securing farm data and IoT devices, ensuring the resilience and integrity of agricultural systems against potential cyber threats. By amalgamating data from diverse sources such as sensor networks, farm management software, and cloud-based platforms, interpretable AI models scrutinize potential vulnerabilities and anomalous patterns within agricultural networks. These models expose the rationale behind security protocols and identify potential risks or cyber threats. This transparency assists farmers and agricultural technologists in comprehending the factors influencing data security, enabling proactive measures to fortify network defenses, detect intrusions, and mitigate potential cyberattacks. However, challenges persist in accurately interpreting intricate data patterns, ensuring real-time adaptability to evolving cyber threats, and integrating XAI-driven insights into existing agricultural systems while ensuring minimal disruptions to farming operations. Moreover, XAI in securing farm data and IoT devices focuses on enhancing cybersecurity resilience by providing insights into the underlying causes of potential security breaches or system vulnerabilities. These interpretable AI models not only detect threats but also assist in understanding how various factors, such as network configurations or device interactions, impact overall system security. Challenges encompass the need for interpretability in cybersecurity alerts and interventions, accurate analysis of diverse data sources, and the seamless integration of XAI-driven insights into existing agricultural infrastructures.

15.6 Smart grid: enhancing efficiency through XAI

In today's rapidly evolving energy landscape, the integration of XAI within smart grid systems stands as a pivotal juncture in advancing the efficiency, reliability, and sustainability of power distribution networks. This section delves into the realm of smart grid technologies, where the fusion of advanced AI methodologies with power grid infrastructure promises to optimize energy utilization and grid management and enhance overall system performance. XAI in smart grids represents a groundbreaking approach, offering transparent and interpretable insights into power distribution, demand forecasting, and grid stability.

In the landscape of modern energy systems, the convergence of XAI with smart grid technologies stands as a beacon of innovation, poised to revolutionize the efficiency, adaptability, and resilience of power networks. This section delves into the dynamic domain of smart grids, illuminating how the integration of sophisticated AI methodologies within grid infrastructures promises to optimize energy distribution, enhance grid stability, and transform the energy sector's operational landscape. XAI in smart grids epitomizes a groundbreaking paradigm, providing transparent and interpretable insights into power management, demand forecasting, and system optimization. From harnessing data-driven precision to address grid complexities, enhance load balancing, and mitigate power disruptions, XAI emerges as a transformative catalyst reshaping the trajectory of energy distribution. Join us in exploring the multifaceted applications and the promising potential of XAI within smart grid systems, where transparency, reliability, and sustainability converge to meet the evolving demands and complexities of the modern energy ecosystem [10].

Table 15.5 provides a comprehensive summary of the diverse applications where XAI intersects with smart grid technologies. It delineates the crucial roles of XAI across 10 key areas within smart grids, elucidating the types of data employed, the varying degrees of opaqueness associated with each application, the essential need for integrating XAI, and the primary challenges encountered in implementing XAI-driven solutions.

XAI revolutionizes energy demand prediction and load balancing within smart grids by leveraging intricate data analytics to forecast energy consumption patterns and optimize grid performance. Through the analysis of historical consumption data, weather forecasts, and evolving consumer behavior, interpretable AI models decode intricate relationships between various factors influencing energy demand. These models shed light on the reasoning behind demand predictions and load-balancing strategies. Such transparency assists grid operators and utility providers in understanding the variables impacting energy demand, facilitating proactive adjustments in energy distribution, and optimizing load-balancing strategies for more efficient grid operations. Challenges remain in accurately interpreting fluctuating consumer behaviors, ensuring real-time adaptability to changing demand dynamics, and integrating XAI-driven insights seamlessly into existing grid infrastructure, demanding adaptable and interpretable models capable of dynamic load balancing across varying consumption patterns and grid conditions.

Table 15.5 Overview of XAI applications in smart grids: data sources, opaqueness levels, necessity, and challenges

Applications	**Data**	**Level of opaqueness**	**Need for XAI**	**Challenges**
Demand forecasting	Historical consumption and weather data	Low–medium	High	Accuracy and variability in consumer behavior
Fault detection and diagnosis	Sensor data and grid performance metrics	Medium	Critical	Data quality and diverse fault patterns
Energy theft detection	Consumption patterns and anomaly detection	Low–medium	High	Privacy concerns and detection accuracy
Grid optimization and control	Real-time grid data and load balancing	Medium	Critical	Scalability and real-time adaptability
Renewable energy integration	Solar/wind generation data and weather forecasts	Low–medium	High	Integration complexities and variability in renewable sources
Predictive maintenance	Equipment sensor data and historical records	Medium	Critical	Data integration and real-time monitoring
Grid stability monitoring	Voltage and frequency measurements	Low–medium	High	Sensitivity to grid fluctuations and interpretability
Outage prediction and management	Historical outage data and weather forecasts	Medium	Critical	Data accuracy and prediction timelines
Grid resilience improvement	Grid stress tests and risk analysis	Low–medium	High	Scenario variability and adaptability to unforeseen events
Smart meter analytics	Metering data and consumer behavior analysis	Low–medium	High	Data privacy and analyzing large-scale metering information

Moreover, XAI in energy demand prediction and load balancing aims to enhance grid efficiency by providing insights into the underlying factors influencing energy consumption and grid stability. These interpretable AI models not only forecast demand but also assist in comprehending how various elements, such as weather variations or emerging consumption trends, impact energy needs. Challenges encompass the need for real-time adaptability to fluctuating demand, accurate interpretation of diverse data sources, and ensuring interpretability while handling large volumes of energy consumption and grid performance data [11].

XAI is pivotal in transforming energy demand prediction and load balancing within smart grids by harnessing sophisticated analytics to anticipate energy consumption patterns and optimize grid efficiency. By analyzing a myriad of data sources including historical consumption records, weather patterns, and consumer behavior, interpretable AI models unveil intricate relationships among various factors influencing energy demand. These models provide transparent insights into the decision-making process, shedding light on the rationale behind demand forecasts and load-balancing strategies. This transparency assists grid operators and utility companies in comprehending the variables shaping energy consumption, enabling proactive adjustments in energy distribution and refined load-balancing strategies for heightened grid operational efficiency. Challenges persist in accurately interpreting dynamic consumer behaviors, ensuring real-time adaptability to fluctuating demand patterns, and integrating XAI-driven insights seamlessly into existing grid infrastructure, necessitating versatile and interpretable models capable of dynamic load balancing across diverse consumption scenarios and grid conditions.

Moreover, XAI in energy demand prediction and load balancing aims to optimize grid stability by delving into the underlying factors influencing energy consumption and grid performance. These interpretable AI models not only predict demand but also facilitate a deeper understanding of how various elements, such as weather fluctuations or emerging consumption trends, impact energy needs. Challenges include the need for real-time adaptability to changing demand, precise interpretation of diverse data sources, and ensuring interpretability while handling vast amounts of energy consumption and grid performance data.

XAI is a transformative force in renewable energy integration and optimization within smart grids, enabling the seamless assimilation of renewable energy sources and enhancing their operational efficiency. Through the analysis of solar and wind generation data, weather forecasts, and grid dynamics, interpretable AI models decipher intricate patterns crucial for integrating renewable energy sources into the grid. These models offer transparent insights into decision-making processes, elucidating the rationale behind optimal renewable energy utilization and grid integration strategies. This transparency assists grid operators in comprehending the multifaceted factors impacting renewable energy integration, facilitating proactive adjustments in energy distribution, and enabling effective optimization strategies to maximize the benefits of renewable sources. Challenges persist in accurately interpreting variable energy generation patterns, ensuring adaptability to diverse renewable sources, and integrating XAI-driven insights into existing grid infrastructure, necessitating adaptable and interpretable models capable of dynamic renewable energy integration across varying environmental conditions and grid configurations.

Moreover, XAI in renewable energy integration and optimization aims to bolster grid sustainability by providing insights into the underlying factors shaping the integration and efficient utilization of renewable resources. These interpretable AI models not only optimize renewable energy usage but also assist in understanding how various elements, such as weather variability or intermittency of renewable sources, impact grid stability and energy distribution. Challenges encompass the need for real-time adaptability to changing renewable energy outputs, precise interpretation of diverse data sources, and ensuring interpretability while handling vast volumes of renewable energy and grid performance data.

XAI plays a pivotal role in grid fault detection and self-healing systems within smart grids by leveraging advanced analytics to detect anomalies and enable autonomous responses to grid disturbances. Through the analysis of sensor data, grid performance metrics, and historical fault patterns, interpretable AI models identify irregularities indicative of potential faults or grid disturbances. These models provide transparent insights into decision-making processes, elucidating the rationale behind fault detection, and the activation of self-healing mechanisms. This transparency aids grid operators in comprehending the factors influencing grid stability, enabling prompt responses to faults and facilitating the implementation of self-healing strategies to restore normal operations swiftly. Challenges persist in accurately interpreting diverse fault patterns, ensuring real-time adaptability to evolving grid conditions, and integrating XAI-driven insights into existing grid infrastructure, requiring robust models capable of dynamic fault detection and autonomous system healing across varying grid scenarios and disturbances.

Moreover, XAI in grid fault detection and self-healing systems aims to enhance grid resilience by providing insights into the underlying causes of potential grid disruptions and facilitating autonomous responses to mitigate the impact of faults. These interpretable AI models not only detect faults but also assist in understanding how various elements, such as grid topology or equipment failures, influence grid stability and the efficacy of self-healing mechanisms. Challenges encompass the need for real-time adaptability to diverse fault scenarios, accurate interpretation of complex grid data, and ensuring interpretability while handling large volumes of grid performance and fault-related data.

XAI assumes a crucial role in energy theft detection and prevention within smart grids, employing sophisticated analytics to uncover irregular consumption patterns and mitigate unauthorized energy usage. Through the analysis of consumption data and anomaly detection techniques, interpretable AI models identify deviations indicative of potential energy theft or unauthorized usage. These models shed light on the rationale behind identifying suspicious energy consumption patterns and initiating preventive measures. This transparency assists utility providers and grid operators in understanding the variables influencing energy usage irregularities, enabling prompt interventions to prevent losses and ensure fair and efficient energy distribution. Challenges persist in accurately discerning subtle anomalies, ensuring real-time adaptability to changing consumption patterns, and integrating XAI-driven insights into existing grid infrastructure, requiring adaptable and interpretable models capable of dynamic energy theft detection across diverse consumer behaviors and grid conditions.

Furthermore, XAI in energy theft detection and prevention aims to safeguard grid integrity by providing insights into the underlying causes of potential energy theft and unauthorized usage. These interpretable AI models not only detect anomalies but also assist in comprehending how various factors, such as consumer behavior or metering inaccuracies, impact irregular energy consumption. Challenges encompass the need for real-time adaptability to evolving theft methods, precise interpretation of diverse data sources, and ensuring interpretability while handling vast amounts of consumption data [12].

XAI plays a crucial role in enhancing grid stability and resilience within smart grids, employing sophisticated algorithms to analyze grid conditions and fortify its capacity to withstand disruptions. Through the analysis of voltage, frequency measurements, and risk assessments, interpretable AI models identify potential stressors that may compromise grid stability. These models reveal the rationale behind grid stability assessments and strategies to reinforce its resilience. This transparency aids grid operators in understanding the factors influencing grid stability, enabling proactive measures to enhance its resilience against potential disturbances and ensuring the continuity of power supply. Challenges persist in accurately interpreting fluctuations in grid conditions, ensuring real-time adaptability to varying stressors, and integrating XAI-driven insights into existing grid infrastructure, necessitating adaptable and interpretable models capable of dynamic stability assessment and resilience enhancement across diverse grid scenarios and disruptions.

Moreover, XAI in grid stability and resilience enhancement aims to fortify grid infrastructure by providing insights into the underlying causes of potential disturbances and vulnerabilities. These interpretable AI models not only predict stressors but also assist in understanding how various factors, such as grid configuration or external disturbances, impact grid resilience. Challenges encompass the need for real-time adaptability to unforeseen disruptions, precise interpretation of diverse grid data sources, and ensuring interpretability while handling extensive volumes of grid performance data. Ultimately, the integration of XAI in grid stability and resilience enhancement fortifies grid reliability, ensures continuity of service, and contributes to the development of a more robust and adaptive energy infrastructure, crucial for meeting evolving energy demands and ensuring grid stability in the face of adversities.

XAI emerges as a critical ally in combating cyber threats and fortifying grid protection within smart grids, employing sophisticated analytics to detect and prevent potential cyber intrusions. Through the analysis of network data, system vulnerabilities, and anomalous patterns, interpretable AI models discern potential cyber threats or intrusions targeting the grid infrastructure. These models provide a clear and comprehensive understanding of decision-making processes, shedding light on the logic underlying threat identification and taking proactive actions to strengthen grid cybersecurity. The openness provided helps grid operators and cybersecurity professionals understand the factors that contribute to cyber threats. This allows for quick responses to suspected breaches and the installation of preventive measures to protect key grid assets. There are ongoing difficulties in accurately identifying complex cyber threats, ensuring the ability to quickly adapt to evolving attack methods, and seamlessly integrating insights from XAI into existing grid cybersecurity frameworks. This requires strong and interpretable models that can dynamically detect threats across different cybersecurity scenarios and intrusions [13].

Furthermore, XAI in cyber threats and grid protection aims to fortify grid resilience by providing insights into the underlying causes of potential cyber vulnerabilities and cyberattack strategies. These interpretable AI models not only detect threats but also assist in understanding how various elements, such as system configurations or emerging attack vectors, impact grid cybersecurity. Challenges encompass the need for real-time adaptability to emerging cyber threats, accurate interpretation of diverse cybersecurity data sources, and ensuring interpretability while handling extensive volumes of network and security-related data.

15.7 Smart retail: personalization and efficiency with XAI

In the contemporary retail landscape, the integration of XAI is a transformative force, heralding a new era of personalized shopping experiences and operational efficiency within the industry. This section delves into smart retail, where the fusion of advanced AI methodologies with retail operations reshapes how consumers engage with products and services while optimizing business processes. XAI in smart retail represents an innovative approach, offering transparent and interpretable insights into consumer behaviors, preferences, and operational dynamics. From enhancing personalized recommendations and optimizing inventory management to streamlining supply chains and improving customer service, utilizing XAI facilitates retailers in making data-driven decisions while unveiling the reasoning behind these decisions. Join us in exploring the multifaceted applications and the promising advancements of XAI within smart retail, where transparency, personalization, and operational efficiency converge to address the ever-evolving landscape of retail commerce.

In Table 15.6, we encapsulate a comprehensive overview of the multifaceted applications where XAI intersects with smart retail, delineating the types of data employed, the varying degrees of opaqueness associated with each application, the critical necessity for XAI integration, and the primary challenges inherent in implementing XAI-driven solutions. This table serves as a valuable reference, shedding light on the intricate requirements and obstacles involved in utilizing XAI across diverse facets of the retail industry.

Applying complex data analytics, XAI is crucial in smart retail customer behavior analysis and predictive analytics in understanding customer behavior patterns and predicting future shopping trends. Interpretable AI models reveal insights about customers' interests, preferences, and purchasing habits by analyzing data from multiple sources, including demographics, browsing routines, and purchase histories. These models explain the rationale behind predictive analytics and behavior analysis strategies. This transparency assists retailers in understanding the multifaceted factors influencing consumer behaviors, enabling proactive responses to market shifts and facilitating the development of targeted marketing campaigns. Challenges persist in accurately interpreting diverse consumer behaviors, ensuring real-time adaptability to evolving preferences, and integrating XAI-driven insights into retail strategies, necessitating robust and interpretable models capable of dynamic behavior analysis across varying consumer segments and market dynamics [14].

Moreover, XAI in customer behavior analysis and predictive analytics aims to enhance retailers' understanding of their clientele by providing insights into the underlying factors shaping consumer preferences and future shopping trends. These

Table 15.6 Overview of XAI applications in smart retail: data utilization, opaqueness levels, necessity, and challenges

Applications	Data	Level of opaqueness	Need for XAI	Challenges
Personalized recommendations	Customer purchase history and preferences	Low–medium	High	Data privacy and accuracy in recommendation algorithms
Inventory management	Sales records and supply chain data	Medium	Critical	Data integration and real-time inventory tracking
Customer segmentation	Demographic data and purchase behavior	Low–medium	High	Data accuracy and interpreting diverse customer profiles
Price optimization	Sales data and market trends	Low–medium	High	Dynamic pricing models and understanding market shifts
Demand forecasting	Historical sales and seasonality	Medium	Critical	Accuracy in prediction and adapting to changing trends
Supply chain optimization	Logistics data and supplier information	Medium	High	Complex supply networks and real-time adaptability
Fraud detection and prevention	Transaction data and behavioral patterns	Medium	Critical	Identifying anomalies and real-time fraud prevention
Customer service enhancement	Customer interactions and feedback	Low–medium	High	Natural language understanding and personalized service
Store layout optimization	Customer movement and heatmaps	Low–medium	High	Privacy concerns and optimizing customer experience
Shelf stocking and product placement	Shelf sensors and sales data	Low–medium	High	Real-time stocking and consumer behavior analysis

interpretable AI models predict behaviors and understand how various elements, such as changing market dynamics or emerging consumer preferences, influence retail strategies. Challenges encompass the need for real-time adaptability to evolving consumer behaviors, accurate interpretation of diverse data sources, and ensuring interpretability while handling vast amounts of customer data.

XAI serves as a powerful tool in driving personalized recommendations and marketing strategies within smart retail, leveraging advanced analytics to comprehend individual consumer preferences and offer tailored experiences. Through the analysis of customer purchase history, browsing behavior, and demographic details, interpretable AI models decode intricate patterns that shape individual preferences.

These models provide transparent insights into decision-making processes, elucidating the reasoning behind personalized recommendations and targeted marketing approaches. Such transparency aids retailers in understanding diverse consumer preferences, facilitating the delivery of tailored recommendations and marketing campaigns that align with individual needs. Challenges persist in accurately interpreting nuanced consumer behaviors, ensuring real-time adaptability to evolving preferences, and integrating XAI-driven insights into retail strategies, necessitating robust and interpretable models capable of dynamic personalization across various customer segments and market contexts.

Moreover, XAI in personalized recommendations and marketing endeavors to enhance customer engagement by offering insights into the underlying factors influencing consumer choices and preferences. These interpretable AI models not only predict individual preferences but also assist in understanding how various elements, such as seasonal trends or changing buying patterns, influence marketing strategies.

XAI is a cornerstone in revolutionizing inventory management and demand forecasting within smart retail by harnessing intricate data analysis to optimize inventory levels and anticipate consumer needs. By analyzing of historical sales data, market trends, and seasonal patterns, interpretable AI models decipher complex demand patterns and fluctuations. These help in interpreting the rationale behind inventory optimization and demand forecasting strategies. This transparency aids retailers in comprehending the multifaceted factors influencing consumer demand, enabling proactive adjustments in inventory levels and precise forecasts of product demand. Challenges persist in accurately interpreting diverse consumer behaviors, ensuring real-time adaptability to fluctuating demand, and integrating XAI-driven insights into inventory management systems, necessitating robust models capable of dynamic inventory optimization across various product categories and market conditions.

Moreover, XAI in inventory management and demand forecasting aims to enhance retailers' understanding of market dynamics by providing insights into the underlying factors shaping consumer preferences and purchasing trends. These interpretable AI models not only forecast demand but also assist in understanding how various elements, such as promotional activities or external market influences, impact inventory needs. Challenges encompass the need for real-time adaptability to evolving demand patterns, accurate interpretation of diverse data sources, and ensuring interpretability while handling vast amounts of sales and market data [15].

XAI is essential to smart retail's dynamic price and promotions optimization, which uses advanced analytics to improve promotional campaigns and make real-time adjustments to price tactics. Analyzing competition pricing strategies, consumer behaviors, and sales data from the past, interpretable AI models can uncover complex patterns that impact promotional tactics and pricing decisions. These models provide transparent insights into decision-making processes, highlighting the rationale behind dynamic pricing and targeted promotions. This transparency aids retailers in understanding market dynamics, enabling the implementation of dynamic pricing strategies that align with demand fluctuations and adjusting promotional campaigns to maximize their impact. Challenges persist in accurately interpreting diverse consumer behaviors, ensuring real-time adaptability to market shifts, and integrating XAI-driven insights into pricing and promotional strategies, necessitating adaptable and interpretable models capable of dynamic optimization across varying market contexts.

Moreover, XAI in dynamic pricing and promotions optimization aims to enhance retailers' strategies by providing insights into the underlying factors, that influence consumer response to pricing changes and promotional efforts. These interpretable AI models not only optimize pricing but also assist in understanding how various elements, such as seasonal trends or competitive market moves, impact pricing strategies and promotional effectiveness. Challenges encompass the need for real-time adaptability to evolving market dynamics, accurate interpretation of diverse data sources, and ensuring interpretability while handling extensive volumes of pricing and promotional data.

XAI is essential in optimizing supply chain and managing vendors in the context of smart retail. It utilizes advanced analytics to streamline operations and improve efficiencies throughout the supply chain. Interpretable AI models utilize logistics data, inventory levels, and supplier performance analysis to uncover complex patterns crucial for managing the supply chain. Challenges persist in accurately interpreting diverse supply chain behaviors, ensuring real-time adaptability to changes, and integrating XAI-driven insights into supply chain systems, necessitating robust and interpretable models capable of dynamic optimization across varying supply chain scenarios and market conditions.

Moreover, XAI in supply chain optimization and vendor management aims to enhance retailers' understanding of vendor relationships and supply chain inefficiencies by providing insights into the underlying factors shaping vendor performance and supply chain bottlenecks. These interpretable AI models not only optimize inventory levels but also assist in understanding how various elements, such as supplier reliability or market demand, impact supply chain operations. Challenges encompass the need for real-time adaptability to changing supplier behaviors, accurate interpretation of diverse data sources, and ensuring interpretability while handling extensive volumes of supply chain data.

XAI serves as a crucial asset in retail data protection and fraud detection within smart retail, employing advanced analytics to safeguard sensitive information and detect fraudulent activities. By analyzing transaction data, customer behavior patterns, and cybersecurity trends, interpretable AI models discern anomalies indicative of potential security breaches or fraudulent transactions. These models provide transparent insights into decision-making processes, shedding light on the rationale behind data protection strategies and fraud detection methodologies. This transparency assists retailers in comprehending the diverse factors influencing data security and fraud risks, enabling proactive measures to secure sensitive data and detect anomalies in real-time. Challenges persist in accurately identifying sophisticated fraud patterns, ensuring real-time adaptability to evolving threats, and integrating XAI-driven insights into cybersecurity frameworks, necessitating adaptable and interpretable models capable of dynamic fraud detection across diverse retail scenarios and security landscapes.

Moreover, XAI in retail data protection and fraud detection aims to fortify retailers' defense against cyber threats and fraudulent activities by providing insights into the underlying factors influencing security vulnerabilities and fraudulent behaviors. These interpretable AI models not only detect anomalies but also assist in understanding how various elements, such as transaction patterns or emerging cybersecurity risks, impact retail security frameworks. Challenges encompass the need for real-time adaptability to evolving fraud techniques, accurate interpretation of diverse data sources, and ensuring interpretability while handling vast amounts of transaction and security-related data [16].

15.8 Smart building: optimizing spaces with XAI

In the realm of modern architecture and infrastructure, the integration of XAI marks a significant paradigm shift in optimizing spaces within smart buildings. This section delves into the critical application of XAI within the domain of smart buildings, where the fusion of sophisticated AI methodologies with architectural designs revolutionizes how we understand, manage, and utilize built environments. XAI in smart buildings represents an innovative approach, offering transparent and interpretable insights into space optimization, energy efficiency, and occupant comfort.

Table 15.7 serves as an extensive resource, encapsulating diverse applications where XAI intersects with smart buildings, elucidating the types of data employed, the varying degrees of opaqueness associated with each application, the critical necessity for XAI integration, and the primary challenges inherent in implementing XAI-driven solutions.

XAI is essential for smart building energy efficiency and HVAC control optimization because it uses advanced analytics to improve building energy management and HVAC unit regulation. In order to optimize energy usage and HVAC settings, interpretable AI models examine building energy usage data, temperature variables, and occupancy patterns to uncover complex consumption patterns. These models provide transparent insights into decision-making processes, shedding light on the rationale behind energy efficiency strategies and HVAC control methodologies. This transparency aids building managers in understanding the multifaceted factors influencing energy consumption, enabling proactive adjustments in HVAC settings and precise control over energy consumption. Challenges persist in accurately predicting occupancy patterns, ensuring real-time adaptability to fluctuating climate conditions, and integrating XAI-driven insights into HVAC systems, requiring robust models capable of dynamic optimization across varying environmental contexts and occupant behaviors.

XAI is a crucial tool in monitoring occupancy and space utilization within smart buildings, utilizing advanced analytics to track occupancy patterns and optimize space allocation efficiently. By analyzing occupancy sensor data and space utilization metrics, interpretable AI models decipher intricate patterns crucial for understanding how spaces are used and adapting them accordingly. These

Table 15.7 Overview of XAI applications in smart buildings: data usage, opaqueness, necessity, and challenges

Applications	Data	Level of opaqueness	Need for XAI	Challenges
Energy management	Building energy usage and sensor data	Low–medium	High	Real-time adaptation and data accuracy
Predictive maintenance	Equipment sensor data and maintenance logs	Low–medium	Critical	Data integration and predicting failure patterns
Indoor air quality monitoring	Air quality sensors and occupancy data	Low–medium	High	Sensor accuracy and occupant behavior
Occupancy and space utilization	Occupancy sensors and space utilization data	Low–medium	High	Data privacy and accuracy in space usage
Smart lighting control	Lighting usage data and occupancy patterns	Low–medium	High	Sensor accuracy and personalized settings
HVAC optimization	HVAC system data and climate conditions	Low–medium	Critical	Real-time adaptability and optimizing comfort levels
Security and access control	Surveillance data and access logs	Low–medium	High	Data privacy and false positives in threat detection
Waste management and recycling	Waste bin sensors and recycling data	Low–medium	High	Sensor accuracy and optimizing waste separation
Personalized environment	Occupant preferences and behavior patterns	Low–medium	Critical	Privacy concerns and adapting to diverse preferences
Building automation and control	Control system data and automation logs	Low–medium	High	Integration complexities and system scalability

models provide transparent insights into decision-making processes, elucidating the rationale behind occupancy strategies and space utilization methodologies. This transparency aids building managers in comprehending the complex factors influencing space usage, enabling proactive adjustments in building layouts, optimizing room allocations, and enhancing the overall efficiency of space utilization. Challenges persist in accurately interpreting diverse occupancy behaviors, ensuring real-time adaptability to changing occupancy patterns, and integrating XAI-driven insights into space allocation systems, necessitating adaptable and interpretable models capable of dynamic analysis across varying occupancy scenarios and spatial configurations.

Moreover, XAI in occupancy monitoring and space utilization aims to augment the understanding of space usage by providing insights into the underlying factors shaping occupancy trends and space optimization opportunities. These interpretable AI models monitor occupancy and assist in understanding how various elements, such as work patterns or organizational changes, impact space utilization strategies and efficiency. Challenges encompass the need for real-time adaptability to evolving occupancy trends, accurate interpretation of diverse data sources, and ensuring interpretability while handling extensive volumes of occupancy and space utilization data.

XAI is crucial in the field of smart buildings, specifically in predictive maintenance for building systems. It utilizes sophisticated data analysis techniques to predict equipment failures and improve maintenance schedules. Interpretable AI models utilize data from equipment sensors, maintenance logs, and historical performance metrics to identify detailed patterns essential for anticipating possible system breakdowns or malfunctions. Moreover, XAI in predictive maintenance for building systems aims to augment the understanding of equipment health by providing insights into the underlying factors shaping equipment failure patterns and maintenance requirements. These interpretable AI models not only forecast failures but also assist in understanding how various elements, such as usage patterns or environmental conditions, impact system reliability and maintenance needs. Challenges encompass the need for real-time adaptability to evolving equipment conditions, accurate interpretation of diverse data sources, and ensuring interpretability while handling extensive volumes of equipment performance and maintenance data. Ultimately, the integration of XAI in predictive maintenance enhances building systems' reliability, reduces operational costs, and ensures optimal performance by proactively addressing potential equipment issues within smart building infrastructures [17].

In smart buildings, XAI is a crucial asset in fortifying security and access control systems, utilizing advanced analytics to bolster surveillance, threat detection, and access management. By analyzing surveillance data, access logs, and behavioral patterns, interpretable AI models discern intricate patterns critical for identifying potential security threats or unauthorized access.

Moreover, XAI in security and access control aims to augment the understanding of security vulnerabilities by providing insights into the underlying factors shaping security breaches and access control protocols. These interpretable AI models not only detect anomalies but also assist in understanding how various elements, such as behavioral patterns or emerging threats, impact security frameworks and access management strategies.

In smart buildings, XAI is a critical component in optimizing smart lighting and environmental comfort, employing sophisticated analytics to personalize lighting settings and regulate environmental conditions efficiently. Through the analysis of lighting usage data, occupancy patterns, and climate variations, interpretable AI models discern intricate patterns crucial for adjusting lighting configurations and environmental parameters to enhance occupant comfort. These models provide transparent insights into decision-making processes, elucidating the reasoning behind personalized lighting strategies and environmental adjustments. This transparency aids building managers in comprehending the complex factors influencing occupant comfort, enabling proactive adjustments in lighting schemes and climate control to cater to individual preferences. Challenges persist in accurately predicting occupants' comfort needs, ensuring real-time adaptability to changing environmental conditions, and integrating XAI-driven insights into smart building systems, requiring adaptable and interpretable models capable of dynamic analysis across varying environmental contexts and occupant behaviors.

Moreover, XAI in smart lighting and environmental comfort aims to augment the understanding of occupant preferences by providing insights into the underlying factors shaping comfort levels and lighting preferences. These interpretable AI models not only optimize lighting but also assist in understanding how various elements, such as personal preferences or external factors, impact occupant comfort and lighting control strategies. Challenges encompass the need for real-time adaptability to evolving occupant preferences, accurate interpretation of diverse data sources, and ensuring interpretability while handling extensive volumes of environmental and occupant-related data.

In smart buildings, XAI plays a crucial role in fortifying building automation security measures, employing sophisticated analytics to safeguard automated systems and control mechanisms. By analyzing control system data, automation logs, and network behavior, interpretable AI models decipher intricate patterns essential for identifying potential vulnerabilities or security breaches in automated building systems. These models provide transparent insights into decision-making processes, elucidating the rationale behind security protocols and automation control strategies. This transparency aids building security teams in comprehending multifaceted factors influencing system vulnerabilities, enabling proactive responses to potential threats, and facilitating robust security measures to safeguard automated building functions. Challenges persist in accurately detecting anomalies, ensuring real-time adaptability to evolving security risks, and integrating XAI-driven insights into building automation frameworks, necessitating adaptable and interpretable models capable of dynamic analysis across varying automation scenarios and potential threat vectors.

Moreover, XAI in building automation security measures aims to augment the understanding of system vulnerabilities by providing insights into the underlying factors shaping potential security risks and automation control protocols. These interpretable AI models not only detect anomalies but also assist in understanding how various elements, such as system configurations or emerging threats, impact automation security frameworks and control strategies. Challenges encompass the need for real-time adaptability to evolving security risks, accurate interpretation of diverse data sources, and ensuring interpretability while handling extensive volumes of automation-related data [18].

15.9 Nanotechnology: advancements through XAI

In cutting-edge technological advancements, the convergence of nanotechnology and XAI marks a pivotal leap forward in scientific exploration and innovation. This section delves into the transformative role of XAI within nanotechnology, where the integration of interpretable AI methodologies reshapes how we understand, analyze, and manipulate matter at the nanoscale. XAI in nanotechnology is innovative, offering transparent insights into intricate molecular behaviors, material properties, and nanoscale phenomena. From enhancing material design to accelerating drug delivery and optimizing manufacturing processes, XAI facilitates interpretable analysis and decision-making, shedding light on the reasoning behind these scientific advancements within the complex and dynamic realm of nanotechnology. Join us in exploring the diverse applications and promising advancements at the intersection of XAI and nanotechnology, where transparency, precision, and unprecedented control intersect to redefine the future of scientific exploration and nanoscale engineering.

Table 15.8 serves as a comprehensive guide delineating the intricate applications of XAI within the dynamic field of nanotechnology. It provides a succinct overview of ten diverse applications where XAI intersects with nanotechnology, outlining the types of data employed, the varying levels of opaqueness associated

Table 15.8 Summary of applications of XAI in nanotechnology

Applications	**Data**	**Level of opaqueness**	**Need for XAI**	**Challenges**
Drug delivery	Molecular structure and drug interactions	Medium	High	Precise targeting and biological complexity
Material design	Material properties and atomic structure	Low–medium	Critical	Predictive accuracy and complexity of materials
Nanoscale imaging	Nanoparticle behavior and imaging data	Low–medium	High	Resolution limits and imaging techniques
Nanorobotics	Nanomachine control and sensor data	Low–medium	Critical	Precision control and limited physical manipulation
Cancer therapeutics	Cell behavior and drug response	Medium	High	Personalized treatments and biological variability
Nanoelectronics	Quantum behavior and electrical data	High	Critical	Quantum effects and miniaturization challenges
Energy storage	Nanomaterials properties and energy data	Medium	High	Scalability and durability of nanomaterials
Environmental remediation	Nanoparticle interactions and pollutants	Low–medium	High	Site-specific treatment and toxicity concerns
Nanomanufacturing	Nanoparticle assembly and manufacturing	Low–medium	Critical	Scalability and reproducibility in mass production
Tissue engineering	Cell–nanomaterial interaction and data	Medium	High	Biocompatibility and tissue regeneration

with each application, the necessity for integrating XAI, and the primary challenges inherent in implementing XAI-driven solutions, from enhancing material design and drug delivery to optimizing manufacturing processes and enabling nanoscale imaging. It is a valuable resource, offering insights into the multifaceted requirements and obstacles involved in leveraging XAI across various facets of nanoscale exploration and engineering [19].

XAI is crucial in the field of nanomaterials design and optimization, since it employs sophisticated analytics to fundamentally transform the process of creating and improving materials at the nanoscale. Interpretable AI models utilize molecular structures, material properties, and computer simulations to comprehend complex patterns that are crucial for predicting and customizing nanomaterial characteristics. These models provide transparent insights into the decision-making processes underlying material design and optimization strategies. This transparency aids researchers and material scientists in comprehend the complex factors influencing nanomaterial behavior, enabling proactive adjustments in material composition and structure to achieve desired properties. Challenges persist in accurately predicting material properties at the nanoscale, ensuring real-time adaptability to changing design requirements, and integrating XAI-driven insights into material engineering frameworks, necessitating robust models capable of dynamic analysis across diverse nanomaterial compositions and properties.

Moreover, XAI in nanomaterials design and optimization aims to augment the understanding of material behavior by providing insights into the underlying factors shaping nanomaterial properties and design parameters. These interpretable AI models not only optimize material properties but also assist in understanding how various elements, such as atomic configurations or surface interactions, impact nanomaterial performance and functionality. Challenges encompass the need for real-time adaptability to evolving material requirements, accurate interpretation of diverse data sources, and ensuring interpretability while handling extensive volumes of nanomaterial-related data.

In the realm of pharmaceutical sciences, XAI stands as a transformative tool in developing and optimizing of drug delivery systems and targeted therapies at the nanoscale. Utilizing complex analytics and molecular simulations, interpretable AI models decipher intricate patterns crucial for predicting drug behavior, interactions, and optimizing delivery mechanisms. These models provide transparent insights into the decision-making processes underlying drug delivery strategies, shedding light on the rationale behind targeted therapies and nanoscale drug delivery systems. This transparency aids pharmaceutical researchers and biomedical engineers in comprehending the multifaceted factors influencing drug behavior within the body, enabling proactive adjustments in drug formulation and delivery methods to enhance therapeutic efficacy. Challenges persist in accurately predicting drug behavior in vivo, ensuring real-time adaptability to varying biological contexts, and integrating XAI-driven insights into drug development frameworks, necessitating robust models capable of dynamic analysis across diverse drug compounds and biological environments. Moreover, XAI in drug delivery systems and targeted therapies aims to augment the understanding of drug interactions by providing insights into the underlying factors shaping drug bioavailability and therapeutic responses. These interpretable AI models optimize drug delivery and also assist in understanding how various elements, such as molecular structures or physiological responses, impact drug efficacy and targeted treatment strategies. Challenges encompass the need for real-time adaptability to evolving biological conditions, accurate interpretation of diverse data sources, and ensuring interpretability while handling extensive volumes of drug-related data.

XAI contributes significantly to the domain of nanoscale imaging and diagnostics, leveraging advanced analytics to enhance imaging techniques and diagnostic tools at the molecular level. By analyzing imaging data, molecular structures, and diagnostic patterns, interpretable AI models discern intricate details critical for improving resolution, accuracy, and interpretability in nanoscale imaging technologies. Moreover, XAI in nanoscale imaging and diagnostics aims to augment the understanding of cellular structures and disease markers by providing insights into the underlying factors shaping imaging outcomes and diagnostic interpretations. These interpretable AI models enhance imaging resolution and assist in understanding how various cellular elements or pathological changes impact diagnostic accuracy and disease characterization. In the domain of nanomanufacturing and fabrication, XAI emerges as a critical driver, leveraging advanced analytics to revolutionize precision, efficiency, and scalability in nanoscale manufacturing processes. Through the analysis of fabrication data, material properties, and manufacturing parameters, interpretable AI models decode intricate patterns essential for optimizing fabrication techniques and enhancing material synthesis at the nanoscale.

These models provide transparent insights into decision-making processes, elucidating the rationale behind manufacturing strategies and process optimizations. This transparency aids engineers and researchers in comprehending the multifaceted factors influencing nanomanufacturing, enabling proactive adjustments in fabrication methodologies and material design to achieve desired properties. Challenges persist in accurately predicting fabrication outcomes, ensuring real-time adaptability to changing fabrication requirements, and integrating XAI-driven insights into manufacturing frameworks, necessitating adaptable and interpretable models capable of dynamic analysis across varying material compositions and fabrication processes [20]. Moreover, XAI in nanomanufacturing and fabrication aims to augment the understanding of material synthesis by providing insights into the underlying factors shaping fabrication results and manufacturing efficiencies. These interpretable AI models not only optimize fabrication processes but also assist in understanding how various fabrication parameters or material compositions impact manufacturing outcomes and nanomaterial properties. Challenges encompass the need for real-time adaptability to evolving fabrication demands, accurate interpretation of diverse fabrication data sources, and ensuring interpretability while handling extensive volumes of manufacturing-related data.

XAI is crucial in promoting the use of nanotechnology for environmental applications. By utilizing advanced data analysis techniques, XAI helps tackle environmental issues and improve sustainability by implementing solutions at the nanoscale. Interpretable AI models analyze environmental data, material interactions, and pollutant behaviors to uncover complex patterns that are essential for optimizing nanomaterials and technologies in order to reduce pollution, cleanse water, and restore polluted habitats. These models provide transparent insights into decision-making processes, shedding light on the rationale behind environmental applications of nanotechnology and the design of innovative solutions. This transparency aids environmental scientists and engineers in comprehending the complex factors influencing nanomaterial interactions with pollutants or environmental matrices, enabling proactive adjustments in nanomaterial design and deployment strategies to achieve eco-friendly solutions. Challenges persist in accurately predicting nanomaterial performance in diverse environmental conditions, ensuring safety and minimizing unintended consequences, and integrating XAI-driven insights into environmental frameworks, necessitating robust models capable of dynamic analysis across varying environmental contexts and nanomaterial functionalities.

Moreover, XAI in environmental applications of nanotechnology aims to augment the understanding of nanomaterial–environment interactions by providing insights into the underlying factors shaping environmental remediation and pollution control strategies. These interpretable AI models not only optimize nanomaterial performance but also assist in understanding how various environmental parameters or nanoparticle characteristics impact environmental cleanup and sustainability efforts. Challenges encompass the need for real-time adaptability to evolving environmental conditions, accurate interpretation of diverse environmental data sources, and ensuring interpretability while handling extensive volumes of environmental and nanotechnology-related data [21,22].

Data security and privacy are crucial in the field of nanotechnology, and XAI plays a vital role in strengthening these features. XAI facilitates the development of transparent and secure frameworks for handling sensitive nanotechnology-related data, such as molecular structures, research findings, and proprietary information. By employing advanced encryption algorithms, interpretable AI models enable secure data transmission and storage, ensuring confidentiality and integrity in nanotechnology research and development. Moreover, XAI-driven approaches allow for interpretable security measures, providing insights into the decision-making processes behind data protection strategies and enhancing the traceability of data access and usage. Challenges persist in ensuring robust cybersecurity measures to prevent unauthorized access, maintaining data integrity amidst evolving threats, and integrating XAI-driven security protocols seamlessly into nanotechnology data infrastructure, necessitating adaptive and interpretable security models capable of dynamic analysis to counter diverse cybersecurity threats [21].

Furthermore, XAI in nanotechnology data security and privacy aims to augment the understanding of data vulnerabilities by providing insights into the underlying factors shaping data protection strategies and privacy controls. These interpretable AI models not only reinforce data security measures but also assist in comprehending how various cyber threats or data access patterns impact security frameworks and privacy regulations. Challenges encompass the need for real-time adaptability to emerging cyber threats, accurate interpretation of diverse security data sources, and ensuring interpretability while handling extensive volumes of sensitive nanotechnology-related data. Ultimately, the integration of XAI in nanotechnology data security ensures the safeguarding of intellectual property, research findings, and confidential information, bolstering trust and compliance in nanotechnology research while navigating the complexities of data security and privacy in this rapidly evolving field.

References

[1] Bharati, S., Mondal, M. R. H., Podder, P., and Kose, U. (2023). Explainable artificial intelligence (XAI) with IoHT for smart healthcare: A review. *Interpretable Cognitive Internet of Things for Healthcare*, 1–24.

[2] Ahmad, I., Zhu, M., Li, G., Javeed, D., Kumar, P., and Chen, S. (2024). A secure and interpretable AI for smart healthcare system: A case study on epilepsy diagnosis using EEG signals. *IEEE Journal of Biomedical and Health Informatics*, DOI:10.1109/JBHI.2024.3366341.

[3] Payrovnaziri, S. N., Chen, Z., Rengifo-Moreno, P., *et al.* (2020). Explainable artificial intelligence models using real-world electronic health record data: a systematic scoping review. *Journal of the American Medical Informatics Association*, *27*(7), 1173–1185.

[4] Atakishiyev, S., Salameh, M., Yao, H., and Goebel, R. (2024). Explainable artificial intelligence for autonomous driving: A comprehensive overview and field guide for future research directions. *IEEE Access*, 12.

[5] Nwakanma, C. I., Ahakonye, L. A. C., Njoku, J. N., *et al.* (2023). Explainable artificial intelligence (xai) for intrusion detection and mitigation in intelligent connected vehicles: A review. *Applied Sciences*, *13*(3), 1252.

[6] Park, S., Kim, D., Kwon, S., and Kang, S. (2023, October). Routability prediction and optimization using explainable AI. In *2023 IEEE/ACM International Conference on Computer Aided Design (ICCAD)* (pp. 1–8). IEEE.

[7] Mugurusi, G., and Oluka, P. N. (2021, August). Towards explainable artificial intelligence (XAI) in supply chain management: A typology and research agenda. In *IFIP International Conference on Advances in Production Management Systems* (pp. 32–38). Cham: Springer International Publishing.

[8] de Benito Fernández, M., Martínez, D. L., González-Briones, A., Chamoso, P., and Corchado, E. S. (2023, June). Evaluation of XAI Models for Interpretation of Deep Learning Techniques' Results in Automated Plant Disease Diagnosis. In *Sustainable Smart Cities and Territories International Conference* (pp. 417–428). Cham: Springer Nature Switzerland.

[9] Aqib, A. I., Fatima, M., Muneer, A., *et al.* (2023). Explainable Artificial Intelligence (XAI) in the veterinary and animal sciences field. *Explainable Artificial Intelligence for Biomedical Applications*, 33–56.

[10] Machlev, R., Heistrene, L., Perl, M., *et al.* (2022). Explainable Artificial Intelligence (XAI) techniques for energy and power systems: Review, challenges and opportunities. *Energy and AI*, *9*, 100169.

[11] Ozdemir, G., Kuzlu, M., Sarp, S., Catak, F. O., Dimd, B. D., and Cali, U. (2024). The role of eXplainable Artificial Intelligence (XAI) in smart grids. In *Big Data Application in Power Systems* (pp. 349–370). Elsevier Science.

[12] Abraham, O. A., Ochiai, H., Hossain, M. D., Taenaka, Y., and Kadobayashi, Y. (2024). Electricity theft detection for smart homes: Harnessing the power of machine learning eith real and synthetic attacks. *IEEE Access*, *12*, 26023–26045.

[13] Sarker, M. A. A., Shanmugam, B., Azam, S., and Thennadil, S. (2024). Enhancing smart grid load forecasting: An attention-based deep learning model integrated with federated learning and XAI for security and interpretability. *Intelligent Systems with Applications*, *23*, 200422.

[14] Behera, R. K., Bala, P. K., and Rana, N. P. (2023). Creation of sustainable growth with explainable artificial intelligence: An empirical insight from consumer packaged goods retailers. *Journal of Cleaner Production*, *399*, 136605.

[15] Albayrak Ünal, Ö., Erkayman, B., and Usanmaz, B. (2023). Applications of artificial intelligence in inventory management: A systematic review of the literature. *Archives of Computational Methods in Engineering*, *30*(4), 2605–2625.

[16] Mill, E. R., Garn, W., Ryman-Tubb, N. F., and Turner, C. (2023). Opportunities in real time fraud detection: An explainable artificial intelligence (XAI) research agenda. *International Journal of Advanced Computer Science and Applications*, *14*(5), 1172–1186.

[17] Dounis, A. I. (2022). Machine intelligence in smart buildings. *Energies*, *16* (1), 22.

[18] Himeur, Y., Elnour, M., Fadli, F., *et al.* (2023). AI-big data analytics for building automation and management systems: A survey, actual challenges and future perspectives. *Artificial Intelligence Review*, *56*(6), 4929–5021.

[19] Ieracitano, C., Mammone, N., Paviglianiti, A., and Morabito, F. C. (2021, June). Toward an augmented and explainable machine learning approach for classification of defective nanomaterial patches. In *International Conference on Engineering Applications of Neural Networks* (pp. 244–255). Cham: Springer International Publishing.

[20] Rajagopal, A., Nirmala, V., Andrew, J., and Vedamanickam, A. M. (2021, December). AI Visualization in Nanoscale Microscopy. In *International Conference on Big Data, Machine Learning, and Applications* (pp. 707–719). Singapore: Springer Nature Singapore.

[21] Rana, A., Gautam, D., Kumar, P., and Das, A. K. (2024). Architectures, benefits, security and privacy issues of Internet of Nano Things: A comprehensive survey, opportunities and research challenges. *IEEE Communications Surveys & Tutorials*. DOI: 10.1109/COMST.2024.3423477.

[22] Linley, S., and Thomson, N. R. (2021). Environmental applications of nanotechnology: Nano-enabled remediation processes in water, soil and air treatment. *Water, Air, & Soil Pollution*, *232*(2), 59.

Index

accountability
  defined 136
  in federated learning (FL) 345–6
accumulated local effect (ALE) plots 202–4
action–value function: *see* Q-value
AdaBoost 100
advanced message queuing protocol (AMQP) 17
adversarial attacks 278, 295, 301–3
  and defense in explainable security intelligence 277
    evaluating adversarial attacks 291–3
    explainability and robustness 303–5
    explainable adversarial attacks 305–8
    explainable defense mechanisms 308–9
    zeroth-order optimization (ZOO) attacks 290–1
  knowledge of 279–81
  one-time versus iterative attacks 281–2
  targeted versus non-targeted attacks 284–7
  training versus inference attacks 283–4
adversarial autoencoders 115
adversarial defense 293
  gradient masking and obscuring 300–1
  preprocessing techniques
    data augmentation 294
    feature squeezing 296–7
    input normalization 294–5
    randomization techniques 295–6
  robust model architectures
    adversarial training 297–8
    certified defense methods 299–300
    defensive distillation 298
    ensemble methods 298–9
adversarial perturbation techniques 288
  Carlini–Wagner (CW) attack 289–90
  fast gradient sign method (FGSM) 288
  projected gradient descent (PGD) 289
adversarial training 297–8
aggregation methods 329–33
AIX360 145
algorithmic explainability 140–1
Alibi 145
Anchor method 217, 232–3
application PEPs (A-PEPs) 368
Application Programming Interfaces (APIs) 76, 83, 364–5
architecture, IoT 14
  five-layer IoT architecture 23
    application layer 26
    middleware layer 24–5

network layer 24
perception layer 23
processing layer 26
four-layer IoT architecture 21
application layer 23
network layer 22
perception layer 21–2
processing layer 22–3
seven-layer IoT architecture 26
application layer 30
business layer 30–1
connectivity layer 26
edge/fog computing layer 26–8
physical layer 26
processing layer 29–30
security layer 31–2
three-layer IoT architecture 14
application layer 20–1
network layer 16–19
perception layer 14–16
ARP spoofing 82
artificial intelligence (AI) emergence in IoT 3–4
asset-centric security 364
asynchronous FL 338
attack surfaces in IoT 60
cloud and backend attack surfaces 73
backend system vulnerabilities 76
cloud infrastructure vulnerabilities 76
data transmission security 76–7
distributed denial of service (DDoS) and service disruption 77–8
third-party service vulnerabilities 76
data attack surfaces 72–3
network-based attack surfaces 60
Bluetooth vulnerabilities 61
fifth generation (5G) networks 64–5
fourth generation (4G) networks 64–5
second generation (2G) networks 61–5
third generation (3G) networks 64–5
Wi-Fi vulnerabilities 60–1
Zigbee vulnerabilities 61
physical attack surfaces 64–9
software attack surfaces 69–71
autoencoders 114–16
automated dependence plots (ADPs) 201–2
average sensitivity metric 260–1
axiomatic metrics 270
completeness metric 270–1
input invariance metric 272–3
non-sensitivity metric 271–2

backdoor exploitation 83
backend system vulnerabilities 76
bandwidth-efficient FL methodologies 340
Bayesian networks 181–2
belief networks: *see* Bayesian networks
Bellman equation 123
BeyondCorp 359
bias, defined 136
bidirectional encoder representations from transformers (BERT) 117
black-box machine learning for IoT security 95
classical ML 98–104
deep learning 105
autoencoders 114–16
convolutional neural networks (CNNs) 110–12

feedforward neural networks (FNNs) 105–6
generative adversarial networks (GANs) 112–14
graph neural networks (GNNs) 122
memory-augmented neural networks (MANNs) 119–20
recurrent neural networks (RNNs) 106–10
spiking neural networks (SNNs) 120–1
transformers 116–19
fundamentals of 96–8
meta-learning 126–9
reinforcement learning (RL) 122–6
see also security vulnerabilities in IoTs
black-box ML 95
black-box models
integration within IoT applications 137
and interpretability 320
Bluetooth 62
Bluetooth Low Energy (BLE) technology 12
Bluetooth vulnerabilities 61
bricking attacks 83
brute force attacks 82
buffer overflow 83

California Consumer Privacy Act (CCPA) 89
Carlini–Wagner (CW) attack 289–90
CatBoost 102
causability, defined 134
causal probabilistic networks: *see* Bayesian networks
cellular (3G/4G/5G) 63
centralized computing 40–1
centralized FL 321–3
certified defense methods 299–300
circuit tracing 79
classification and regression trees (CART) 153
client–server architectures 39, 42–3
cloud and backend attack surfaces 73
backend system vulnerabilities 76
cloud infrastructure vulnerabilities 76
data transmission security 76–7
distributed denial of service (DDoS) and service disruption 77–8
third-party service vulnerabilities 76
cloud computing 43
cloud infrastructure vulnerabilities 76
cloudlet computing 48–9
Cloud of Things (CoT) 54
cloud PEPs (C-PEP) 368
code injection 83
common IoT attacks 78
data privacy attacks 80–1
firmware and software attacks 84
network attacks 81–3
physical security attacks 79–80
communication efficiency 320
communication mechanism 325–9
completeness metric 270–1
complexity and granularity 157–9
computation-efficient FL methodologies 340
computing technologies, convergence of IoTs and 39
advanced distributed computing paradigms 47
cloudlet computing 48–9
dew computing 50
mist computing 50–2
swarm computing 49–50

centralized computing 40–1
distributed computing 41
client–server architecture 42–3
cloud computing 43
edge computing 43–5
fog computing 45–6
quantum computing 46–7
emerging technologies and innovations
Cloud of Things (CoT) 54
fabric computing 52–3
mobile computing 55
multi-access edge computing (MEC) 53
conditional generative adversarial network (cGAN) 113
conditional probability distribution (CPD) 182
consistency metric 262–3
constrained application protocol (CoAP) 17
contextual explainability 142–3
continuity test 257–8
Continuous Adaptive Risk and Trust Assessment (CARTA) framework 359
continuous diagnostics and mitigation (CDM) framework 367
continuous monitoring and incident response 92
contractive autoencoders 115
contrastive explanations 216, 227–9
convergence techniques 336–8
convolutional neural networks (CNNs) 110–12
cost-complexity pruning 184–5
counterfactual explanations 229–32
credential attacks 81
cross-site scripting (XSS) 81, 83
cryptojacking 81, 83
customer relationship management (CRM) systems 29
CycleGAN 114

DALEX 145–6
data attack surfaces 72–3
data augmentation 294
data breaches 80
data distribution service (DDS) 18
data explainability 141–2
datagram transport layer security (DTLS) 18
data handling processes 76
data integrity vulnerabilities 73
data interception 81
data manipulation 81
data PEPs (D-PEP) 369
data privacy attacks 80–1
data privacy preservation 319
data privacy vulnerabilities 72
data transmission security 76–7
decentralized FL 323–4
decision rules 186–9
decision trees 182–6
deep convolutional generative adversarial network (DCGAN) 113
deep learning 105
autoencoders 114–16
convolutional neural networks (CNNs) 110–12
feedforward neural networks (FNNs) 105–6
generative adversarial networks (GANs) 112–14
graph neural networks (GNNs) 122
memory-augmented neural networks (MANNs) 119–20
recurrent neural networks (RNNs) 106–10

spiking neural networks (SNNs) 120–1
transformers 116–19
deep neural networks (DNNs) 97, 105
deep Q-networks (DQNs) 124
defense-in-depth strategy 91
defensive distillation 298
defensive strategies and mitigation techniques 84
encryption and authentication 85–6
monitoring and incident response 88–9
network security measures 86–7
secure design principles 84–5
secure software development 87–8
delayering 79
demographic parity metric 242–3
denial-of-service (DoS) 83
device authentication and identity management 91
device relocation 80
dew computing 50
diameter-based attacks 64
dice coefficient 266
differentiable neural computers (DNCs) 119–20
differential privacy 350
differentiated services (DiffServ) 19
dImensions of explainability in IoT security 140
algorithmic explainability 140–1
contextual explainability 142–3
data explainability 141–2
human-centric explainability 143–4
directed acyclic graphs (DAGs) 181
disparate impact metric 243–4
distributed computing 41
client–server architecture 42–3
cloud computing 43
edge computing 43–5
fog computing 45–6
quantum computing 46–7
distributed computing paradigms 47
cloudlet computing 48–9
dew computing 50
mist computing 50–2
swarm computing 49–50
distributed denial of service (DDoS) attacks 4, 77–8, 82
distributed FL 324
DNS spoofing/poisoning 82
domain-specific XAI methods 160–1

eavesdropping/sniffing 81–2
edge computing 40, 43–5
"edge of things" paradigm 314
emerging technologies and innovations
fabric computing 52–3
mobile computing 55
multi-access edge computing (MEC) 53
Cloud of Things (CoT) 54
encryption and authentication 85–6
end-to-end encryption (E2EE) 86
energy-efficient FL methodologies 341
ensemble methods 298–9
enterprise resource management (ERM) 29
enterprise resource planning (ERP) 29
entropy 183–4
environmental attacks 80
equal odds metric 245
equal opportunity metric 244
Ethernet 63
EthicalML-XAI 146
evaluating adversarial attacks 291–3
explainability
defined 134

in IoT systems 136–40
and robustness 303–5
explainable adversarial attacks 305–8
explainable defense mechanisms 308–9
explainable security intelligence 151, 237
axiomatic metrics 270
completeness metric 270–1
input invariance metric 272–3
non-sensitivity metric 271–2
complexity and granularity 157–9
domain-specific XAI methods 160–1
evaluation metrics, foundations of 238–40
fairness metrics 241
demographic parity metric 242–3
disparate impact metric 243–4
equal odds metric 245
equal opportunity metric 244
false positive rates (FPR)-parity metric 246–7
group unawareness metric 241–2
PPV-parity metric 245–6
faithfulness metrics 247
faithfulness correlation metric 248
faithfulness estimate metric 249
infidelity metric 256
monotonicity correlation metric 252–3
monotonicity metric 251–2
pixel-flipping metric 250–1
selectivity metric 253–4
SensitivityN metric 254–5
global versus local interpretability 155–7
human-understandable versus machine-oriented explanations 159–60
interpretability type 163–7
localisation metrics 265
pointing game metric 265–6
relevance mass accuracy metric 267–8
relevance rank accuracy metric 269–70
top-K intersection metric 266–7
model-agnostic versus model-specific XAI methods 152–4
post-hoc versus intrinsic XAI methods 154–5
quantitative versus qualitative explanation 161–3
robustness metrics 257
average sensitivity (Avg Sensitivity) metric 260–1
consistency metric 262–3
continuity test 257–8
local Lipschitz estimate 258–9
max-sensitivity metric 259–60
relative input stability metric 263–4
relative output stability metric 264–5
explainable security intelligence in IoT applications 377
nanotechnology 407–12
smart agriculture 390–4
smart building 404–7
smart grid 395–400
smart healthcare 378–81
smart manufacturing 387–90
smart retail 400–4
smart transportation 381–6
eXtensible Markup Language (XML) 16
extensible messaging and presence protocol (XMPP) 18

fabric computing 52–3
failure tolerance 339–40
fairness, defined 136
fairness metrics 241
 demographic parity metric 242–3
 disparate impact metric 243–4
 equal odds metric 245
 equal opportunity metric 244
 false positive rates (FPR)-parity metric 246–7
 group unawareness metric 241–2
 PPV-parity metric 245–6
faithfulness metrics 247
 faithfulness correlation metric 248
 faithfulness estimate metric 249
 infidelity metric 256
 monotonicity correlation metric 252–3
 monotonicity metric 251–2
 pixel-flipping metric 250–1
 selectivity metric 253–4
 SensitivityN metric 254–5
false positive rates (FPR) 245–7
fast gradient sign method (FGSM) 288
feature squeezing 296–7
federated averaging (FedAvg) 316, 318, 323, 327, 329
federated learning (FL) methods 313
 accountable FL 344
  accountability in FL 345–6
  regulatory compliance 346–8
 advantages and challenges 319
  black-box models and interpretability 320
  communication efficiency 320
  data privacy preservation 319
  model divergence and convergence 321
  non-IID data distribution 320
  reduced communication overhead 319–20
 definition and motivation 315–16
 intersection of FL and XAI 341
  challenges in achieving explainability in FL 341–2
  techniques for explainable FL 342–4
 key concepts in 316–19
 taxonomy of 321
  aggregation methods 329–33
  communication mechanism 325–9
  convergence techniques 336–8
  failure tolerance 339–40
  model architecture 333–4
  network topology 321–5
  participant selection 334–6
  resource allocation 340–1
 trustworthy FL 348
  security and privacy measures 349–52
  trust challenges in FL 348–9
federated model explanations (FME) 343
federated training 317
federated transfer learning 326, 328–9
feedforward neural networks (FNNs) 105–6
FFNN layers 117
fifth generation (5G) networks 64–5
firewalls and IDSs 86
firmware and software attacks 84
firmware attacks 81
firmware rollback 83
firmware tampering 83
5G mobile network 28
fog computing 45–6
fog-to-edge computing: *see* mist computing

fourth generation (4G) networks 64–5
frameworks and tools for implementing explainable AI in IoT security 145–7
fundamentals of IoT 12–14

gated recurrent units (GRUs) 107
  architecture of 109
Gaussian distributions 295
General Data Protection Regulation (GDPR) 89, 144, 346
generalized additive models (GAMs) 178–9
generalized linear model (GLM) 177–9
generative adversarial networks (GANs) 112–14
generative pre-trained transformer (GPT) 117
Gini impurity 183
globally interpretable machine learning, model-agnostic methods for 195
  accumulated local effect (ALE) plots 202–4
  automated dependence plots (ADPs) 201–2
  feature interaction 205–6
  functional decomposition 206–8
  global surrogate models 209–11
  individual conditional expectation (ICE) plots 198–201
  partial dependence plot (PDP) 196–8
  permutation feature importance 208–9
  prototypes and criticisms 211–12
global versus local interpretability 155–7
GoogLeNet 112
gradient-boosting machines (GBM) 98
gradient masking and obscuring 300–1
Gramm-Leach-Bliley Act (GLBA) 346
graphical models 179
  Bayesian network 181–2
  Naive Bayes classifier 179–81
graph neural networks (GNNs) 122
group unawareness metric 241–2

H2O 146
hardware vulnerabilities in IoT devices 64
Health Insurance Portability and Accountability Act (HIPAA) 89, 346
heterogeneous FL 334
hierarchical aggregation 332–3
hierarchical FL 324–5
homogeneous FL 333–4
horizontal FL 325–7
host-based PEPs (H-PEPs) 368
H-statistic 205–6
human-centric explainability 143–4
human–machine interface 7
human-understandable versus machine-oriented explanations 159–60
Hypertext Transfer Protocol (HTTP) 12

IBM System/360 41
IEC 62443 Series 90
ImageNet large-scale visual recognition challenge (ILSVRC) 111
Inception Networks 112
individual conditional expectation (ICE) plots 198–201, 216–19
Industrial IoT (IIoT) 3
infidelity metric 256
InfoGAN 114
information gain 184

information security management system (ISMS) 90
Infrastructure as a Service (IaaS) 42
input invariance metric 272–3
input normalization 294–5
Institute of Electrical and Electronics Engineers (IEEE)
P7003 Standard for Algorithmic Bias Considerations 144
interactive AI, defined 135
interactive explanations within the realm of XAI 165
interconnected nature of IoT ecosystems 73
inter-device communication and data sharing within IoT networks 72–3
international mobile equipment identity (IMEI) system 64
International Organization for Standardization (ISO) 89
Internet of Things (IoT) revolution 2–3
internet protocol (IP) 17
interoperability 13
interpretability, defined 134
interpretability type 163–7
InterpretML 146
intersection of FL and XAI 341
  challenges in achieving explainability in FL 341–2
  techniques for explainable FL 342–4
intrinsically explainable security intelligence 169
  graphical models 179
    Bayesian network 181–2
    Naive Bayes classifier 179–81
  K-nearest neighbors (KNN) 189–92
  linear models 170
    generalized linear model (GLM) 177–9
    linear regression 170–4
    logistic regression 174–6
  tree-based models 182
    decision rules 186–9
    decision tree 183–6
intrusion detection systems (IDSs) 66
  firewalls and 86
IoT botnets 81–2
IoT Security Foundation (IoTSF) 90–1
IoT worms 82
ISO/IEC 27000 90
ISO/IEC 27001 90, 347

Jaccard index 266
JavaScript Object Notation (JSON) 16
Jericho Forum Commandments 358

kernel methods 103–4
K-nearest neighbors (KNN) 189–92
Kolmogorov–Smirnov test 171

layer-wise relevance propagation (LRP) 153
learning to learn: *see* meta-learning
least privilege, principle of 364
least squares generative adversarial network (LSGAN) 113–14
LightGBM 101
lightweight machine-to-machine (LwM2M) 18
linear models 170
  generalized linear model (GLM) 177–9
  linear regression 170–4
  logistic regression 174–6
linear regression 170–4
Lipschitz continuity 257
liquid state machines (LSM) 120
local interpretable model-agnostic explanations (LIMEs) 216, 343

localisation metrics 265
pointing game metric 265–6
relevance mass accuracy metric 267–8
relevance rank accuracy metric 269–70
top-K intersection metric 266–7
local Lipschitz estimate 258–9
locally explainable AI, model-agnostic methods for 215
Anchor 232–3
contrastive explanations 227–9
counterfactual explanations 229–32
individual conditional expectation (ICE) plots 217–19
local interpretable model-agnostic explanations 220–2
PyExplainer 224–7
SHapley Additive exPlanations 222–4
location tracking exploitation 81
logistic regression 174–6
logit: *see* log-odds
log-odds 175
long range (LoRa) 19
long short-term memory networks (LSTMs) 107–8
LoRaWAN 63
low-power wide area network (LPWAN) technology 12, 19

machine learning (ML) 32, 95
see also black-box machine learning for IoT security
machine-to-machine (M2M) communication 18
mainframe computing 40
malware infections 83
man-in-the-middle (MitM) attacks 61, 80, 82
max-sensitivity metric 259–60
memory-augmented neural networks (MANNs) 119–20
Message Queuing Telemetry Transport (MQTT) 12
message queuing telemetry transport (MQTT) 17, 63
meta-learning algorithm 126–9
Meta-SGD (Meta-stochastic gradient descent) 128
microprobing 79
minimal cost-complexity pruning: *see* cost-complexity pruning
Mirai botnet attack in 2016 4
mist computing 50–2
mobile computing 55
mobile equipment identity (MEID) system 64
model-agnostic meta-learning (MAML) 126
model-agnostic versus model-specific XAI methods 152–4
model-agnostic XAI methods 153
model architecture 333–4
model divergence and convergence 321
monotonicity correlation metric 252–3
monotonicity metric 251–2
multi-access edge computing (MEC) 53
Cloud of Things (CoT) 54
multifaceted nature of XAI 138
multi-party computation (MPC) 349–50

Naive Bayes classifier 179–81
nanotechnology 407–12
narrowband IoT (NB-IoT) 19
National Institute of Standards and Technology (NIST) 89, 359–61

- NIST cybersecurity framework 90
- NIST Special Publication 800-183 90

natural language processing (NLP) 116, 160
near-field communication (NFC) 62
network attacks 81–3
network-based attack surfaces 60
- Bluetooth vulnerabilities 61
- fifth generation (5G) networks 64–5
- fourth generation (4G) networks 64–5
- second generation (2G) networks 61–5
- third generation (3G) networks 64–5
- Wi-Fi vulnerabilities 60–1
- Zigbee vulnerabilities 61

network function virtualization (NFV) 64
network layer attacks 82
network PEPs (N-PEPs) 368
network security measures 86–7
network sniffing 82
network topology 321–5
network traffic analysis 82
neural turing machines (NTMs) 119
neuromorphic computing models 121
node capture 79
node destruction 79
node malfunction 79
node outage 79
node replication 80
non-IID data distribution 320
non-sensitivity metric 271–2

one-time versus iterative attacks 281–2
opaqueness, defined 133
OpenAI 117
Open Connectivity Foundation (OCF) 91
ordinary least squares (OLS) 171–2
overfitting 97

partial dependence plot (PDP) 196–8
participant selection 334–6
Payment Card Industry Data Security Standard (PCI DSS) 347
personalized/adaptive aggregation 332
phishing 81
physical access exploitation 80
physical attack surfaces 64–9
physical security attacks 79–80
physical tampering 80
physical vulnerabilities in IoT devices 66
pixel-flipping metric 250–1
Platform as a Service (PaaS) 42
pointing game metric 265–6
policy decision point (PDP) 367
policy enforcement point (PEP) 367
post-hoc versus intrinsic XAI methods 154–5
post-pruning 185
PPV-parity metric 245–6
preprocessing techniques
- data augmentation 294
- feature squeezing 296–7
- input normalization 294–5
- randomization techniques 295–6

pre-pruning 185
principle of least privilege 91
principles of XAI 132–6
privacy policy violations 81
privilege escalation 83
projected gradient descent (PGD) 289
protocol interoperability 18
protocols, coexistence of 17
public key infrastructure (PKI) 367
PyExplainer 216, 224–7

Q-learning algorithm 123–4
quality of service (QoS) 17–19
quantitative versus qualitative explanation 161–3
quantization-based aggregation 330–1
quantum computing 46–7
Quantus 146
Q-value 123

radio access network (RAN) 53
randomization techniques 295–6
random selection FL 335
recurrent neural networks (RNNs) 106–10
reduced communication overhead 319–20
reduced error pruning 185
regular updates and patch management 91–2
regulatory frameworks and standards 89–92
regulatory landscape and standards 144–5
reinforcement learning (RL) 122–6
relative input stability metric 263–4
relative output stability metric 264–5
relevance mass accuracy metric 267–8
relevance rank accuracy metric 269–70
reproducible AI, defined 135
Reptile 127–8
resilient FL methodologies 339
resource allocation 340–1
resource reservation protocol (RSVP) 19
responsibility, defined 136
RoBERTa (robustly optimized BERT pretraining approach) 117
robust model architectures
  adversarial training 297–8
  certified defense methods 299–300
  defensive distillation 298
  ensemble methods 298–9
robustness metrics 257
  average sensitivity metric 260–1
  consistency metric 262–3
  continuity test 257–8
  local Lipschitz estimate 258–9
  max-sensitivity metric 259–60
  relative input stability metric 263–4
  relative output stability metric 264–5
rogue devices 82
rootkit installation 83
R-squared statistic 172–3
rule-based systems leverage decision rules 187

safeguarding and security, need for 4
safeguarding IoT systems, XAI for 131
  dimensions of explainability in IoT security 140
    algorithmic explainability 140–1
    contextual explainability 142–3
    data explainability 141–2
    human-centric explainability 143–4
  frameworks and tools for implementing explainable AI in IoT security 145–7
  need for explainability in IoT systems 136–40
  principles of XAI 132–6
  regulatory landscape and standards 144–5
second generation (2G) networks 61–5
secure design principles 84–5
secure development lifecycle (SDL) 91
secure elements (SEs) 85
secure software development 87–8
securing the Internet of Things 11
  architecture, IoT 14

five-layer IoT architecture 23–6
four-layer IoT architecture 21–3
seven-layer IoT architecture 26–32
three-layer IoT architecture 14–21
fundamentals of IoT 12–14
security challenges in IoT 32–6
security, enhancing 5–6
security challenges in IoT 32–6
security considerations 17
security vulnerabilities in IoTs 59
attack surfaces in IoT 60
cloud and backend attack surfaces 73–8
data attack surfaces 72–3
network-based attack surfaces 60–4
physical attack surfaces 64–9
software attack surfaces 69–71
common IoT attacks 78
data privacy attacks 80–1
firmware and software attacks 84
network attacks 81–3
physical security attacks 79–80
defensive strategies and mitigation techniques 84
encryption and authentication 85–6
monitoring and incident response 88–9
network security measures 86–7
secure design principles 84–5
secure software development 87–8
regulatory frameworks and standards 89–92
see also black-box machine learning for IoT security
selectivity metric 253–4
SensitivityN metric 254–5
session hijacking 82
Shapiro–Wilk test 171
SHapley Additive exPlanations (SHAP) method 216, 222–4, 343
Shor's algorithm 46–7
Siamese networks 127
side-channel attacks 80
SigFox 19
Signaling System 7 (SS7) protocol 61
signal interception 80
Skater 147
smart agriculture 390–4
smart building 404–7
smart grid 395–400
smart healthcare 378–81
smart manufacturing 387–90
smart retail 400–4
smart transportation 381–6
social engineering 81
Software as a Service (SaaS) 42
software attack surfaces 69–71
software-defined network (SDN) 64
software vulnerabilities in IoT devices 70
sparse autoencoders 115
spike timing-dependent plasticity (STDP) 120
spiking neural networks (SNNs) 120–1
spoofing attacks 82
SQL injection 81, 83
strategic selection FL 335–6
stress-testing devices 68
Stuxnet worm 4

supply chain attacks 80
support vector machines (SVMs) 103–4
swarm computing 49–50
symbolic explanations 164–5
synchronous FL 337–8

targeted versus non-targeted attacks 284–7
temporal difference (TD) error 125
text-to-text transfer transformer (T5) 117
textual explanations in the field of XAI 163
Tf-explain 147
The Explainable AI Toolkit (XAITK) 147
third generation (3G) networks 64–5
third-party manufacturing risks 68
third-party service vulnerabilities 76
thread 63
time-sensitive networking (TSN) 18
top-K intersection metric 266–7
training versus inference attacks 283–4
transformer networks 116–19
transparency, defined 133
transport layer security (TLS) 18
tree-based models 98–103, 182
  decision rules 186–9
  decision tree 183–6
tree pruning 184
trust, building 6–7
Trusted Computing Group (TCG) 91
trusted execution environments (TEE) 349–51
trusted platform modules (TPMs) 85
trustworthiness, defined 135
trustworthy FL 348
  security and privacy measures 349–52
  trust challenges in FL 348–9
t-statistic 173
two-factor authentication (2FA) 86

understanding XAI 4–5
universal subscriber identity (USIM) cards 64
user datagram protocol (UDP) 17

variance 97
variational autoencoders (VAEs) 115
vendor risk assessment and supply chain security 92
vertical FL 325, 327–8
VGG networks 111
virtual private networks (VPNs) 357

Wasserstein GAN (WGAN) 114
weakest link pruning: *see* cost-complexity pruning
What-If Tool 147
wide area networks (WANs) 19, 26
Wi-Fi 60–1

XLNet 118

zero-day exploits 82–3
zeroth-order optimization (ZOO) attacks 290–1
zero-trust architecture (ZTA) 359, 361, 366–9
zero-trust eXtended (ZTX) framework 359
zero-trust IoT security 355
  history of 356–60

primary principles 360–4
secondary principles 364–6
synergies between zero trust and XAI
challenges and opportunities 372–4
enhancing trust through XAI 371–2
explainability in zero-trust systems 369
integration of XAI in zero trust 369–71
Zigbee 61–2
Z-Wave 62

www.ingramcontent.com/pod-product-compliance
Lightning Source LLC
LaVergne TN
LVHW011559200926
848484LV00001B/331
* 9 7 8 1 8 3 9 5 3 8 0 2 5 *